Coding for Chemists

Coding for Chemists

Saving Time and Solving Problems with Python

Christopher J. Johnson

and

Benjamin J. Lear

A University Science Book

The MIT Press

Cambridge, Massachusetts • London, England

The MIT Press
Massachusetts Institute of Technology
77 Massachusetts Avenue
Cambridge, MA 02139
mitpress.mit.edu

The MIT Press would like to thank the anonymous peer reviewers who provided comments on drafts of this book. The generous work of academic experts is essential for establishing the authority and quality of our publications. We acknowledge with gratitude the contributions of these otherwise uncredited readers.

This book was set in UtopiaStd-Regular by Nova Techset Private Limited, Bangalore and Chennai, India. Printed and bound in the United States of America.

Library of Congress Cataloging-in-Publication Data is available.

ISBN: 9780262054485 (Paperback)
ISBN: 9780262054577 (PDF)

10 9 8 7 6 5 4 3 2 1

EU Authorised Representative: Easy Access System Europe, Mustamäe tee 50, 10621 Tallinn, Estonia | Email: gpsr.requests@easproject.com

Contents

Who This Book Is For

Are you a researcher aware that programming exists, but are unsure of how it might be applied to your research? Do you have a reasonable understanding of how to code, but struggle to find applications in your research? If the answers to either of these question is yes, then you are in the core audience that inspired this book.

You may or may not know the exact problems you wish to solve, but you likely wish to avoid manually completing boring repetitive tasks, add new capabilities for your chemistry toolset, and learn a skill that will set you apart in the job market. This book aims to help you find these applications.

Although there is a overwhelming amount of free information available for learning how to code, it can prove hard to know how to apply this material in the context of chemical research. You probably don't have a large amount of time learn core programming like a computer scientist would before you get to apply your new skills to accomplish your chemistry goals. This book aims to help you apply coding as quickly as you can.

In short, this book aims to reduce the time to application to zero. You will immediately gain new skills and see how to apply them to automate time-consuming day-to-day chemistry tasks and do things you couldn't have done before. Welcome to the exciting world of programming!

How To Read This Book

This book is structured differently than most books on coding. Whereas most books on coding aim to teach you coding first—focusing on developing a base of core coding concepts and then shifting to problems—this book aims to get you solving problems as quickly as possible, using that problem solving to teach coding concepts along the way.

The chapters are built around a narrative structure that follows a student from the start of their coding journey—a hazy awareness that coding may save them time and frustration—to their competence in creating sophisticated tools that they would not have access to otherwise. Each narrative chapter presents a problem (or set of related problems) that might be encountered by someone performing experimental chemical research. The coding concepts necessary to solve the problem are then introduced via very quick examples along the way.

Each chapter ends with a fully functional code that demonstrates a complete solution to the problem introduced in the chapter. The design process to arrive at the code is discussed in the text, a flowchart graphically outlining the code is provided, and the working of the code itself is explained line by line.

To support this approach to learning coding, we also provide a series of technical chapters that address select coding concepts in more detail or provide important background information that is not strictly necessary to simply solve the problem presented. Although not needed to complete the work addressed in the narrative chapters, they are provided for students who desire a deeper understanding of these concepts. The principal goal of the book is to get you to the point that you can solve problems encountered in real-world research and that you have a good enough understanding of Python that more standard online resources are a viable means by which you can further your coding skills.

Because of this utility-forward approach, coding concepts that are typically taught together are instead introduced only when needed to solve the problem at hand, potentially many chapters apart. To help parse this structure, each time we introduce a new concept, it is given a subsection. The heading of this subsection will state what the concept is and what it will be used for. Additionally, at the start of each chapter, we provide a list of concepts introduced in that chapter, paired with any concepts we believe are prerequisites for understanding the new concept. Although the book is designed to be read in order, starting at Chapter 0, the headings and prerequisites provide an entry point for readers who wish to begin by tackling problems solved in latter chapters.

We are also committed to supporting a community of coding chemists by maintaining a companion website to this book (https://www.codingforchemistsbook.com/). This website will house tools to save chemists time that were developed in the course of writing this book, extra content on topics that are evolving too rapidly to fossilize on paper, guidance for setting up your Python installation, and additional exercises to hone your chemistry problem-solving acumen. These tools are contained in a Python library called codechembook, and which can be found on PyPi and conda-forge.

Finally, we acknowledge that are many guides and options for installing Python, adding libraries to Python, and coding Python within integrated development environments, all with their relative strengths and weaknesses. There are also many guides available online for setting up a coding environment for science. Some of these guides will cause you frustration or limit the tools you have to analyze your code and its inevitable bugs. We **strongly** encourage you to start out by following our installation guide included on the companion website for this book. We feel that this guide ensures that you have the smoothest experience learning Python—instead of learning to fix problems with your Python installation.

TYPOGRAPHIC CONVENTIONS

There is a fairly large amount of terminology that needs to be introduced and kept track of. Thus, to some extent, learning to program involves learning to "speak" programming, and this can present a significant barrier to those learning programming for the first time. Our solution to this challenge is to provide a glossary at the end of the book. The first time a bit of terminology is encountered in a subsection, we **bold** it. This indicates you can find a corresponding entry in the glossary. The glossary contains definitions of these terms, as well as the pages were these terms appear in the book—in case you wish to see them used in different contexts. Of course, we also encourage the interested reader to search these terms on the internet, or to ask artificial intelligence (AI) tools about them.

FOR STUDENTS STUDYING INDEPENDENTLY

This book and its companion website contain a number of resources to help you teach yourself problem solving using programming. We include exercises at the end of each chapter and solutions to the exercises online. We also encourage you to use the included content as additional instructional materials.

Because you won't have the built-in support structure of a course, when issues arise, you will rely on the wealth of online resources for coding, most of what are not meant for chemists. Artificial intelligence will also be helpful for overcoming barriers—at the time of writing this book, ChatGPT, Antropic's Claude, Google's Gemini, and Microsoft's Copilot are currently (in the spring of 2024) reasonable options.

FOR EDUCATORS

The book was written at a level that targets advanced undergraduate and graduate students, although we believe *any* student who has taken a laboratory course that requires analysis and reporting of data will appreciate the material presented.

The companion website for this book includes resources that can aid in the design and implementation of a course on coding. We provide proposed course syllabi for both 15-week and 10-week semesters and additional problems that could supplement the ones included in the book. We also discuss approaches to homework, projects, and exams—including ideas and examples for each.

FOR ADVISORS AND LABORATORY MANAGERS

We believe that students could gain good coding proficiency in 8 weeks. Specifically, working through Chapters 0–7 (one per week) will provide a large number of general tools. The rest of the book is also useful and may help in data analysis. Thus, devoting 8–16 weeks of group meetings could be a way to introduce coding to your research groups.

NOTE ON PYTHON VERSION USED

The code examples for this book were largely written using Python version 3.11, and so many of the conventions and usages or Python reflect those required or enabled by that particular version. We note this here, simply to explain differences you may find and you use newer version of Python that allow for different syntax.

Acknowledgments

We thank our wives, Diana Schlamadinger and Shana Lear, for their patience during the writing of this book. Three years ago, when this book was just a random idea we discussed on a hike, we had no idea how much working on it would come to dominate our shared time together.

We also wish to thank several of our colleagues who helped us think through the presentation and scope of the material. Although we chatted with dozens of people about the book, we particularly thank Peter Tonge, Peter Khalifah, Ben Levine, Tom Allison, Michael White, and Trevor Sears.

Our research groups also deserve our gratitude. Firstly, because writing this book took significant time away from our work with them, we thank them for their understanding. Additionally, several students provided consistent and pointed feedback that was invaluable for crafting the book. We are grateful to Annapoorani Hariharan, Jake Bordenca, Kristen Aviles, Anthony Katona, and Ben Kaercher for that feedback.

Students in the Stony Brook CHE358/534 course in the last two years have been guinea pigs for much of the content of this book. In particular, Madeeha Ali and Indrajeet Kumar gave valuable feedback to help us develop this book.

Finally, we wish to thank the editors we worked with at University Science Books: Barbara Dickson and Katerina Heidhausen. This is our first book and they answered many questions and guided us over many hurdles.

PART ONE

SAVING TIME AND AGGRAVATION

CHAPTER 0

Simplifying Repetitive Calculations

- ☐ **New concepts**
  — *and their prerequisites*

- ☐ **Giving your computer instructions using Python commands**
- ☐ **Having Python do basic math by evaluating mathematical expressions**
- ☐ **Accessing capabilities beyond basic math using functions**
- ☐ **Clarifying how your code works with comments**
- ☐ **Temporarily storing data using variables**
- ☐ **Accessing a history of code you have executed in the console**
- ☐ **Expressing data as readable text using strings**
- ☐ **Avoiding repetitive typing by making your own functions**

0.0 PROBLEM

You have recently developed a new aqueous catalyst that relies on a proton transfer step. Obviously, the efficacy of the catalyst will depend on the pH of the solution relative to the pK_a of the catalyst. However, you are also concerned that the pK_a of the catalyst will, in turn, depend on the ionic strength of the solution. This suggests you need to run multiple titration experiments, each conducted under a different (but constant) ionic strength.

Although you are excited to run these experiments and lay the groundwork for your first paper, it will take some work. Fortunately, your lab has a 24-plate reader that can measure the pH of the solutions. This means that you could run multiple experiments at once, where you could use the four rows for different constant ionic strengths and the six wells in each row for different pHs. Each well's condition could be made by mixing together different amounts of stock solutions. Deciding you will titrate with HCl, you produce the sketch of the experiment shown in Figure 0.1:

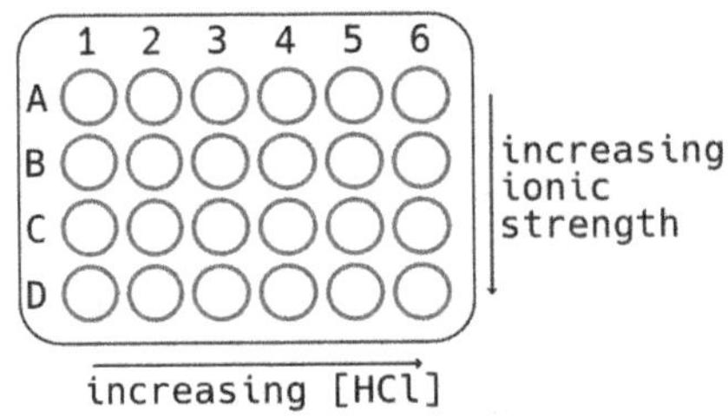

FIGURE 0.1. Schematic of a 24-well plate, showing the desired changes in chemical composition.

Although the plate reader will save you measurement time, it also means that you need to calculate all 24 conditions and keep track of them. You mention to your labmate that you are excited to do the experiment but are not looking forward to doing these tedious calculations and double checking them. This is particularly true because your advisor has a habit of changing their mind regarding the exact limits of reaction—and each time the limits change, you will need to recalculate the conditions.

You know that you could set up a spreadsheet to do this in Microsoft Excel, but the fact that you wanted to maintain a constant volume across the experiments means that the formulas in Excel become cumbersome, and it will be hard to use the spreadsheet while you're actually making the solutions. You'd rather just have a checklist of what goes in each well so that you can be sure you won't make a mistake and have to repeat the experiment.

You are talking about this in lab on morning, and your labmate mentions that this really hasn't been a problem for them since they learned to code. Intrigued, you ask what they mean. They say that coding is a perfect tool for these repetitive tasks—it has reduced their mistakes *and* improved their life in lab. They say that you could write a simple program that would calculate everything you need and could adapt to whatever limits your advisor decides they want that day. They say that it should take only two lunch hours to get you to the point you can solve this problem.

They point out that the very *first* time you implement this solution might take longer than calculating by hand once, but that you will have generated a general tool that can be used again and again. So, the more frequently your advisor changes their mind, the more time you will save in the long run.

You realize that this could reclaim *years* of your life. You agree to meet over lunch, and your labmate tells you to bring your laptop with you.

0.1 SOLUTION

When the first lunchtime comes, you walk into the break room, and your labmate is waiting. The solution to your problem, they explain, will be achieved by outlining the constraints of the experiment, expressing them in terms of calculations that you would normally do on a calculator, and then writing a bit of code that will do these calculations automatically for you. You will do so by storing information about your stock solutions and required final concentrations in the computer and then telling it how to calculate the volumes of each stock solution to add to the well. This first lunch hour, they want to get you to the point you can do this for a single well. Even this will show you how using the code can save you time over using a calculator or Microsoft Excel. Then, in the next lunch hour, they will show you how to adapt this to the rest of the wells in a straightforward manner. This all sounds great to you, so you sit down and open your laptop.

0.1.0 GIVING YOUR COMPUTER INSTRUCTIONS USING PYTHON COMMANDS

"The most basic thing to understand," your labmate starts, "is that programming is a set of instructions for the computer and, at its heart, a computer is not much more more than a calculator with memory. This means your code is going to be a set of instructions, but that these instructions are ultimately turned into numbers. So, you need a way to take language that *humans* understand (i.e., words) and translate them into language that your *computer* understands (i.e., numbers). When someone talks about programming languages, they are referring to how this translation is accomplished."

"Each coding language includes a program that performs this translation. The most basic task in coding is to write your instructions in a way that the translating program can understand." The structure and terminology required for these instructions to be understood by the translating program is called the **syntax**, and the appearance of this syntax is one of the most obvious differences between languages. That is, the exact same instructions written in two languages will look very different.

"I like the Python language," your labmate says, "because its syntax is easy to learn and understand and there are a huge number of tools for chemistry that can be used with it. There are a lot of ways to install Python on your computer, but I recommend starting by installing basic Python from https://www.python.org/."[1]

"Once Python is installed, the simplest thing you can do is to open Python's Integrated Development and Learning Environment (IDLE),[2] which allows you to type Python commands and start calculating things."

Opening IDLE, you are presented with a window that looks something like what is shown in Figure 0.2 Your labmate explains that this box is sometimes called a **shell** or a **console**. You can enter commands and then press enter and the computer will read them and try to execute them. In this case, you are going to input Python commands, and the **Python interpreter** program is going to try to execute them.

[1] We strongly recommend following the procedure outlined on the companion website for this book, https://www.codingfor-chemistsbook.com/, to obtain the full functionality that will be covered in this book. There are many ways to get something that will work, but you will avoid annoyance and get some extra capability installing Python our way!

[2] If you are having trouble, visit the website for this book (https://www.codingforchemists-book.com/) for help.

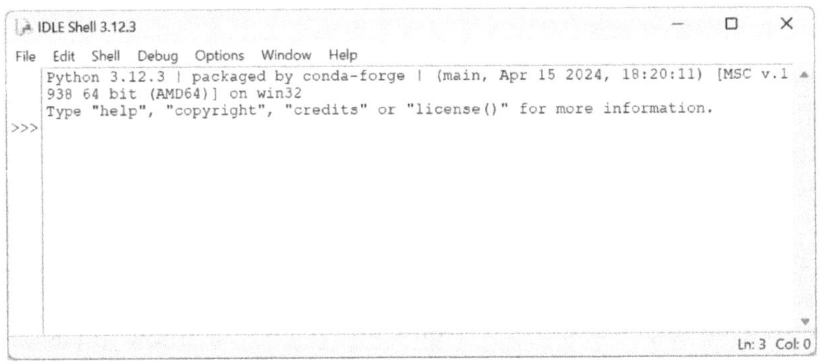

FIGURE 0.2. Screenshot of the IDLE interface for Python in the Windows operating system. This is one way to use Python interactively. You can enter commands with proper Python **syntax**, and they will be executed.

0.1.1 HAVING PYTHON DO BASIC MATH BY EVALUATING MATHEMATICAL EXPRESSIONS

"Remember how I said the computer is a calculator?" Your labmate adds, "Try adding 2 and 3. You can type it into the command prompt, denoted by >>>."

You look at your labmate, but they just raise their eyebrows and wait. So, you tentatively type:

```
>>> 2+3
```

and press enter. Immediately, a new line appears with:

```
5
```

"Great! This is the first line of code that you have ever written! What you typed into the command prompt is a mathematical **expression**, using the **operator** +. Python reads the expression on the input line (represented in this book by light gray and preceded by >>>), carries out all the operations specified by operators, and prints out the result on an output line (represented in this book by darker gray). Now what if you wanted to multiply them?" Here, you think a bit, but remember that other programs, like Excel, use * to indicate multiplication and so you try that.

```
>>> 2*3
6
```

"Ok, now, how about division? What if you divide 2 by 3?"

```
>>> 2/3
0.6666666666666666
```

"Okay," your labmate says. "This is working just like a calculator. You know that 2/3 is not a rational number—that it will keep repeating forever. But a *calculator* has a limited display, so it just writes 0.666... until it runs out of display room. The computer is similar to this. It has a lot more display room than your typical calculator, but each number does require some storage in the computer, and the computer does not have infinite storage. So it has to eventually stop writing numbers. We will talk about how to control the representation of numbers later. Another thing you will run into is this," they say while typing:

```
>>> .01/.1
0.09999999999999999
```

"We know that the answer should be 0.1, but actually, the computer doesn't know what 0.1 is. The problem is, again, that the computer has a limited amount of storage, and so it cannot store all of the real numbers—which would take an infinite amount of storage. Thus, the computer works with only a limited subset of real numbers and 0.1 isn't one of them. It is up to you to realize that 0.09999999999999999 is as close as Python will come to the correct answer. This concept is called **numerical precision** and in almost all cases, you will find that the precision of Python is close enough for your needs. You might have a strong desire to make the output look nicer, but if you can wait just a bit, we will get to that. For now, realize that as soon as you move away from situations where all the numbers involved are whole numbers, your answers will typically be *very* close to correct but not *exactly* correct."

"Now, there will be other things that are a bit different from a calculator or Excel. For instance, what if you want to calculate two to the third power? In Excel, the **operator** for exponent is ^, and many calculators use the same symbol. In Python, this operator is **:"

```
>>> 2**3
8
```

You can see that the output is what we expect for raising two to the third power. Thus, this is the same operator as ^, Python just uses a different **syntax** than Microsoft Excel or your calculator.

0.1.2 Accessing Capabilities Beyond Basic Math Using Functions

"Now," your labmate says, "you can probably think of lots of different symbols for doing math, but sometimes programs don't have them all. Because we are considering similarities and differences with Microsoft Excel, consider taking the absolute value of a number. If you wanted to take the absolute value of a number in Excel, you use the absolute value **function** using the command `abs(-2)`. It turns out that, Python happens to use the exact same idea and notation: `abs()`."

```
>>> abs(-2)
2
```

"And here, we have introduced a new bit of **syntax**. We have taken a value we want to operate on and placed it between two parenthesis, preceded by a name."

"In Python, the term for the command with the parentheses is a **function**, while the number that we place between the parentheses is an **argument**. Functions can evaluate to values in the same way as mathematical expressions can be evaluated to yield a single value. In this case, it evaluated to the value of 2, but not all functions evaluate to values. Functions can also be written to accept any number of arguments—including no arguments. The `abs()` function expects a single argument. But we can consider a different function—one that can accept either one or two arguments: `round()`."

"Invoking the code in a function is referred to as **call**ing the function. To start with, we include, only a single argument in the function call: the number to be rounded. Once called, the function evaluates to the argument value, rounded to a whole number."

```
>>> round(3.14159)
3
```

```
>>> round(3.5)
4
```

```
>>> round(4.5)
4
```

"Take a close look at these last two examples," your labmate says, pointing at their computer's monitor. "You might be surprised that both numbers were rounded to four. This is probably not how you were taught to round. Most people in the USA were taught to round up, when presented with a five at the decimal place where they are rounding. However, Python uses a different rule, known as 'bankers' rounding. Because always rounding a five upwards will tend to bias in reported statistics for measurements, a rule is created to eliminate this bias. If the digit before the five is odd, we round up. If the data before the five is even, we round down. Although this may look strange, it produces a result that, in the long term, has less bias."

"We can exercise greater control over the behavior of the `round()` function by using another argument. This second argument comes after the number we wish to round, has a specific name, and in this case, tells the function the number of decimal places to keep after rounding."

```
>>> round(3.14159, ndigits = 2)
3.14

>>> round(1.61803, ndigits = 3)
1.618
```

"This could be used to solve the issue we encountered above with **numerical precision** and the answer to 0.01 divided by 0.1. For instance, if we wanted we could use the following:"

```
>>> round(0.01/0.1, ndigits = 1)
0.1
```

"And so now we are getting the answer *displayed* as we expect, but there are two things to realize here. First, the answer the computer calculated for 0.01/0.1 was not 0.1. We saw that before (Section 0.1.1). We have asked it to round to give the answer we expect... So really we are just asking for the answer to a precision of one decimal place. This is a technical point, and it is, but it is worth keeping in mind. Second, you will notice that we did math *within* an argument passed to a function. That is, we specified a mathematical operation, and this was evaluated, and then the result of this evaluation was used as the argument. This is our first hint that, when coding, we are going to be able to combine many different operations in a single line of code!"

"One more important thing about function arguments—you notice that we always need to supply a value for the number that we are rounding, but we don't always have to supply the digits to round to. We say that the former is an required argument that is often referred to as a **positional argument**, whereas the latter is an optional argument that is often called a **keyword argument**. Positional arguments are arguments without which the function cannot run. When you call the function, all required arguments must have values. However, for keyword arguments, the function has a default value that it will use if you do not pass a value for it. In the case of round, `ndigits` uses the default value that essentially acts as if it were zero—meaning that the number will be rounded to its integer value."

"Another thing that we see above is that we don't always need to use the construction `ndigits = <value>`. In the last input line, we simply gave the value we wanted. This is the behavior of positional arguments, and the order in which they are passed determines how they are used in the function. However, we can also use the name of the keyword argument that we wish to assign the value to, and this is what we were doing when we used `ndigits = <value>`. All things being equal, it is much clearer to just use the keyword, so that's what I always do."

"All of this might seem a bit overwhelming at first, and there is certainly a lot of terminology to learn. But Python also has some built-in functions to help. If you want to know what the required and optional arguments of a function are, the easiest way to do it is to type `help(<name of function>)`, where `<name of function>` is just the name of the function. In this case, we could type:"

```
>>> help(round)
1 Help on built-in function round in module builtins:
2
3 round(number, ndigits=None)
4     Round a number to a given precision in decimal digits.
5
6     The return value is an integer if ndigits is omitted or None. ↩
7     ↪Otherwise the return value has the same type as the number. ↩
      ↪ndigits may be negative.
```

"Let's have a look at this output."

"Line 1 just tells you what you asked for. It gives you the name of the function, and where it is located. In this case, `builtins` indicates this is part of the base Python."

"Line 3 is the start of the function—what is called the function 'signature.' We see the name of the function, and then information between parentheses. These are the things that the arguments get passed to—basically **variables** within the function. The technical name for this is **parameters**. So, parameters exist in the function (as variable names) and arguments are the values that are passed (assigned) to them, when you **call** a function. In this case, we see that there is a parameter `number`, and a parameter `ndigits`. The `ndigits` parameter is also followed by an assignment (`=None`)[3], which is the default value used, if no value is passed as an argument. The parameter `number` does not have this assignment, and so has no default value. This is why `number` is a required argument and `ndigits` is an optional argument in the function call."

[3] A thing to note here: the default value is actually `None`, not 0. `None` is the value specific to a **data type** in Python. You don't need to know more about this yet, but if you are curious, this is discussed in the chapter on data types (Chapter B).

"The rest of the lines gives details about how the function works and what values it will **return**. For instance, look at the very end of Line 7. This states that `ndigits` can accept a negative number! But it doesn't say what this does, so let us see!"

```
>>> round(42, ndigits = -1)
40
```

```
>>> round(180.156, ndigits = -2)
200.0
```

"So, you can see that negative values passed to `ndigits` will result in integers that are rounded to the specified number of digits before the decimal."

"This could be a useful way to show significant figures, but be careful—as you can see in the last example, when rounding a number with a decimal place to negative decimal places, the number will still contain a decimal place. This might lead to confusion regarding significant figures. There are ways to solve this," your labmate says. "But that is going to have to wait until our next lunch."

"For now, let's start using this cool calculator for some chemistry!"

0.1.3 CLARIFYING HOW YOUR CODE WORKS WITH COMMENTS

"Even now, you already know enough to calculate chemically-relevant things like the number of molecules present in 1 g of water. You also already know how to do this on paper; it would look like:"

$$\text{number of molecules of water} = 1\,\text{g}\left(\frac{1\ \text{mole}}{18\,\text{g}}\right)\left(\frac{6.022\times10^{23}\ \text{molecules}}{1\ \text{mole}}\right)$$

"We can translate these numbers and mathematical operations directly to code, with one important difference: *Python does not know about units by default.* Like your calculator, it just does math on numbers and you will have to keep track of units yourself. Keeping this in mind, we can calculate the number of molecules in 1 g of water:"

```
>>> 1*1/18.02*6.022*10**23
3.341842397336293e+22
```

"It is important to talk about how the answer is formatted. We input Avogadro's number using the sort of scientific notation that most scientists are taught—the general format <#>$\times 10^{<\#>}$. However, even though we might expect our final answer to also be formatted using this same notation, Python represented it differently. Instead of using *10**<##> for the order of magnitude, it used e<##>, which saved some room, and might even be easier to read.[4] A *critical* point here is that there are no spaces between the digits and the 'e.'"

[4] This is sometimes referred to as **e notation**.

"And, remember, Python doesn't include any units in the answer. But *we* know units are important. Rather than needing to always remember what the units are, we can we can leave ourselves notes about this using **comments**. In Python, we can add a comment to any line using # and then adding text after it. Once a comment is started, that will be the last thing on the line. In other words, if you want the line to have both **executable instructions** and comments, the comments need to come last." Your labmate leans over and does the typing this time:

```
>>> 1*1/18.02 # final answer in mols
0.05549389567147614
```

"You see three things here. This is the same format that Python output last time. First, the comment is a different color from the numbers, which are also different colors from the mathematical operator symbols you supplied (like multiplication, ∗). They say this is called **syntax highlighting** and is meant to make it easier to read code. Second, I have added a comment to the end of the expression. Third, the output is the same, but we now have a record that the units were molecules."

"You could also make a line entirely a comment." They also lean over and type:

```
>>> # We want to calculate number of molecules in 1 g water.
>>> 1 * 1 / 18.02 * 6.022e23 # final answer in units of molecules
3.341842397336293e+22
```

Your labmate stops here and says that there are again three things they want you to notice. "First, the line that was entirely a comment was run, but it didn't result in an output. Second, I used the e notation for the order of magnitude." Finally, they point out that although you may not have noticed, they added spaces between several of the numbers and the operations. They say that Python largely ignores spaces *between* numbers and operators in an expression and so you can add in these space to help increase the readability of the code—but it is *critical* to not add a space between the number and 'e' when you are using this to add an order of magnitude, because this represents a single number only when they are given together without spaces.

"There are other rules for using spaces, but I will help you learn them as you go."

"Before we move on, there's one more thing," your labmate says. "Everything we have been doing so far is evaluating **expressions**, where expressions are basically a complete set of instructions. In our case, the result the evaluation was printed below our input line, and this is the default behavior in Python. In other words, unless we go out of our way to direct the result to something else, it just goes to the command line output.[5] However, there are other options. Above (Section 0.1.2), we passed the result of the expression 0.01/0.1 to the `round()` **function**. Another option is that we can transfer the result to the computer's **memory**, so that we can use it later."

[5] This is not actually always the case, but for this moment in time, we will pretend it is. Later in this chapter, we will learn a different way to ensure this always happens—the `print()` function.

0.1.4 Temporarily Storing Data Using Variables

"Up to this point," your labmate says, "you may not really see the major advantage of using coding. You may be asking yourself why not just use a calculator?" However, they point out that the computer has better **memory** than a calculator, and this can be very useful. For instance, what if you didn't want to type out the value of Avogadro's number (N_A) all the time? You can give it a name, and then use that name. This is known as creating a **variable**, which is a name that you can use in your code to avoid having to keep track of the value yourself. You can pick almost any name you want, as long as you follow a few rules, such as the name cannot start with a number.[6] For our variable, `N_A` might be a nice choice, because it looks a bit like N_A. Using sensible and descriptive names is good practice. Your future self will definitely be grateful that your past self did so.

[6] If you want to know the full set of rules, you can read technical Section A.1.1.1.

```
>>> N_A = 6.022e23
>>> N_A
6.022e+23
```

"On the first input line, you created a **variable** called '`N_A`' and **assigned** it a value of `6.022e23` using the assignment **operator** (=). The number `6.022e23` is not a variable, it is the actual value that you want to use in the expression, and this turns out to be known as a **literal**. Literals are not stored and have to be retyped every time."

"This assignment **statement** does not result in an output, because the assignment operator (=) redirects the result of evaluating the **expression** (on the right side of the equals sign) from the output to instead be stored by the `N_A`. However, in the next input line, you ask for the value of the variable N_A. Now, Python evaluates the expression 'N_A' and passes the result (6.022e23) to the output."

"Rather than passing the value of the variable to the output, we can pass it to an operator, like this:"

```
>>> N_A * 1 * 1 / 18.02
3.3418423973362933e+22
```

"Here, multiple things happened in a single line. First, N_A was evaluated, effectively replacing the variable with its value. Then, the leftmost multiplication was evaluated. The value resulting from that evaluation was used in the next multiplication. That multiplication was evaluated and *its* values was used in the division. Finally, the division operation was evaluated and the result was displayed in the output line. We can also take the result of expressions with variables in them and assigned the result to a variable. For instance:"

```
>>> N_water = N_A * 1 * 1 / 18.02
```

"Then, asking for the value of N_water will return the value we saw above:"

```
>>> N_water
3.3418423973362933e+22
```

"It is worth noting something about the order of all these operations. Essentially, the math follows standard order of operations.[7] But in Python, the part of the expression to the right of the equals sign is evaluated first, and then passed to the assignment operator (=), which then assigns the value to the variable N_water."

[7] You probably learned this as PEMDAS (parentheses, exponent, multiplication, division, addition, subtraction).

"Whew! That is a lot of detail," your labmate says, "and not anything we need to talk about over and over again, but it does show that all Python is doing is evaluating expressions and then passing results along. Coding is basically planning out expressions and describing how their results are passed around. And variables are going to play a key role in how we manage all of that."

"You can use variables for anything you would like the computer to remember, like holding the value of the number of grams you have or the molecular mass of water."

```
>>> g_water = 1
>>> MW_water = 18.01528 # g/mol
>>> g_water / MW_water * N_A
3.3427179594211138e+22
```

"There is a point worth making here. The number 18.01528 might look funny as the mass of water, but it shows yet another advantage of coding and computer **memory**: because we can save a value and use it over and over again, we don't waste time typing all those decimal places over and over, letting use as high precision as we like quite easily."

"Now imagine we had a different amount of water. We can simply change the value of the g_water variable."

```
>>> g_water = 4.3121
```

"We could just type out the whole expression again, but we can actually do something much faster, as we will see next!"

0.1.5 Accessing a History of Code You Have Executed in the Console

"So far, we have just been typing out everything we want to write. But the **console** keeps a record of what we have already typed, so that if we want to repeat a previous command—or modify a small part of a previous command—we can save some time by accessing this record."

"If you are ready to type a command into the console, you can instead press the up arrow key on your keyboard. The cursor will then move through commands you already entered, from the most recent to the farthest back—moving up through the IDLE display. If you move the cursor to the line that reads `g_water / MW_water * N_A` and press enter, this will place those commands on the input line, and if you press enter again, it will repeat the calculation. However, between the first time we entered these commands, and now, we have changed the value of the `g_water` **variable**. This time, the expression will use the new value for `g_water`, and will give you the following result:"

```
1.4414134112819785e+23
```

"A variable can only have a single value at a time. If we reassign the value of a variable, we can only access the current value."

"Okay," your labmate says. "We are not *quite* in a place to completely solve your original problem, but we are in a place to start making progress. For instance, we can create variables to store your stock concentrations so that you don't have to remember them, like this:"

```
>>> conc_stock_cat = 0.1 # M
>>> conc_stock_HCl = 6.0 # M
>>> conc_stock_NaCl = 3.0 # M
```

"We can also store the final volume in a different variable:"

```
>>> vol_final = 1.0 # mL
```

"Now you never have to think about that again! Next, we need to figure out the math that we need to do and translate that to code. We can calculate how much we will need using the $M_1V_1 = M_2V_2$ approach:"

$$V_2 = \frac{M_1V_1}{M_2}$$

"If you were to do this on a calculator and you wanted the answer in mL, you would compute:"

$$V_{\text{stock catalyst}} = \frac{0.01 \text{ mol catalyst}}{1 \text{ L}} (0.001 \text{ L final volume}) \frac{1 \text{ L stock solution}}{0.1 \text{ mole catalyst}} \left(\frac{1000 \text{ mL}}{1 \text{ L}}\right)$$

"We can translate this equation into Python code by replacing the math variables with Python variables. Let's say you want the final catalyst concentration to be 0.01 M, the final HCl concentration to be 0.006 M, and the ionic strength to be 0.02 M. Let's go ahead and make those variables too."

```
>>> conc_cat = 0.01 # M
>>> conc_HCl_final = 0.006 # M
>>> I = 0.02 # M
```

"You could get the volume of stock catalyst solution needed by:"

```
>>> conc_cat * vol_final / conc_stock_cat # mL
0.09999999999999999
```

"If you are paying close attention, you will notice that the code we typed in does not exactly map onto the solution we outlined above. In particular, we are missing the part where we multiply by 0.001 L for the final volume and then the final conversion between L and mL. The reason for this is that these two steps cancel out, so as long as we leave them *both* out we will still get the correct final answer. This also means that we had three calculations that we did not ask Python to perform, and this speeds up the calculation. Of course, here, the difference would not be noticeable. But if you were going to repeat simple calculations a billion times or more (as you will find yourself doing surprisingly often), you might start to notice the difference. So, simplifying the number of steps can be good, as long as you can justify it."

"Because the calculation is so similar for HCl, we'll just use the up arrow to select the previous line, press Enter, and then edit the line to replace `con_cat` and `conc_stock_cat` with `conc_HCl_final` and `conc_stock_HCl`, respectively:"

```
>>> conc_HCl_final * vol_final / conc_stock_HCl # mL
0.001
```

"Ionic strength is only a little more complicated. You calculated the ionic strength as follows:"

$$I = \frac{1}{2}\left([H^+] + [Cl^-] + [Na^+] + [Cl^-]\right)$$

"So you can get the volume of NaCl stock solution needed by:"

$$V_{NaCl}^{\text{stock}} = \frac{(I - [HCl]) * V_{\text{sol'n}}}{[NaCl]_{\text{stock}}}$$

"So we just type:"

```
>>> (I - conc_HCl_final) * vol_final / conc_stock_NaCl # mL
```

"The last question is how much water you need to make up the total volume. We need to calculate:"

$$V_{\text{water}} = V_{\text{sol'n}} - V_{\text{cat}} - V_{\text{HCl}} - V_{\text{NaCl}}$$

"This would be easy if we stored all of the results we already got as variables; otherwise, we have to either write them down or recalculate them. We can go back and type:"

```
>>> vol_cat = conc_cat_final * vol_final / conc_stock_cat # mL
>>> vol_HCl = conc_HCl_final * vol_final / conc_stock_HCl # mL
```

"which then allows us to do this:"

```
>>> vol_NaCl = (I - conc_HCl_final) * vol_final / conc_stock_NaCl # mL
>>> vol_water = vol_final - vol_cat - vol_HCl - vol_NaCl # mL
>>> vol_water
0.8943333333333333
```

"This is convenient, but it would still be nice to be able to output numbers in context, with units attached to them, so let's talk about using text in Python."

0.1.6 Expressing Data as Readable Text Using Strings

"Even though Python doesn't know the units associated with the numbers, *we* do. So we can add them to the answer if we like. When we want to give Python text (like letters, words, and numerals) instead of code and numbers, we place the text between apostrophes ('), quotation marks ("), triple apostrophes ('''), or triple sets of quotation marks """). This lets Python know that we are dealing with text, rather than numbers or code. From Python's perspective, numbers and text are different kinds of data, or different **data types**. The technical term for textual data is a **string**. We can create these strings and assign them to **variables**:"

```
>>> greeting = 'Hello there.'
>>> greeting
'Hello there.'
```

"Here, you can see that Python indicates that the value output is a string by placing it between apostrophes. Of course, you might not want to see the apostrophes when Python writes out the string, and this can be done by using the `print()` **function**. This function takes one or more arguments and outputs the value of the **argument** to the standard output, which is in this case the output in our **console** window (the thing we have been seeing as the output)."

```
>>> print(greeting)
Hello there.
```

"You can see the difference is that Python has output the value assigned to the variable `greeting` (Hello there.), without an indication of the type of data this is (i.e., there are no bracketing apostrophes)."

"Of course, we can also provide data types besides strings as arguments to the `print()` function. For instance, we could use one of the variables that has numbers associated with it."

```
>>> print(MW_water)
18.01528
```

"Another thing that is useful about `print()` is that it can accept multiple arguments, as long as we separate them by commas. It will then display each one sequentially, separated by a space."

```
>>> water_molecules = g_water / MW_water * N_A
>>> print(water_molecules, 'molecules of water.')
1.4414134112819785e+23 molecules of water.
```

"Suppose we wanted this output to be on two different lines. If we type them separately, then this happens:"

```
>>> print('Molecules of water:')
Molecules of water:
```

```
>>> print(water_molecules)
1.4414134112819785e+23
```

"But the outputs are now broken by a line of input, which isn't great. Luckily, we can make one print statement that prints multiple lines of text:"

```
>>> print('Molecules of water:\n', water_molecules)
Molecules of water:
1.4414134112819785e+23
```

"Here, we used `\n` in our string, and this really gets to the heart of how computers see text.[8] The computer has no concept of paragraphs. Even though it is automatic for us to end a paragraph and start a new line, we have to *tell* the computer where to do this. We do this by using a symbol to indicate this. The symbol is the **newline character** and in Python, it is `\n`. So, we are using this character to make the output more readable. There are many other such characters that specify formatting, with the other one you will likely need being tab: `\t`."

"So you can see that we get the output we wanted. The next time we meet, we will talk about how to format this even more. For today, let's end by looking at everything we have done so far and talk about how to tie it all up so it is easy to use."

[8] If you want to read more about how text works in detail, see Technical Chapter C. The information is not needed for this chapter, but we point it out in case you are interested.

0.1.7 AVOIDING REPETITIVE TYPING BY MAKING YOUR OWN FUNCTIONS

"You have already used a couple of **functions** that are built in to Python: `print( )` and `round( )`. These functions allow you to easily accomplish common tasks. Essentially, you can think of them as taking a series of instructions that accomplish this task and bundling them together, so that you only need to write out the name of the function, rather than all the instructions. This is certainly very convenient! And wouldn't it be nice if you had a function that did this for the amounts of stock solution you need to add for your experiment?"

"Well, you are in luck! Although Python may not automatically include such a function, you can make your own functions that do basically anything that you can code up! To do so, you simply take code you write and then define a function to use it. Then, you can just **call** the function, and the code you wrote is hidden away inside of it so that only Python has to deal with it. Functions have four basic parts:"[9]

- A name—function names need to follow the same conventions as variable names.[10]
- Zero or more **parameters** that define the data that the function will use. These hold that values that are passed as arguments.
- The code in the function—this needs to be indented to make it clear that the code belongs to the function instead of the rest of the code. Indentation is one type of **white space**, used to tell Python how code should be executed. For any function, all of the contiguous lines of code after a function definition that have the same indentation are interpreted to be part of the code inside the function. The first line of code where there is no longer indentation is outside of the function and is the signal to the **Python interpreter** that the code is no longer part of the function. As you will see later, there are reasons why you will need to use additional indents to specify more complex code inside of functions, but this is enough for now.
- The value(s) that the function will **return** when executed. Sometimes, there are no values returned, and this is fine. Remember when we used `round( )`? The rounded value was the value that was returned from that function when it was evaluated.

"When Python evaluates the line of code with the function call, it will evaluate the code in the function and then replace the function call with the returned value."

"Using the IDLE (as we have so far), you *could* choose to create a function by typing the function into the **console**, line-by-line. However, a *much* simpler way is to create a file that contains the code we want to use and then read the code from this file. If you click File->New File in IDLE, you will get a new window to define your function."

[9] A more complete discussion of how functions work can be found in the technical chapter on functions (Chapter G).

[10] See Section A.1.1.1.

Your labmate types into the IDLE text editor:

```
1  # A function to calculate the volumes of stock solutions needed to create solutions with
       given catalyst, HCl, and ionic strength
2  # All concentrations are in M
3  # The final volume is in mL and assumed to be 1 unless provided
4
5  def calcPlateVols(conc_cat, conc_HCl, I, vol_final = 1):
6
7      # Define our stock solutions
8      conc_stock_cat = 0.1 # M
9      conc_stock_HCl = 6 # M
10     conc_stock_NaCl = 3 # M
11
12     # Calculate the volumes needed
13     vol_cat = conc_cat / conc_stock_cat * vol_final
14     vol_HCl = conc_HCl / conc_stock_HCl * vol_final
15     vol_NaCl = (I - conc_HCl) / conc_stock_NaCl * vol_final
16
17     # Calculate the water needed to make up 1 mL
18     vol_water = vol_final - vol_cat - vol_HCl - vol_NaCl
19
20     print('[ ] catalyst solution (mL)\n', vol_cat)
21     print('[ ] HCl (mL)\n', vol_HCl)
22     print('[ ] NaCl (mL)\n', vol_NaCl)
23     print('[ ] water (mL)\n', vol_water)
```

and saves the file as calcPlateVols.py.

"Let's talk through this code line by line."

1–3: We use **comments** to let people know what this code does and how to use it. The first comment explains what the code does. The next two give the assumptions the code makes.

4: This line in blank. Though not required, using blank lines can increase the readability of the code. We often use them to indicate separation of tasks.

5: The definition of the **function** starts with the `def` **keyword**, which tells Python that what follows is the name of the function and that the lines that follow will have the instructions that comprise the working of the function. The name of the function is the name that you will need to type in Python when you want to use the function, like "print" for the `print()` function or "round" for the `round()` function. After that, you have parentheses, and everything between the parentheses is a **parameter**... the names of the **variables** inside the function that **arguments** will be **passed** to when the function is **called**. This is like how you provided values to the `round()` function. In the case of this function, you can type `calcPlateVols(.01, .006, .02)` and in the function, there will be three variables: `conc_cat` (holding the value 0.01), `conc_HCl` (holding the value 0.006), and `conc_NaCl` (holding the value 0.02). We will set the final volume of the wells, vol_final, to be a **keyword argument**. This allows us to set a default value for the volume (in this case 1 mL), but also allow the user to easily change the value when the function is called.

7: Another comment to tell us what this part of the function does.

8–10: Here, we are defining variables to hold the concentrations of our stock solutions, just like we did when using the interpreter. This type of usage is known as **hard coding**, because we cannot change the values without changing the code. A more general usage would be to add these as keyword parameters in the function definition, with these default values assigned. However, because this was a quick tool for a specific situation, and you commonly use these stock concentrations, they are hard-coded.

12: Another comment to tell us what this part of the function does. It is good practice to leave comments that help a user understand the steps you are trying to accomplish.

13–15: Now we can compute the volumes of stock solutions needed the same way as we did earlier in this chapter.

17: Another comment, because we are now moving on to a new step in our solution.

18: We're calculating the volume of water needed.

20–23: This is a series of print statements to output the resulting arrays in human-readable form. Each one prints two lines, one that has a checkbox and a description of the solution to be added, and a second that prints the volume that needs to be added. Even though the print statement is on a single line, it will output two lines, because we are using the newline character, \n.

"Now, when you run this code by clicking Run->Run Module, Python will step through this code line by line and run each line in the interpreter. When it is finished, Python will now know that there is a new function and how it should behave. So, we can use it like any other function:"

```
>>> calcPlateVols(.01, .006, .02)
[ ] catalyst solution (mL)
 0.09999999999999999
[ ] HCl (mL)
 0.001
[ ] NaCl (mL)
 0.004666666666666667
[ ] water (mL)
 0.8943333333333333
```

> **The Python Kernel**
> Each Python session involves the "kernel." The Python Kernel is the core execution environment that runs Python code in interactive computing environments like the IDLE. The kernel is what keeps track of variable names, variable values, defined functions, and so on. If you close IDLE, this closes the kernel. The reason to mention this is that, if you close IDLE and then open it and immediately try to run `calcPlateVols()`, you will get an **error** message:
> ```
> NameError: name 'calcPlateVols' is not defined
> ```
> which indicates that Python does not know what this function is. You must first run the code that defines the function before you can use it.

"Now, you just have to press the up key and enter to copy the last line into the input, change whichever concentration you want to, press enter again to run it, and get a new calculation! You can use this to calculate the value for all the wells in your experiment."

"In my experience," your labmate says, "this is *already* better than using a calculator or Excel, but if we learn one more concept, we can make these calculations even easier. Like, we can apply this function to every well of your 24-well plate. And if we plan it out correctly, we could use the same approach for a 96-well plate, or a plate of any size!" Your labmate glances at their watch and says, "But I have to get back to the lab to finish up an experiment, so we can pick this up tomorrow at lunch!"

Given the amount you have learned in only this hour, you can hardly wait.

CHAPTER 0 EXERCISES

TARGETED EXERCISES

Exercises for Section 0.1.0

0. Using pip (Windows) or pip3 (Macs), verify that the following Python **libraries** are installed, and if not, install them. If you want directions you can use the resources at the book's website (https://www.codingforchemistsbook.com/).

a) numpy
b) scipy
c) lmfit
d) plotly
e) kaleido
f) pillow
g) openpyxl

h) python-docx

i) rdkit

j) nbformat

k) codechembook

1. Run plotly-test.py from the codechembook library. If you need more directions, refer to the resources at https://www.codingforchemistsbook.com/

Exercises for Section 0.1.1

2. Compute the following outside of Python, and then have Python compute them. Compare the answers:

   a) $9^{1/2}$

   b) $9^{0.5}$

   c) $5 + \frac{7}{8}$

   d) $5 + 7 \div 8$

   e) $(5 + 7) \div 8$

   f) $(4 + 5)^{1/2}$

   g) $9 \div 3^2$

   h) 2^{3+1}

Exercises for Section 0.1.2

3. Compute the following outside of Python, and then have Python compute them. Compare the answers:

   a) $2 - 3$

   b) $|3 - 2|$

   c) $4.5 + 3.2 \times 7.8 - 12.1$

   d) $(4.5 + 3.2) \times (7.8 - 12.1)$

   e) $(4.5 + 3.2) \times (|7.8 - 12.1|)$

4. An arithmetic operator that is included in Python which you haven't used yet in this chapter is the modulo operator (%). This produces the remainder of a division operation. Calculate the following outside of Python and using Python, and then compare the answers:

   a) $13 \div 10$

   b) $10 \div 13$

c) $7 \div 2$

d) $6 \div 2$

5. A function included in Python that you haven't yet seen is `pow( )`, which raises one number to the power of another and accepts two **positional arguments** and one **keyword argument**. Use the `help( )` function to get information on how to use `pow( )`, and then use it to compute the following:

   a) $2^{3.14}$

   b) $\sqrt{2}$

   c) $7^{3/4}$

   d) a rendering of Avogadro's number

6. In Exercise 4, you learned modulo, and in Exercise 5, you learned to use the `pow( )` function. Using the `pow( )` function and modulo operator only, check to see if 7^3 is a factor of 3.

7. Compute the following answers, rounded to the specified number of digits, using the `round( )` function.

   a) $2 \div 3$ rounded to the nearest thousandth.

   b) π rounded to the nearest hundredth.

   c) $5.6^{3.2}$ rounded to the nearest tens digit.

   d) $1 \div 3.3$ rounded to the nearest billionth.

Exercises for Section 0.1.3

8. Produce comments for the code you used to solve Exercises 3b, 4a, 5a, and 6.

Exercises for Section 0.1.4

9. Which of the following are valid variable names in Python? For invalid variable names, propose a new name that is acceptable.

   a) `2nd_trial`

   b) `length*width`

   c) `^volume`

   d) **`def`**

   e) `__hidden_variable`

   f) `_`

   g) `fast component`

10. Create valid Python variables for the masses of carbon, hydrogen, oxygen, and nitrogen. Then using these variables, compute the molar masses of:

    a) NO_2
    b) water
    c) glucose
    d) benzene
    e) pyridine
    f) glycine

11. Using only the mass of hydrogen, carbon, nitrogen, and oxygen, compute the mass of hexaglycine simply by adding up all the masses of the elements.

Exercises for Section 0.1.5

12. Solve Exercise 11 and then, without retyping any code, press up until you highlight the line for calculating the mass of a single glycine. Modify this line to store the mass as a variable. Then, use this new variable to calculate the mass of hexaglycine without using the atomic masses directly

Exercises for Section 0.1.6

13. Enter each of the following into the console and press return, then also put the value inside a `print()` statement and compare the outputs.

    a) 2.0
    b) '2.0'
    c) 2/2
    d) '2/2'
    e) 'test'
    f) test

14. Using a single line of code with a single print command, output a grammatically correct sentence that answers the following questions (perform the math, if necessary, and output with correct units):

    a) What is the number of moles of water in 14.7 g of water?
    b) How many mL are in 1.35 gallons?
    c) What is 72.5°F in °C?

Exercises for Section 0.1.7

15. Create functions that do the following tasks:

    a) Print the number of moles of water for a given mass of water. Here, the mass of water should be a **positional argument**. Confirm that your function works by reproducing the result of Exercise 14a.

    b) Convert a temperature in degrees Fahrenheit to degrees Celsius and print the result. Confirm that your function works by reproducing the result of Exercise 14c.

    c) Solve dilution problems ($M_1V_1 = M_2V_2$) for M_1. Show two examples that prove that it works correctly.

    d) Solve dilution problems ($M_1V_1 = M_2V_2$) for V_1. Show two examples that prove that it works correctly.

COMPREHENSIVE EXERCISES

16. Write a function that will take two masses of samples of water and compute the difference in the *number* of molecules of water of the two samples, but rounded to the nearest thousandth of a mole. Write your function so that the order of masses supplied as arguments to the function does not matter.

17. You have a spectroscopic detector that you need to cool to 6 K using liquid He. You know the heat capacity of the detector is 105 $Jkg^{-1}K^{-1}$. Write a function that will take the current (room) temperature of the detector and print the number of liters of liquid He it will take to cool down the detector, rounded to the nearest tenth of a liter. Assume that it requires 5 L to simply cool down the transfer line, before you start cooling the detector. Also, print a statement that converts the function's output to a cost in dollars to cool down the detector, rounded to the nearest dollar. At the time of this writing, the cost of liquid helium is $40/L. Make this the default value, but write the function so that the user can specify any price.

CHAPTER 1

Automating Repetitive Calculations

- ☐ **New concepts**
  - *— and their prerequisites*

- ☐ **Importing libraries to add capabilities to Python**
- ☐ **Operating on collections of data efficiently with Numpy arrays**
  - *— Section 0.1.1: Having Python do basic math by evaluating mathematical expressions*
- ☐ **Interpreting Python error messages to fix mistakes**
- ☐ **Structuring data using arrays of arrays**
- ☐ **Generating arrays of all ones or zeros using Numpy**
- ☐ **Generating an array of equally spaced numbers using Numpy**
- ☐ **Planning a multi-step solution to a complex problem**
- ☐ **Writing scripts to execute complete solutions in one click**
- ☐ **Keeping track of changes to your code using versioning practices**

1.0 PROBLEM

You are still working to plan out the experiment in the 24 well plate, as shown in Figure 1.1.

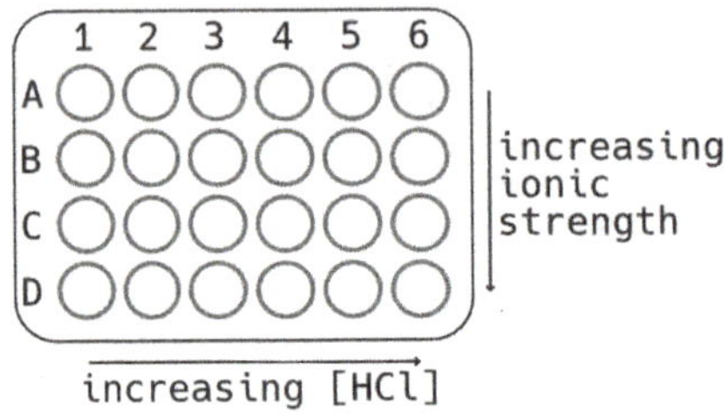

FIGURE 1.1. Schematic representation of a 24 well plate, with the desired changes in chemical composition noted. This is the same diagram as shown in Figure 0.1. In the last chapter, we learned how to calculate the conditions for any single well. In this chapter, we will learn how to calculate for all wells at once.

At this point, you have written a function so that you can calculate the volumes of each stock solution needed for a single well using single command in Python. However, you still need to input the concentrations of HCl and ionic strengths for each well, and you have to run the function once for every well. This might not be a problem for a single 24-well plate, but you think that doing this multiple times (or extending to a 384-well plate) might still be a big problem. What you really want is to calculate all of the wells at once, regardless of the number of wells or the shape of the plate, which your labmate promised you would accomplish over lunch today!

1.1 SOLUTION

When you arrive at the break room, your labmate is waiting and, as you sit down, says, "One nice thing about Python is that the **function** that we wrote yesterday to calculate what to do for a single well in the plate can be just as easily used to calculate any number of wells. Just like we stored single values in variables, we can store many values as a collection in a single variable, and Python will perform the same operation on each value without you having to tell it to do it. Today, we will see how this works, and then

> **Installing new libraries in Python**
> At the time of this writing, there are two dominant means by which to install new libraries in Python: *pip* and *conda*. The most important thing to know is that you should **not** mix these two. So, if you started using pip, please continue using it. If you started using conda or Anaconda, then continue to use it. One thing to note is that conda installations can be *slow*, sometimes taking hours. If you wish to understand more, please visit the website to read about the reasons we selected conda.

write a simple program that can handle any number of wells and any range of concentrations."

"In order to accomplish what we want to do, we need to use capabilities beyond standard Python. In fact, what you can do easily with the core Python programming language is relatively limited. Fortunately, there is a huge range of other code written in Python that you can access. This significantly expands the things that we can do without having to reinvent the wheel. I'll show you how we can use these additional capabilities to take the function you already made and simply adapt it to work with any size well plate."

1.1.0 IMPORTING LIBRARIES TO ADD CAPABILITIES TO PYTHON

"To understand why we might want to expand the capabilities of Python, consider this: the base Python knows how to calculate exponents, but it doesn't know how to do logarithms. So, if we want to calculate logarithms, we need to either write a function to do this, or add that capability from some external source."

Your labmate says the latter can be done by importing a **library** and then using the capabilities of this library. You will use **Numpy**,[1] which is the dominant library for doing the sorts of mathematics that are common in science. They show you how to quickly install Numpy to work with your Python installation.[2] Also, because we have just seen that a feature of coding is making it so that we don't have to write out long lines of code over and over, we will also import the Numpy library and assign it a shorter name, called an **alias**. In this case, we will assign it the alias `np`.

```
>>> import numpy as np
```

We refer to the capabilities of this library by using the alias, adding a period, and then naming the capability we wish to use:

```
>>> np.log(10)
2.302585092994046
```

> **Numpy 1 versus Numpy 2**
> At the time this book was written, Numpy 2 had just been released. This book was written using Numpy 1, and Anaconda currently still ships Numpy 1, but users of `pip` probably have Numpy 2, and Anaconda will eventually update as well. For our use in this book, there is little difference between the two versions. One annoyance is that Numpy2 will print the results of expression evaluations in a way that explicitly denotes the numerical **data type** along with the value: `np.float64(3.14)` rather than simply `3.14`. Here, `float64` indicates a 64-bit floating point number. To disable this behavior, you can run the command `np.set_printoptions(legacy='1.25')` at the beginning of your Python session or script (but after importing Numpy).

[1] If you would like more information beyond what is in this book, check out https://numpy.org/devdocs/user/absolute_beginners.html.

[2] See the companion website for this book (https://www.codingforchemistsbook.com/) for detailed instructions to install Numpy and all the other libraries you will use in this book.

From this value, you can see that `log()` is a function in the Numpy library and calling `np.log(<number>)` performs a natural logarithm. Your labmate says that the base can be changed, but we will not need to do so right now.

1.1.1 Operating on Collections of Data Efficiently with Numpy Arrays

Another advantage of **Numpy** is that this library includes a **data type** known as **Numpy array**. The Numpy array holds a collection of values that can all be acted on using a single **operator**. For instance, imagine you had three different samples of water, with masses of 1 g, 4.3121 g, and 3 g. If you wanted to calculate the number of molecules for each, you *could* do this three separate times, or you can place the masses in a Numpy array, and then do math on the array:

```
>>> MW_water = 18.01528
>>> N_A = 6.022e23
>>> masses = np.array([3, 4.3121, 1]) # units of g
>>> molecules = masses / MW_water * N_A
>>> molecules
array([1.00281539e+23, 1.44141341e+23, 3.34271796e+22])
```

"In the fourth, we solved three different math problems at once! In this example, we have first put the values in a Numpy array and **assign**ed it to the **variable** `masses`. This is called array **creation** or array **instantiation**. The latter is term is used because we have created an **instance** of a Numpy array. You will notice that you didn't need to import Numpy each time. Once a library is imported (as was done in Section 1.1.0), you can keep using it."[3]

[3] This was a bit more explicitly addressed in the final box at the end of Chapter 0.

"Next, we do some math on the array, using its variable name. We call this 'operating on the array.' We assign the result of these operations to a new variable. Then, we ask for the value of the new variable by typing its name. In the output, we can see that the new variable is also an array, and that it contains the results associated with running the calculations on each of the values in the array. Rather than needing to specify that the operations should be done on each component, one trait of arrays is that the operations are naturally applied to all elements. In this case, we have one array variable (`masses`) and two variables that have only single values (`MW_water` and `N_A`). When Python evaluates this **expression**, it breaks the problem up into three parts, each corresponding with one element of the array, and yields a new array that is the same size. Variables that only hold single values are just applied the same in each part. This behavior, where single-value variables are applied the same to each element of the array, is an example of the concept of **broadcasting**."

"Of course, one can **pass** the array as an **argument** to the `print()` **function**:"

```
>>> print(molecules)
[1.00281539e+23 1.44141341e+23 3.34271796e+22]
```

"The difference in the output is that we no longer have the explicit indication this is an array, and the numbers are no longer separated by commas. Using the `print()` function simply makes the output more readable."

> **Why not Pandas?**
> If you spend anytime searching the internet regarding working with data in Python, you will encounter the Pandas library. This is an extremely popular library for data science. However, we do not use it here for two reasons. First, it is not needed, and so we did not think that the added complexity of working with Pandas was worth introducing. Second, Pandas is written for working with data that has a particular structure—a structure that is not always met by experimental chemistry data. Thus, we think the flexibility and relative simplicity of using only Numpy is worth missing out on the few conveniences of Pandas. Of course, as you learn Python, you can also elect to use it in your own workflows—but neither of the authors of this book currently use Pandas regularly.

"When expressions contain multiple arrays that are the same size, operations involving them are performed in an element-by-element manner. In other words, if we multiply two arrays that are the same size, the result will be a new array in which the first element is the product of the first two elements of the existing arrays, the second is the product of the second two, and so on. This process is called **vectorization**.[4] For instance, imagine we wanted to know the difference in the number of molecules between a sample that is 100% D_2O and the 'normal' sample. We can do the following:"

[4] Even though this term suggests a relationship to mathematical vectors, arrays are not vectors!

```
>>> MW_d2o = 20.0276 # g/mol
>>> molecules_d2o = masses / MW_d2o * N_A
>>> molecules_d2o - molecules
array([-1.00760224e+22, -1.44829387e+22, -3.35867413e+21])
```

The second input created a new array that had the molecules of D_2O that was the same size as the `molecules` array and the third subtracted the numbers from the `molecules` array from this one, on an array-element by array-element basis.

"The last thing that you need know is that, although you have just seen that you can use operators like `+`, `-`, `*`, `/`, `**` on arrays, not everything works this way. For instance, if we try to round our array, we need to use a rounding function in Numpy instead of the normal Python one that we used before:"

```
>>> np.round(123.456, decimals = 1)
123.5
```

"For `np.round()`, you see that the keyword for digits to round is `decimals` rather than `ndigits`. If we try to use the 'normal' `round()` function, we get this:"

```
>>> round(123.456, decimals = 1)
Traceback (most recent call last):
  File "<pyshell#8>", line 1, in <module>
```

```
    round(123.456, decimals=1)
TypeError: 'decimals' is an invalid keyword argument for round()
```

"Let's talk about this error."

1.1.2 Interpreting Python Error Messages to Fix Mistakes

"Error messages in any programming language are irritating, but they usually contain helpful information that you can use to figure out what went wrong and how to fix it. A few things to know:"

- When Python encounters an error, the line of code that it was running when the error happened does not finish and, if it was supposed to assign a value to a **variable**, that doesn't happen.
- If the error happens inside of a **function**, it just stops. The rest of the function does not execute, it does not return a value, no other code executes, and nothing else happens.
- Just because you don't get an error message doesn't mean that your code ran as you expected it, or that your answers are correct. It just means that nothing literally stopped Python from being able to process the code.

There are three basic pieces of information in an error message:

1. The file in which you can find the code that was running when the error happened. In our case, the "file" was actually the IDLE Shell (`<pyshell#8>`).
2. The actual line of code that was running when the error happened. Here, the line of code was the one that we tried to run containing the `round()` function.
3. A description of the error including a standard error type (here, a "TypeError," which denotes that an operation was being done on a **data type** that was not supported by the operation.

"The first two parts are just there to help you figure out what was happening when the error occurred. Sometimes, the problem isn't specific to the code that was executing, but rather is caused by some previous code that executed fine but set up a situation in which the current line could not run correctly. The last part will help you understand what the problem was. Python has standard types of errors that you can look up when you come across them. In fact, just copying the last two lines of the error message and searching (e.g., using Google or AI tools like ChatGPT) will often get you the solution to your problem quickly!"

"Let's look at another example of an error. We know that, because Numpy arrays are vectorized, we can subtract one array from another, and it will do this on an element-by-element basis. However, you have to ensure that the lengths of the arrays are the same, or you will get an error. For instance:"

```
>>> molecules - np.array([1, 1, 1, 1]) # trying to subtract 1 from each element
Traceback (most recent call last):

  Cell In[4], line 6
    molecules - np.array([1, 1, 1, 1])

ValueError: operands could not be broadcast together with shapes (3,) (4,)
```

"In this case, Python is telling us that we cannot **broadcast** or vectorize these arrays because they do not have the correct **shapes** that would allow this. The error is a ValueError, which usually signifies that some variable is the right data type but has some other property that makes the operation unable to finish. In this case, the arrays are different lengths, so we could neither broadcast nor vectorize them and Python didn't know what else to do."

"In both of these cases, the problem was with the line of code that we wrote. Most of the time, as you write more complex code, the error will actually occur in some function that is called by some other function that is called by some other function. When that occurs, the **traceback** will be very helpful because it will let you track back through the nested series of functions that got from the main code to the line that actually caused the issue."

1.1.3 STRUCTURING DATA USING ARRAYS OF ARRAYS

"One convenient thing about **Numpy arrays** is that one array can hold other arrays. This is called **nesting**. For instance, we can have an array that holds four arrays, each of which hold six values. We call the arrays holding the other four arrays the **outer array**, and the four arrays are called the **inner arrays**."

```
>>> array1 = np.array([ 1,  2,  3,  4,  5,  6])
>>> array2 = np.array([ 2,  4,  6,  8, 10, 12])
>>> array3 = np.array([ 3,  6,  9, 12, 15, 18])
>>> array4 = np.array([ 4,  8, 12, 16, 20, 24])
>>> outer_array = np.array([array1, array2, array3, array4])
>>> outer_array
array([[ 1,  2,  3,  4,  5,  6],
       [ 2,  4,  6,  8, 10, 12],
       [ 3,  6,  9, 12, 15, 18],
       [ 4,  8, 12, 16, 20, 24]])

>>> outer_array3 = outer_array * 3
>>> outer_array3
array([[ 3,  6,  9, 12, 15, 18],
       [ 6, 12, 18, 24, 30, 36],
       [ 9, 18, 27, 36, 45, 54],
       [12, 24, 36, 48, 60, 72]])
```

"From this example, although we operated on the outer array, you can see that the **broadcasting** applies to all inner arrays, which is convenient for doing math on all the arrays.

This sort of behavior happened because we were operating on an array using a single value. However, as we saw in Section 1.1.1, there is another option: we can apply an operation using another array of exactly the same **shape**—where shape refers to the size of the inner and outer arrays. In our case, we say the array `outer_array3` is of [4, 6], meaning that the outer array has four elements, each of which has six elements. If we choose another array that has this same dimensions, we can perform operations with it, and it will match up the corresponding elements of the array."

"For instance, what if we subtracted our array from itself?"

```
>>> outer_array3 - outer_array3
array([[0, 0, 0, 0, 0, 0],
       [0, 0, 0, 0, 0, 0],
       [0, 0, 0, 0, 0, 0],
       [0, 0, 0, 0, 0, 0]])
```

"You can see that each element has been subtracted from itself, which is why the result is all zeros."

At this point your labmate smiles at you. "This [4, 6] array matches the format of 24-well plate we have been considering" They point out the sketch you made earlier (Figure 1.1). "So, if we can figure out a way to map our calculations onto these arrays, we will solve our problem! We are almost there, but it is going to be convenient to know a few other things about using Numpy arrays."

1.1.4 Generating Arrays of All Ones or Zeros Using Numpy

"It can be irritating to have to make array **literal**s all the time. Fortunately, Numpy includes a bunch of functions that create **Numpy arrays** for us. You will often want to make arrays in which each element is one, or arrays in which each element is zero. There are several functions that do this automatically for us. Let's take making ones as example. If we know the size of the array we want, we can use `numpy.ones( )` to make it. You just have to tell the function what size to make the array. If you want a one-dimensional (1D) array of six elements, just pass it the argument six."

```
>>> np.ones(6)
array([1., 1., 1., 1., 1., 1.])
```

"If you want a two-dimensional (2D) array like your well plate, 4×6, then you pass the dimensions of the array in parentheses:"

```
>>> np.ones((4,6))
array([[1., 1., 1., 1., 1., 1.],
       [1., 1., 1., 1., 1., 1.],
       [1., 1., 1., 1., 1., 1.],
       [1., 1., 1., 1., 1., 1.]])
```

"The parentheses are important because the `ones( )` function is only expecting one total argument, and it will interpret an argument having two dimensions in parentheses as a single argument specifying (rows, columns)."

"Sometimes, we either don't know or are too lazy[5] to define the dimensions of the array. In that case, there's another function, `numpy.ones_like( )`, that will take as an argument another array and produce an array with the same dimensions but ones for all elements. It doesn't actually change the array that you use as the argument, it just uses it as a template and makes a new array from it. So, we could also make an array of all ones that is the same size as our `outer_array` from before like this:"

[5] This is not necessarily a pejorative. In fact, one of the main motivations for learning to code is that it automates doing stuff for you... so a major motivation is trying to be "lazy."

```
>>> np.ones_like(outer_array)
array([[1., 1., 1., 1., 1., 1.],
       [1., 1., 1., 1., 1., 1.],
       [1., 1., 1., 1., 1., 1.],
       [1., 1., 1., 1., 1., 1.]])
```

"Either way gets us the same array, but sometimes one will be more convenient than the other. For arrays of all zeros, things work the same way, but the functions are `numpy.zeros( )` and `numpy.zeros_like( )`."

1.1.5 GENERATING AN ARRAY OF EQUALLY SPACED NUMBERS USING NUMPY

"Sometimes you will want to make **Numpy arrays** that have equally spaced numbers in them, and Numpy has shortcuts for this, too. The two main ones you will use are `linspace( )` and `arange( )`. They are pretty similar, but you use `linspace( )` when you know how many elements need to be in the array, and `arange( )` when you know what the spacing of the elements should be. In our case, we'll be using `linspace( )`. The arguments you need to pass to it are, in order, the value of the first element, the value of the last element, and the number of elements. For instance:"

```
>>> np.linspace(3.14, 100, 11)
array([ 3.14 , 12.826, 22.512, 32.198, 41.884, 51.57, 61.256, 70.942, ↩
    ↪80.628, 90.314, 100.])
```

"Here, you can see we have 11 equally spaced values, starting at the first (3.14) and ending at the last (100). So, both the upper and lower limit are **inclusive**. Finally, although these **arguments** can involve either whole numbers or numbers with decimal places, `linespace( )` always returns a `numpy array` of numbers with decimal places."[6] This is important because numbers with decimal places are the ones most likely to encounter issues with **numerical precision**, as discussed in the last chapter.[7]

"The function `arange( )` is pretty similar to `linspace( )`, but there is one main difference: the stop value is not inclusive, meaning that your array's last element will always be less than the stop value. So, if we wanted to make an array that goes from zero to one with 0.1 spacing, we would have to do something like:"

[6] This is true even if only integers are used and the requested values could be represented at integers. If you want to see, try `np.linspace(1, 11, 11)`.

[7] One other thing to note here is that the output had a fancy curly arrow at the end of the first line, and one at the start of the second. These arrows indicate that the line was too long to fit on the page, so the output was "wrapped" onto a line. This arrow is not part of the Python output, but is part of the formatting of this specific book, so you will not have these arrows in your outputs.

```
>>> np.arange(0, 1.1, 0.1)
array([0. , 0.1, 0.2, 0.3, 0.4, 0.5, 0.6, 0.7, 0.8, 0.9, 1. ])
```

1.1.6 PLANNING A MULTI-STEP SOLUTION TO A COMPLEX PROBLEM

"Okay, we're ready to solve our original problem! What we want to do is to define the first and last values for HCl concentration and ionic strength, then use Numpy to make **Numpy arrays** based round this for us. Then, we can reuse your **function** from earlier (the end of the last chapter) to operate on each element in the arrays simultaneously and display the results as text. This way, you will only have to do one calculation, and you'll get the volumes for all of the wells in the blink of an eye."

"Start by determining the range of parameters you want to explore, the final conditions you wish to have, and starting materials you are working with."

"Like before, you want each well to have 1 mL final volume. You want to use a constant catalyst concentration of 0.01 M for the titrations, and you have a 0.1 M stock catalyst solution, a 6 M stock HCl solution, and a 3 M stock NaCl solution. For ionic strengths, the catalyst has a single protonation site, and you plan your titration to run to 1 equivalent of acid added, so the lowest possible ionic strength is going to be 0.02 M. You want to run up to a strength of 0.2 M. Finally, you say you want the four different titrations to be at equally spaced ionic strengths."

"This, hopefully, makes sense. However, it can be a bit cumbersome to say or read through. There are other tools that can help us plan out a solution. For instance, many people find flowcharts to be a helpful graphical depiction of what they want to do in code, so they can outline the solution in a flowchart, and then write out the code to accomplish it. Let's put together a flowchart (Figure 1.2) to guide us on this solution."

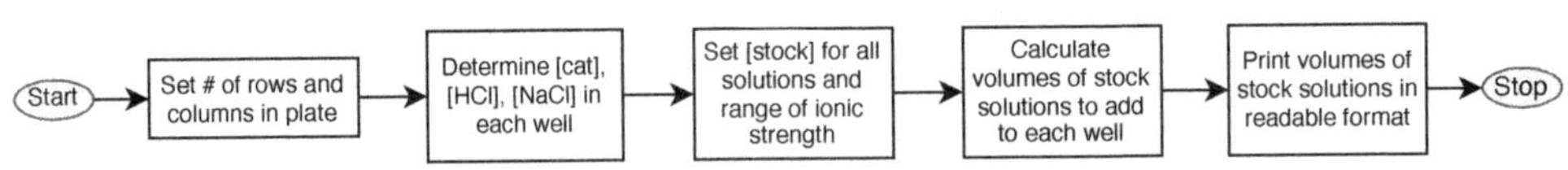

FIGURE 1.2. Flowchart of the approach to calculating the concentrations for the 24-well plate. It can be useful to try to sketch out the steps associated with a problem in this manner to help envision the way the code needs to work. The ability to take the solution for a single well (as seen in the last chapter) and apply it to the entire plate relies on the use of **Numpy arrays**.

"First, let's create arrays holding [HCl] and ionic strengths that we will want to use for each well. For [HCl], we want to span zero to one equivalents of catalyst, so we can define:"

```
>>> innerMHCl = np.linspace(0, 1, 6) * 0.01 # scale the array by [catalyst]
```

"Then, we can make the 2D array like this:"

```
>>> MHCl = np.array([innerMHCl, innerMHCl, innerMHCl, innerMHCl])
```

"We can do the same thing for ionic strength. Each of the titrations are under a constant ionic strength. Running from 0.02 M to 0.2 M, we want them to be equally spaced, and so there should be three steps from 0.02 M to 0.2 M."

```
>>> starting_ionic = 0.02
>>> ending_ionic = 0.2
>>> ionic_step = (ending_ionic - starting_ionic) / 3
```

"Here, we are dividing by 3, even though we have four rows, because there are three steps *between* the rows. Using these **variables**, we can create an `array`."

```
>>> inner_ionic = np.array([0.02, 0.02, 0.02, 0.02, 0.02, 0.02])
>>> ionic = np.array([inner_ionic, inner_ionic + ionic_step, inner_ionic + ↩
    ↪2*ionic_step, inner_ionic + 3*ionic_step])
```

"A fair amount happened in this last step, so let's have a closer look. We created an array and put arrays inside of it. But for each subsequent array we added, we also added a value to that array using our step size we determined earlier. The end result is that we ended up with an array that has the desired ionic strengths, which we can verify by asking for the value."

```
>>> ionic
array([[0.02, 0.02, 0.02, 0.02, 0.02, 0.02],
       [0.08, 0.08, 0.08, 0.08, 0.08, 0.08],
       [0.14, 0.14, 0.14, 0.14, 0.14, 0.14],
       [0.2 , 0.2 , 0.2 , 0.2 , 0.2 , 0.2 ]])
```

"Lastly, we need to make an array to hold the concentration of the catalyst in each well. We have a couple of options: we could define a 4×6 array with all ones, or we can use `np.ones_like()` to create an array of ones that is the same shape as another array. If we base this array on, say, the HCl concentrations, then it will always automatically be the right size in case our well plate changes. So, we can just type: '`vol_final` with' "

```
>>> Mcat = .01 * np.ones_like(MHCl) # M
```

"Now, you can use these three **variables** that hold arrays as arguments to our `calcPlateVols()` function the same way you used single values before, and it will automatically use **broadcasting** and **vectorization** to calculate all the wells at once!"

```
>>> calcPlateVols(Mcat, MHCl, ionic)
[ ] catalyst solution
 [[0.1 0.1 0.1 0.1 0.1 0.1]
 [0.1 0.1 0.1 0.1 0.1 0.1]
 [0.1 0.1 0.1 0.1 0.1 0.1]
 [0.1 0.1 0.1 0.1 0.1 0.1]]
```

```
[ ] HCl
 [[0.         0.00033333 0.00066667 0.001      0.00133333 0.00166667]
 [0.         0.00033333 0.00066667 0.001      0.00133333 0.00166667]
 [0.         0.00033333 0.00066667 0.001      0.00133333 0.00166667]
 [0.         0.00033333 0.00066667 0.001      0.00133333 0.00166667]]
[ ] NaCl
 [[0.00666667 0.006      0.00533333 0.00466667 0.004      0.00333333]
 [0.02666667 0.026      0.02533333 0.02466667 0.024      0.02333333]
 [0.04666667 0.046      0.04533333 0.04466667 0.044      0.04333333]
 [0.06666667 0.066      0.06533333 0.06466667 0.064      0.06333333]]
[ ] water
 [[0.89333333 0.89366667 0.894      0.89433333 0.89466667 0.895     ]
 [0.87333333 0.87366667 0.874      0.87433333 0.87466667 0.875     ]
 [0.85333333 0.85366667 0.854      0.85433333 0.85466667 0.855     ]
 [0.83333333 0.83366667 0.834      0.83433333 0.83466667 0.835     ]]
```

"You can see now how this could save a ton of time. All you have had to do is define two variables, rerun your function, and you have your volumes for all the wells."

1.1.7 WRITING SCRIPTS TO EXECUTE COMPLETE SOLUTIONS IN ONE CLICK

"The solution we just made is good, but we can do even better! It would be time consuming to type this out every time you wanted to change the experimental conditions. One of the benefits of programming is saving the lines of code that you write and running them automatically instead of forcing you to type them in every time. So far, we've used Python interactively, where we give it one instruction at a time. But, we can put all the lines into a text document, called a **script**, that we save and run from the IDLE. When we run it, IDLE will go through the file one line at a time and automatically transfer that line to the **Python interpreter** and run it. Then, it will go to the next line and do the same. That way, you don't have to do all the typing, which saves you a ton of time and prevents a lot of mistakes. To do this, we just add more code to the end of the file where you defined your function (see Section 0.1.7). You didn't know it, but that was a script that you wrote, and you can modify it as you need to. Python will just keep reading the file and will do the instructions that you put in there."

You go back to the IDLE text editor window and make sure the file was saved as `WellPlate.py` so that you don't lose your previous work. "Your Python script file names should always end with '.py' so that your computer knows that they should be run with Python. Other than that, this is just a text file, there is nothing special about it."

You copy and paste the lines of code that you put into the IDLE Shell into the Python script and click Run->Run Module, and you get the exact same result as you did previously without any typing!

"Now that we have this, you could just change the numbers every time our advisor changes their mind, and just run it again by clicking Run->Run Module."

1.1.8 KEEPING TRACK OF CHANGES TO YOUR CODE USING VERSIONING PRACTICES

Your labmate looks over at you, but sees you frowning slightly. They ask what is wrong, and you say that you are bit concerned because you started with a file that was just a single function that you called in the IDLE, and then you added lots of new capabilities to it. Although that was nice, you are thinking about how you can ever hope to keep track of what is going on with code as you build it out. Worse, what happens if you make a change it results in the code not working at all? How could you recover from this?

Your labmate compliments you for your foresight and says if they can just show you how to do this, so you push the laptop over to them.

"What you have brought up is a need for **version control**. The idea behind this is to help you keep track of the changes you have made to your code as you develop it. This is good practice for a lot of reasons: for instance, if you make a change that you decide is bad, you can easily go back to a time you liked the code."

"There are multiple means to handle versioning, and most professional programmers use something like git.[8] We aren't going to worry about this yet because it is a bit much for where we are right now. So, let's talk about a simpler means to do this."

"What I like to do is as follows: each time I am going to change a script, I save the current version of it, with the Script name, and a series of three numbers after it. For a generic Python file, this would look something like: `Name_x.y.z.py`, where x, y, and z are the numbers."

"You may have seen files with version numbers like this before; but what do they mean?" Your labmate looks at you, but you just shake your head. Like, you know that these numbers seem to increase over time, but why does one or another change? That has always been a mystery.

Your labmate explains that each number has a meaning, and you should increment them as follows:

- `z` indicates changes associated with "patches." These are changes to the code that do not alter intended behavior or introduce new features. For instance, perhaps you have a code with a typo in it. Fixing that typo does not change how the code behaves from the user's standpoint.

- `y` indicates minor version changes. This should be used when you change the behavior or introduce new features to your program, but does so in a way that is backward compatible. This means that someone upgrading your program will not need to change how they have been using it—the things they have been doing in the past will continue to work with the new version.

- `x` indicates a major version change. These are changes to the program that break backwards compatibility. Changing this number indicates that someone upgrading your program can expect some things they have been doing will no longer work, and they will have to change how they use your program.

[8] If you want to know more, look at Chapter 17. But we really don't think this knowledge is needed yet.

"Right now, I am going to save a version of your code we just wrote. But this code is already different from just the function. Really, it is trying to solve a related, but different, problem. For this reason, I will save this as a different script. Since this is going to be a script to calculate the volumes in well plates, we can call it `WellPlate.py`."

"But, of course, we already added text to the function, so let us go back to how we started." Your labmate deletes the code that you added, so that it is just the same function that you had at the end of Chapter 0.

"Before we save this, there is one more thing to do: we should **document** these changes in your script using a **comment** at the top. I typically include this for each version, so I can understand what I was doing." Your labmate goes to the top of the script and adds this comment: `# Version 0.1.0 - 240605 - A function that calculates volumes of stock solutions needed for a well in a well plate.` They then save a *copy* of this script as `WellPlate_0.1.0.py`.

"And now we are ready to keep altering the existing script: `WellPlate.py`. When we do this, we can restructure the solution slightly to be better organized and easier to read. We will start by documenting the update, and if we make further changes to this program, we will save this version we are producing now as a different file: `WellPlate_0.2.0.py`."

1.2 CODE

Your labmate then starts typing for a few minutes and shows you the following code:

```
# Version 0.2.0 - 240605 - Took the function and wrote a script to calculate volumes of stock solutions needed for a well plate of arbitrary size.
# Version 0.1.0 - 240605 - A function that calculates volumes of stock solutions needed for a well in a well plate.

import numpy as np

# A function to calculate the volumes of stock solutions needed to create solutions with given catalyst, HCl, and ionic strength
# All concentrations are in M
# The final volume is in mL and assumed to be 1 unless provided

def calcPlateVols(conc_cat, conc_HCl, I, vol_final = 1):

    # Define our stock solutions
    conc_stock_cat = 0.1 # M
    conc_stock_HCl = 6 # M
    conc_stock_NaCl = 3 # M

    # Calculate the volumes needed
    vol_cat = conc_cat / conc_stock_cat * vol_final
    vol_HCl = conc_HCl / conc_stock_HCl * vol_final
    vol_NaCl = (I - conc_HCl) / conc_stock_NaCl * vol_final

    # Calculate the water needed to make up 1 mL
    vol_water = vol_final - vol_cat - vol_HCl - vol_NaCl
```

```
29
30      print('[ ] catalyst solution (mL)\n', vol_cat)
31      print('[ ] HCl (mL)\n', vol_HCl)
32      print('[ ] NaCl (mL)\n', vol_NaCl)
33      print('[ ] water (mL)\n', vol_water)
34  # Define the dimensions of our well plate
35  rows = 4
36  cols = 6
37
38  # Define our experimental concentrations and ionic strengths
39  conc_cat = 0.01 # M
40  conc_HCl_start = 0.0 # M
41  conc_HCl_end = 0.01 # M
42  I_start = 0.02 # M
43  I_end = 0.2 # M
44
45  # Get the concentration of catalyst in each well
46  cat = conc_cat * np.ones((rows, cols))
47
48  # Get the concentration of HCl in each well
49  # We will do this by multiplying two 1D arrays to make a 2D array
50  # First, define the concentration of HCl in each row
51  MHCl_row = np.linspace(conc_HCl_start, conc_HCl_end, cols)
52
53  # Next, each row should be the same, so make an array of all ones
54  MHCl_col = np.ones(rows)
55
56  # Finally, we can outer multiply these two arrays to make a 2D array
57  MHCl = np.outer(MHCl_col, MHCl_row)
58
59  # Now we need to get the ionic strengths in each well by a similar method
60  # Here, the columns are all the same instead of the rows
61  # First, make a row that is all ones
62  ionic_row = np.ones(cols)
63
64  # Next, make an array that represents one column
65  ionic_col = np.linspace(I_start, I_end, rows)
66
67  # Finally, outer multiply to get the 2D array
68  ionic = np.outer(ionic_col, ionic_row)
69
70  # Calculate and print the volumes
71  calcPlateVols(cat, MHCl, ionic)
```

"Let's walk through this again, line by line."

1.2.0 LINE-BY-LINE EXPLANATION

1 and 2: These lines are **comment**s that explain what the **script** does, which is helpful to anyone else reading the code or to you if you forget in a few years. They also include a version number and date, so that you can look back at what you changed over the years. You have a version number for the current version (0.2.0) and one for the original (0.1.0). We increment the second version number

because this change is introducing a new functionality, but not breaking the old functionality.

3: We do not need to add a blank line here; however, from a readability standpoint, it can be a good idea. These blank lines help human being separate out the different parts of the code. We will sometimes use them and sometimes not, depending on exactly how the code is structured.

4: We need to use the **Numpy library**, and we can't assume that we have already imported it, so we will import it here, too, to be sure.

10–33: This is just the same code you developed in the last chapter, with no modifications. See the previous chapter for an explanation.

34–36: We want to be able to use a well plate with any number of rows and columns, so we can define **variables** to hold those numbers here. This means we only have to change them one place in the code, which will make our lives easier and make the code less prone to errors from forgetting to change the number of rows or columns every time we use them.

38–43: Here we are setting up variables for the concentrations and ionic strengths that we will be using in the experiment. Just like the last lines, setting them all here makes it so that we don't have to go through the code and change every value, which would take time and be prone to mistakes.

45 and 46: We are going to make a 2D **Numpy array** of catalyst concentrations here. We can make an array of all ones and multiply it by the catalyst concentration, which is a single value. Numpy will then take care of **broadcasting** the single value across the entire array, multiplying each element by the catalyst concentration and giving us an array where each element is the catalyst concentration that we want.

48–57: This is where we are going to calculate the 2D array representing [HCl]. The basic plan is to make two 1D arrays, one that says how the concentration should change across a row, and one that says how the concentration should change down a column. Then, we can take the "outer product" of the two arrays, which will produce an array where each element is the product of the corresponding elements of the 1D arrays. Here are the details:

51: We will use `numpy.linspace()` to create a six-element Numpy array that goes from the initial to final concentration of HCl. This is how our concentrations are changing.

54: Each row of the plate has the same range of concentrations in it, so we can just make a four-element array that is all ones and that will cause each row to be the same as the row array we just made.

57: Now we have a "trick" to make the 2D array; `numpy.outer()` performs an outer product between two vectors. The outer product between a vector of length four and a vector of length six is a 4 × 6 array. Let's say we have the 1D arrays `a = [a1, a2]` and `b = [b1, b2, b3]`. The outer product of these two vectors is the 2D array `[[a1*b1, a1*b2, a1*b3],[a2*b1, a2*b2, a2*b3]]`. The `numpy.outer()` function accepts two **arguments**: the first is the array whose length is the number of rows we want, and the second is the array whose length is the number of columns we want in our final array of arrays. So, if we pass it the individual column and row arrays that we just made, we get the 2D array that we want in one line.

59–68: Here, we are doing the same thing as for [HCl], and the only difference is that each row is different and each column is the same, rather than the other way around as above.

71: Now, all we have to do is call our function, and we get the printout of the volumes in the IDLE Shell. *Voila*!

"It was convenient to put the script into a file that we could edit, and IDLE had a capability for this, but there are many other programs that you can use that have even more helpful features. These are generally called IDEs (Integrated Development Environments) and I really like using Spyder. Like most things in Python, it is free to install and use, and you can install it the same way you installed Numpy."[9]

[9] See https://www.codingforchemistsbook.com/for detailed instructions.

Why don't we use Jupyter notebooks?

An alternative to using an IDE like Spyder is to use Jupyter notebooks. For sure, you will encounter people using this for solving data analysis workflows. However, there are a few reasons we are not going to use them in this book. First, it is possible to create very bad workflows in Jupyter where your analysis is not valid but you don't realize it. Second, it is challenging to access and code from Jupyter notebooks in external code, which limits the reusability of your code. Thus, while you may eventually decide to use Jupyter as you become more experienced, we think learning to code with an IDE gives you more flexibility and ensures better practices.

Why use Spyder as the IDE?
When reading about IDEs on the web, you will find that VSCode dominates most discussions and recommendations. The reason is that VSCode is free, easy to customize, and can be used with a variety of languages. However, we do not recommend using it for two reasons. First, you are learning Python, and so you don't need a more general tool. Second, Spyder provides some functionalities that are useful for data analysis and for people learning Python that are not present in VSCode. These include a variable explorer, that will give you information on the **objects** you are working with and their current states and a window that keeps a history of the plots and images you make.

Python interpreter vs. iPython
While IDLE uses the standard Python interpreter, Spyder and many other IDEs use an interpreter called IPython. IPython features expanded capability over the standard interpreter: it can run operating system command line commands natively, it can display images, it has helpful input/output line numbering, and autocompletion. If you get used to using IPython in an IDE, you may be surprised that some of the functionality that you have learned is not available when you have to run the standard interpreter (python.exe on Windows and python3 on a Mac).

CHAPTER 1 EXERCISES

TARGETED EXERCISES

Exercises for Section 1.1.0

0. In the IDLE or console of an IDE, import Numpy, and calculate the following. It may be helpful to know the following about Numpy: if you import Numpy as `np` then base-10 logarithms are accessed using `np.log10`, the value of π is accessed using `np.pi`, the value of e is accessed using `np.e`.

a) $\ln(1)$
b) $\ln(10)$
c) $\ln(e)$
d) $\log_{10}(1)$
e) $\log_{10}(10)$
f) $\log_{10}(e)$

1. In this chapter we introduced log operations, but Numpy introduces more mathematical operations. Using the Numpy documentation, a web search, or an AI tool, figure out how to calculate the following using Numpy:

   a) $-e^{\pi}$

   b) $\sqrt{\pi}$

   c) $\sin \frac{\pi}{3}$

   d) Greatest common denominator of 275 and 385

   e) Least common multiple of 12 and 27

2. Using Numpy, calculate the half-life of a material with a decay rate of 1.3×10^{-6} s.

Exercises for Section 1.1.1

3. Use a Numpy array to complete the following steps:

   a) Make an array that holds the following pH values: 1.1, 2.2, 3.3, ..., 9.9.

   b) Use this array to calculate the concentration of protons at each pH.

   c) Dilute all concentrations by a factor of 10.

   d) Convert the diluted concentrations back to pH.

4. You wish to make a simple syrup that contains alcohol. You mix 220 g of glucose with 220 g of ethanol and then add enough water to bring the final volume to 0.75 L. What is the concentration of the glucose and the ethanol? Use a Numpy array to complete this calculation.

5. Numpy has many other functions, which are built to work with Numpy arrays. For instance, `numpy.sum()` can give you the sum of all elements in an array. Use this knowledge to find out the total concentration of solutes in Exercise 4.

6. You are doing a variable-temperature nuclear magnetic resonance (NMR) experiment at eight equally-spaced temperatures from $-10°$C to $60°$C. You have two isomers that exchange and the ΔG of isomerization is 10 kJ/mol. For each species, calculate the concentration/relative population at each temperature.

Exercises for Section 1.1.2

7. Write code that, when run, generates an error of each type below. You might wish to consult the technical chapter on errors (Chapter H) or the internet.

   a) Errors in variable naming.

   b) Errors in making an array.

c) Adding vectors of different shape.

d) Using a function incorrectly.

e) Syntax errors.

f) Indentation error.

g) Forgotten colon (:).

Exercises for Section 1.1.3

8. Make a 3×3 Numpy 2D array, in which each index holds a different number from 1 to 9:

a) Make the numbers count up by one as you go across the "rows."

b) Make the numbers count up by one as you go down the "columns."

9. Imagine you had four solutions that had concentrations of 1.0 M, 0.70 M, 0.55 M, and 0.13 M.

a) Make a Numpy array that holds these concentrations.

b) Create a 2D array that holds three sets of arrays that would represent serial dilution of these concentrations by a factor of 3, performed twice.

c) Subtract the array you just made from a 2D array that represents 12 solutions all of 1.00 M concentration.

Exercises for Section 1.1.4

10. Make a 15 element 2D Numpy array where all elements have a value of 0.33.

a) Make this array, using `numpy.ones()`

b) Make this array, using `numpy.ones_like()`

c) Make this array, using `numpy.zeros()`

d) Make this array, using `numpy.zeros_like()`

Exercises for Section 1.1.5

11. Create the following Numpy arrays. You cannot just type out the values to create the array, you must use at least `numpy.linspace()` or `numpy.arange()` in your solution:

a) An array from 0 to 9, with 10 equally spaced elements, using `numpy.arange()`.

b) An array from 0 to 9, with 10 equally spaced elements, using `numpy.linspace()`.

c) An array spanning 0 to 10, including 10, with 10 elements.

d) An array spanning 0 to 10, including 10, with each element differing by 0.5.

e) An array from 0 to 2π, not including 2π with eight elements.

f) Make an array that, when squared, has elements 0 to 10, equally spaced.

g) Using `numpy.linspace( )` and `numpy.sin( )`, make a Numpy array that has the values $0, \sqrt{2}, 1, \sqrt{2}, 0, -\sqrt{2}, -1, -\sqrt{2}$.

h) An array spanning 1 to 10^{14} (including 10^{14}), where each element is 10 times larger than the element before it.

Exercises for Section 1.1.6

12. You have created a series of molecules that you think will serve as pH sensors, and you need to calibrate their responses to changes in pH. Write a code that will specify how much of a 6 N HCl solution to use for 11 standards, each 10 mL in volume and which area equally spaced between two pH that the user of the code can specify.

13. Draw a flow chart for the function produced at the end of Chapter 0.

14. Create a flowchart for how you might solve an ICE table in general chemistry course.

Exercises for Section 1.1.7

15. Create a file containing the solution to Exercise 2 and save it to a directory on your computer.

Exercises for Section 1.1.8

16. You wrote a Python script. You start at version 0.1.0 and go through the changes described below, in turn. Next to each entry, provide the version number that should be assigned to your script *after* the changes.

a) You add a new **keyword argument** to a function, with a default value that was originally hard coded.

b) You notice a typo in a comment, and fix it

c) You notice a typo in a function that breaks the function, and fix it.

d) You add a new **positional argument** (that does not have a default value) to a function.

e) You notice an mathematical error in a formula, and correct it.

f) You decide to reorder the structure of your code.

h) You have your code output additional text contextualizing results that are printed to the console.

COMPREHENSIVE EXERCISES

17. Take the final code from this chapter and adapt it for a 384-well plate. You should only need to change two numbers to do this. Try to ensure that the printed result is easy to follow.

18. You find that your pipette is only accurate to the nearest 0.1 μL. Modify the final code in this chapter so that it prints out volumes rounded to the nearest 0.1 μL. Looking at your answer, could you carry out this experiment if you had a pipet that was only accurate to 1 μL? How do you know?

19. Your advisor decides that they want you to add a small amount of another solvent to your reaction mixture. Modify the final code in the chapter to allow for the same arbitrary amount of that solvent to be added to each well.

CHAPTER 2

AUTOMATING THE PLOTTING OF DATA

- ☐ **New concepts**
  — *and their prerequisites*

- ☐ **Exploiting built-in methods to manipulate data stored in objects**
- ☐ **Importing subsets of libraries using `from`**
  — *Section 1.1.0: Giving your computer instructions using Python commands*
- ☐ **Performing multiple assignment operations in one line by unpacking**
  — *Section 0.1.4: Temporarily storing data using variables*
- ☐ **Representing answers to yes/no questions using booleans**
- ☐ **Describing the location of files or folders on your computer using Pathlib**
  — *Section 0.1.6: Expressing data as readable text using strings*
- ☐ **Making paths convenient and portable by specifying relative paths**
  — *Section 0.1.6: Expressing data as readable text using strings*
- ☐ **Adding `codechembook` to your Python installation**
- ☐ **Locating files and folders graphically with `codechembook.quickTools.quickOpenFilename()`**
- ☐ **Accessing data in structured text files using Numpy**
  — *Section 1.1.1: Operating on collections of data efficiently with Numpy arrays*
- ☐ **Making appealing and interactive plots using Plotly**
- ☐ **Outlining your solution before coding using comments**
  — *Section 0.1.3: Clarifying how your code works with comments*

2.0 PROBLEM

You have completed the experiment outlined in the prior chapters and have titration curves. You have identified the ionic strength that you wish to work at moving forward. However, you would like to verify that the titration involves only two chemical species: the protonated and deprotonated form of the catalyst. You are lucky that your catalyst has a nice ultraviolet-visible (UV-vis) spectrum, and so you can follow the titration using UV-vis and look for an isosbsetic point. So, you run another experiment in which you obtain the UV-vis spectra at different pH values.

You now have 20 different UV-vis spectra stored on your computer as .csv files. Your advisor has asked that you send them a plot of the *final* spectrum, just so they can verify that the end point makes sense before you work up the rest of the data. Although it is nice they are taking an interest in your training, and you appreciate your advisor wants to verify you are not wasting your time working up nonsensical data, it is also true that your advisor is particular about the appearance of plots. It always takes you some time to get Plots formatted correctly in Microsoft Excel, which seems silly for what should be a simple task. In addition, your advisor likes to have images as png files, but you like to have your own copy as a **scalable vector graphic** image (svg) because you like the fact you can rescale them however you like without a loss in resolution. This disagreement in file preference means you have to run through the image export file twice for each image.

You mention these frustrations to your labmate, and they just say, "Oh I just have a Python script that automatically generates the plot and then saves it as both a svg and a png, formatted correctly. I would be happy to help you make a similar one. It might take a bit longer to learn this than to make a single plot in Excel, but once you have this script, you can make figures in just a few seconds—if not less."

You agree, and then sit down to get started.

2.1 SOLUTION

"You have data," your labmate says. "So, you've already done the hard part! Now you simply need to get this data into Python—what we would call 'loading' or 'reading' the data. Once you have done so, you can plot that data, and then save an example of that plot—what we would call 'outputting' the plot. This input/output is what people call I/O, and we will do it using files on your computer, so people call that 'file I/O.' "

2.1.0 PLANNING

Thinking at a more granular level, you can see that the desired outcome can be broken up into five major steps:

1. *Identify a file that stores the data to be plotted.*
In this case, the file is a **csv** file, which stands for "comma separated values." Essentially, the file contains a series of rows, each with a series of values. Each value is separated by a comma. If you open your file in a text editor,[1] like Notepad (Windows) or Textedit (MacOS), you will see something like what is shown in Figure 2.1.

[1] It is important to note that programs in the Microsoft Office Suite, like Word or Excel, will not render the .csv correctly, so it is best to use a text editor instead.

```
0.999.csv
nm, absorbance
190.0, 1.0322432686044443
191.0, 1.6820443819566664
192.0, 1.9082619467088888
193.0, 1.9882040200855553
194.0, 0.9984422247888888
195.0, 0.9987143425411112
196.0, 1.3122859960822222
197.0, 1.4969219340455557
198.0, 1.1285519645844446
199.0, 2.0405319004255555
200.0, 2.109402121603333
201.0, 2.1053568877788895
202.0, 1.6340021448166664
203.0, 1.68599361894
204.0, 1.7108926628399999
205.0, 1.2511672529266666
```

FIGURE 2.1. Screenshot of a .csv file opened up in a text editor. The .csv file has data separated by commas. This is a common means by which to store data, and in this chapter, you will learn how to read such files in Python.

You can see that each row has two values separated by a comma. You can also see that the top row has text that function as labels for each column—in this case nm (the unit for wavelengths) and absorbance. We call such text rows "**headers**." This is information that is not the data, but intended to help the person looking at the file interpret and understand the data. They can be very useful for documenting what the data actually represents.

We will point to the location of this file on your computer using its **path**, which is the location on your computer. Your labmate points out that there are different **file systems** for different operating systems, but that the solution you will use will work with any of these systems, so we don't really need to worry about it too much.

2. *Read the file and store the data in* ***memory***
Ideally, you would want the data as **Numpy arrays** because you already know how to use these to hold multiple values. This means making one array that holds the x-values (wavelengths) and one array that holds the y-values (intensity) of the UV-vis spectrum.

3. *Construct the plot*
You need to turn the data arrays into a graphical representation of their values. In this case, you would like a line plot because that is the standard choice for UV-vis

spectra. You want to be able to control the presentation of the spectrum, such as the range of data shown and the general appearance of the figure.

4. *Display the figure*
Once you have the figure constructed, you need display it. This step is ultimately optional because you can save the plot without showing it. However, it is nice to have Python display your data so that you can quickly see what it looks like, rather than needing to open up the saved file.

5. *Output the figure to image files*
By saving files, you have something you can send to your advisor or something you can use yourself in other applications. Because you have different preferences, you need to save both .png and .svg files.

You also sketch out a flowchart (Figure 2.2) of what you want to do:

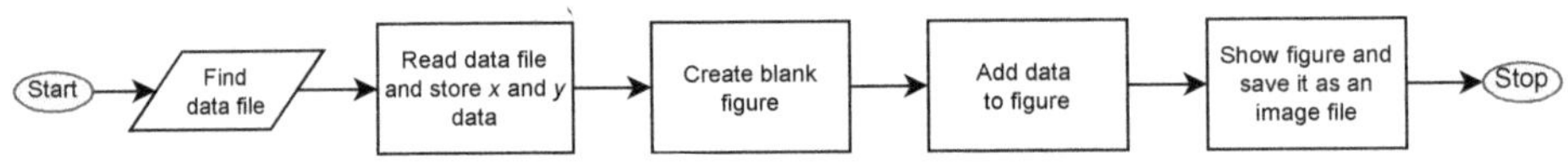

FIGURE 2.2. Flowchart of the solution to plotting a single spectrum, with data read from a .csv file.

Your labmate says that there are Python **libraries** that handle navigating file structure on your computer, libraries that handle importing data, and libraries that handle the plotting and outputting the plots. Although there are many choices for each of these tasks, we are going to use the following:

1: *Navigate the file structure using the Pathlib library*
The reason for using this library is that it does a great job **parsing** and manipulating the components of a file path. That is, it is aware of what folders and filenames are, among other things. This will make it much easier to work with our computer's file structure in Python. Additionally, this library can be used to produce code that automatically works the same on different operating systems with different file systems, like Windows versus MacOS.

2: *Input the data using the* `numpy.genfromtxt()` **function**
We have already used the Numpy library when making Numpy arrays by specifying them directly. However, the library also provides functions for creating Numpy arrays from data read from files and `genfromtxt()` is one such function.

3–5: *Making, displaying, and saving a data plot using the* ***Plotly*** *library*
There are a large number of libraries that handle making, display, and saves of data visualizations in Python, but this one has three advantages for our work. First, it automatically makes interactive plots. Second, the way it composes .svg files is

convenient for later modifications using software (such as PowerPoint, Illustrator, or Inkscape). Third, the way that Plotly makes, combines, and provides access to plot **objects** is very convenient once you understand how to use it.

However, before we can use all of these correctly, there are a few new concepts that you need to know.

2.1.1 IMPORTING SUBSETS OF LIBRARIES USING from

"Some **libraries** are huge, and we don't need or want to import them all. Luckily, we don't have to. Libraries themselves can be broken up into parts. In Python, there are three major components of libraries: **modules**, **classes**, and **functions**, and each can be imported separately when you don't want to import the whole library. We already talked a little about classes and functions, but modules are really just collections of classes and functions that are topically related. We can access modules directly using the syntax <library>.<module>. Modules can also have submodules, which are accessed by <library>.<module>.<submodule> and so on... Similarly, we can access specific classes and functions using <library>.<class> and <library>.<function>, respectively. When a library is very complex, this can lead to rather lengthy constructions to access a particular component."

"But, if you only want to import a component of a library, you can use a different type of import statement that goes like from <library> import <component>, where <component> is the name of the module, class, or function that you want to use. You can keep going, too, with things like from <library>.<component> import <subcomponent>."

"Not only does this save your computer the overhead of having to load the whole library, but it makes your code easier to write and read. Suppose we wanted to use a function that was defined in a submodule of a module of a library. If we wrote:"

```
In: import <library> as <alias>
```

"then every time we wanted to use the function, we would have to type:"

```
In: <alias>.<module>.<submodule>.<function>()
```

"You can image that this would be slow to write and difficult to read. Instead, we could just write:"

```
In: from <library>.<module>.<submodule> import <function>
```

"Now, when we want to use the function, we can just write:"

```
In: <function>()
```

"This is so much nicer! And we will use this kind of importing twice in the code we will produce later on."

2.1.2 PERFORMING MULTIPLE ASSIGNMENT OPERATIONS IN ONE LINE BY UNPACKING

"We have been using the assignment operator (=) in an intuitive way: there has been a single **variable** name on the left and a single **expression** on the right. However, Python allows more complex usage of the assignment **operator**. For instance, you can have more than one variable on the left, as long as there is a matching number of values on the right:"

```
In: one, two = 1, 2
```

"This will create a variable one with the value 1, and a variable two with the value 2."

"You can also take a single object that holds multiple values and pass it to the same number of variables, such as:"

```
In: three, four = np.array([3, 4])
```

"This will create two new variables: three with a value of 3, and four with a value of 4. These behaviors are examples of **unpacking**."

2.1.3 REPRESENTING ANSWERS TO YES/NO QUESTIONS USING BOOLEANS

"So far, we have seen a few different **data types**, such as **strings** or **Numpy arrays**. We are going to need to learn another now: **boolean**. This is a data type with only two possible values: `True` and `False`, which functionally behave a lot like yes and no. These values are often used to tell Python when you want something done or not, or to test whether two things are equal. In Python, capitalization matters, so `TRUE` and `true` are not valid values for booleans. The first letter (and only the first letter) of these values *must* be capitalized."

2.1.4 EXPLOITING BUILT-IN METHODS TO MANIPULATE DATA STORED IN OBJECTS

"One of the nice things about Python is that pretty much everything (**variables**, **functions**, etc.) is an **object**, which makes it so that you have to write much less code yourself. This is because the object contains not just the data that you assign to it, but also a bunch of code that other people wrote that helps you transform the data into something more useful. This means that the object lets you use bunch of complex code without having to write it!"

"Consider an analogy: you could build your own computer. You could figure out what parts you need, make sure that they're all compatible, buy them, put them together, install an operating system, install software, update everything, and you'd have a working computer but you would have also spent a lot of time on it. Alternatively, you could just buy a computer from a company. You don't have to know what is inside the computer, you just have to turn it on and use it and not worry about all the work that went into making it."

"When we assigned a **Numpy array** to a variable, what happened is that the computer made an **instance** of the **class** ndarray (this class is the official **data type** of a Numpy array). Don't worry about what a class is for now; just know that it is the collection of code describing how an object works—in this case stored in the Numpy **library**. You will probably never have to even look at the code describing this class, much less understand in detail how it works. But that code describes two things you need to know about: **methods** and **attributes**. Methods are **functions** that are built into the object, have access to the data in the object, and act on that data. Attributes are variables within the object that provide information about the object."

"Let's just worry about methods for now."

"Methods are not much different than any other function in Python, like `print()`, except that they live within objects and can access data in the object without your intervention. Because a method is not usable without the object in which it resides, you use it a little differently than the functions we have used so far. Instead of `<function>()` like we've already seen, you call a method by first identifying the object containing it, and then giving the method **call**, separated by a period. This construction looks like: `<object>.<method>()`. That way, if we have multiple instances of the same object with different names, Python knows which one we want to operate on. It also means that we do not have to pass the object variable to the method as an argument because its identity is built in. Beyond this, **arguments** and **returns** work pretty similarly to any other function."

You open the Spyder IDE,[2] go to the console window, and type the following:

[2] This was introduced at the end of the last chapter.

```
In: import numpy as np
In: testArray = np.array([1, 5, 3])
```

"You now have created an ndarray **object**." Your labmate continues, "In the ndarray class definition, there are functions that can report statistics of your array: mean(), std(), sum() and so on. Normally, we might think that we need to use a construction like numpy.ndarray.sum(testArray) to access these functions. However, because our testArray is an instance of the ndarray class and contains these functions, we can just use testArray.sum()"

```
In: testArray.sum()
Out: 9
```

You can see that using the sum() method of the Numpy array class returned the sum of all the elements in the Numpy array... all without you having to know how the method worked.

2.1.5 Describing the Location of Files or Folders on Your Computer Using Pathlib

"Files in your computer are organized in a hierarchy of containers intended to make it easy to logically group them and indicate their relationships. The typical representation of this hierarchy is that files are grouped into folders (also known as **directories**). Any folder can also be in another folder, and so on. Eventually, you have one primary folder that is not contained in any other folder, but contains all other folders, subfolders, sub-subfolders, and so on. This primary folder is called the **root** folder, and is typically represented by `C:` on Windows computers and by `/` on Macs and Linux computers."

"We might describe the location of a file in the hierarchy as `<root><folder1><folder2><folder3><file>`. We call this the **absolute path**. The collection of files and folders is called the **file system** and the exact organization (names and locations of the files and folders) that it employs is called the **file structure**."[3]

[3] See Chapter D for more details.

> ****KNOWLEDGE CHECK**: Do you understand the computer's file system?**
> It is *absolutely critical* that you understand file structures when coding. You will be unable to write effective code if you do not know (or learn how) to navigate a file structure. Although we provide tools to try to help make this easy, you cannot avoid it all together. If *any* the following statements describes you, we *strongly* recommend you go read the technical chapter on files (Chapter D) or refer to the tutorials posted on the companion website: https://www.codingforchemistsbook.com/.
>
> - You do not know what a folder on a computer is.
> - You do not know what a file is on a computer is.
> - You do not know what a file extension is.
> - You do not know what a path is.
> - You cannot tell us what the path to your documents folder is.
> - You do not know how to determine the number of folders that lie between your documents folder and downloads folder.
> - You do not know what a directory is.
>
> We also want to stress that it is ok if you don't *yet* know these things—we can help you learn. But you *will* need to learn them.

"Python has a built in **library** for exploring and creating paths called **Pathlib**. It provides a function `Path( )`, which accepts a **string argument** that specifies a path on your computer and converts it into an **object** that Python can work with (a Pathlib Path object)—recognizing things like folder, files, extensions, etc. More details can be found in Chapter E. However, for the solution to our workflow problem we need to understand how to change the **extension** of a file using the `with_suffix( )` **method** of `pathlib`. Now, extensions are the way in which computers know how to treat files, and they are short codes that follow the filename. For instance, we have been considering **.csv** files. The 'csv' that follows the dot is the extension. Likewise, if you have a modern excel file, you might notice that it ends in .xlsx, which is the extension associated with Microsoft Excel. Microsoft Word documents use the extension .docx. In Pathlib, extensions are termed the suffix, and the `with_suffix()` method allows us to control the suffix (extension) associated with a file."

> **Why Pathlib?**
> The "original" way to handle files and file structure in Python was using a **library** called `os`, and which is still included in Python (same as Pathlib). However, Pathlib adds many additional capabilities for handling **file structure** that make it worth using instead. For instance, Pathlib is aware of what a file **extension** is, as well as the file hierarchy. For this reason, we will only use `os` when it can perform a function that Pathlib cannot.

"We can see this behavior by creating a path object, with extension, and then changing this extension."

```
In: from pathlib import Path
In: new_path = Path('C:/Users/me/Documents/Research/fakeFile.csv')
```

"You see that this is a Windows-style path with C: as the root, but we used forward slashes instead of backslashes. This is because the backslash character in strings has special significance: remember that `\n` is the newline character, and Python knows that it isn't just the letter n because of the backslash. So, pathlib knows about this, and can accept forward slashes instead, or if you prefer, two backslashes in a row to signify a backslash. If we then enter this value, it will return a path object:"

```
In: new_path
Out: WindowsPath('C:/Users/me/Documents/Research/fakeFile.csv')
```

"An important thing to remember is that the path is *not* the file or folder. Creating a path does not create a file or folder. And just because you write a path doesn't mean there is a corresponding file or folder. The path is like directions to a location: it will get you to a place, but there may be nothing there when you arrive. You can also write directions to fictitious places—a fact regularly exploited by authors and screenwriters. Paths just specify a location in your file structure. Also, if you are on a Mac, then 'WindowsPath' is replaced by 'PosixPath.' Both are indicating that this is a path object."

"We can then take this path and change the extension, using the `with_suffix()` method:"

```
In: new_path.with_suffix('docx')
Out: WindowsPath('C:/Users/me/Documents/Research/fakeFile.docx')
```

"Here you can see everything about the path is the same, but it ends with docx rather than csv. Now there are two very important things to note about what just happened:

1. Just because a file has an extension, doesn't mean that a program can open it. If you have a random file that ends in docx and try to open it, your computer will try to

open it in Microsoft Word (because the extension is telling it to), but unless the file is formatted correctly, Microsoft Word will fail to make sense of it (or might not even open it).

2. The `with_suffix()` is a method of a path object. You cannot just use it on anything you want. For instance, if you try to use this on a string, you will get an error:"

```
In:  'C:/Users/me/Documents/Research/fakeFile.csv'.with_suffix("docx")
In:
Out: AttributeError: 'str' object has no attribute 'with_suffix'
Out:
```

"This error message is telling you that strings do not have this capability."

2.1.6 Making Paths Convenient and Portable by Specifying Relative Paths

"Paths can get very long and annoying, because your computer can have a lot of folders," your labmate says. "Luckily, there's a shortcut that makes it much easier. You can specify a different folder in which to read or write files one of two ways. First, you can explicitly type out the full path to where you want it saved—often called the **absolute path** You can also type out the path *relative* to the **working directory**. The working directory is the place where the computer is currently active. You can think of it like this: if you were to create a file without specifying any location, the place it would be created is the working directory."

"Specifying the location of a folder or file relative to another is known as a **relative path** and it uses two symbols: a single period (`.`) and a double period (`..`). The single period symbol represents the working directory and the double period indicates the directory containing the working directory. Imagine that our current working directory was `<folder2>` above. Then, `./<folder3><file>` would refer to the file of interest, while `..` would refer to `<folder1>`.[4] If you already have absolute paths, then **Pathlib** makes it easy to automatically find the relative path between them. For example, let's say we have a Python script in directory **A** on our computer, and a data file in directory **B**. We can create absolute paths from **A** to **B** by using the `.relative_to()` **method** of the Path **class** like this:"

[4] You can read more about relative paths in Chapter D.

```
In: relative_path = pathB.relative_to(pathA, walk_up = True)
```

"This yields a new `Path` object that points from your Python script to the data file, which you can use in your code to avoid having to type out the full path. The `walk_up` keyword allows Python to navigate your computer's file structure more completely.[5] Note that `pathA` and `pathB` must both be absolute paths for this to work! You can use the `.resolve()` method to get an absolute path from a relative path:"

[5] Basically, the default behavior is to explore only up to the "base" directory given (i.e., a place to start), but setting `walk_up = True`, Python can go beyond that folder.

```
In: abs_path = relative_path.resolve()
```

"In this case, we are just recreating the absolute path to the data file. A possible usage of `.resolve()` is as an alternate way to get an absolute path to the current working directory or one close by: `path_above = Path('..').resolve()` gets the absolute path of the folder containing the folder we are currently working in."

2.1.7 ADDING codechembook TO YOUR PYTHON INSTALLATION

"Now, it is important to understand how the **file system** works, and how to specify locations in the file system by writing **paths**," your labmate says, adding, but it can also be tedious at times. We are a bit more familiar with finding files using our computers file explorer. Because of this, I wrote a bit of code that I tend to use when I simply want to identify a file to open. This saves me time and effort when trying to specify a specific file location. I put this code in my own **library** named **codechembook**, so that I can call the functions in that library just like any other program. Because I thought others might find some use from it, I have also made it so that you can install it on your computer.[6]

"Nothing we are going to do will *require* that you use this library, but I think you fill find it adds a great deal of functionality for working scientists, while also saving time and reducing frustration—and isn't that why you are learning to code in the first place?"

[6] You learned how to do this in the last chapter, but if you need help, you can refer to our website (https://www.codingforchemists-book.com/) for detailed instructions.

Who, or what, is codechembook for?
Codechembook is a Python library written and maintained by the authors of this book. It originally started as a set of shared tools we used to help us in our own research workflows. For instance, we have modules that automatically format data visualizations for common chemistry journals. When writing this book, we thought we could share these tools with a wider audience. Additionally, we wrote a few tools specifically for this book, which help navigate some of the input/output tasks that might otherwise be frustrating to people new to coding. It is important to note that using codechembook is not strictly required to do any of the work in this book—we have ensured that we provide sufficient information to accomplish any of the tasks without it. However, we also hope you will find this library useful in accomplishing the work in this book, and beyond!

2.1.8 LOCATING FILES AND FOLDERS GRAPHICALLY WITH codechembook.quickTools.quickOpenFilename()

"Now that we have **codechembook** installed, we can access its functionality. In this case, we are going to use a function that I created to identify a single file to open: `quickOpenFilename()`. We can see how this works using the following:"

```
In: from codechembook.quickTools import quickOpenFilename
In: quickOpenFilename()
```

"This will open up a native file explorer that you can use to select any file on your computer. What it will **return** is a **path object** of the **Pathlib library**, which contains the path to the file you selected. Basically we are getting the filename for the file we wish to open. We can take this output and assign to a variable for later use, or pass it to another function, or anything else we do with the output of an **expression** in Python. Also, because this function returns a Pathlib path object, we can work with it using the Pathlib library. For instance, we can use the `with_suffix()` method on this object."

"Another thing to be aware of... even though `quickOpenFilename()` returns a pathlib object does not mean that the script automatically has access to pathlib.Path. All this means is that you will get back a Path object. You can operate on that object using Path methods because the instructions for how to do so are contained in the object. But if you want to use funtions from pathlib.Path, elsewhere in the script, you will still need import it as normal."

"At any rate, the `quickOpenFilename()` function is a way to get the path to a *single* file on your computer. The next time we meet, I can show you how to get the path to *multiple* files at once, but this is sufficient to solve our current workflow problem."

"One more thing! If you click cancel in the file dialog, `quickOpenFilename()` will return the value `None`, which is a special data type that represents literally nothing.[7] Your codes will mostly be expecting strings or Path objects, so getting `None` will probably cause an error and the **Python interpreter** will stop executing your code. For now, we'll just live with this possibility, but later we'll talk about how to handle these situations gracefully."[8]

[7] For more information, see Chapter B.

[8] See Section 7.1.4.

2.1.9 Accessing Data in Structured Text Files Using Numpy

"Now, we need to talk about reading data files," your labmate says. "Structured text files are a convenient way to store human-readable numerical data. You are already familiar with the generic term "**csv** file." This is a very common structured text data file format, in which data (often, but not exclusively, numerical) is presented in rows and columns. Strictly speaking, for comma-separated value files, each column is **delimited** by commas. Delimiting is necessary because computers cannot easily tell the difference between numbers, white space, and other characters, and thus do not necessarily know when one value ends and the next one begins. With a comma **delimiter**, a program can determine that one value should exist between each pair of commas (as well as between the beginning of a line and the first comma, and between the last comma and the end of a line), which makes it much easier to parse the data in the file. Other common delimiters are tab characters `'\t'` and spaces."

"Because this manner of representing is so robust, standard **functions** can be written that automatically load data from structured text files such as a .csv file into **variables** in your code. One very convenient one is built into **Numpy**, the function `genfromtxt()`. This function has only one required **positional argument**, the file path (as a string), but you will generally want to pass several additional optional **keyword arguments**." These are:

- `delimiter`: the delimiter character (as a string). Common values are `','` for commas, `' '` for spaces, and `'\t'` for tabs.

- `skip_header`: the number of rows to skip before reading in values. Often, these files have one or more rows with information in them other than the values that you want to load into arrays. For instance, the headings of each row (wavelength, absorbance for a UV-vis spectrum). In this case, you would want to skip one row because you can't load text into your x and y data arrays. You can optionally attempt to capture these headings into separate variables using the `names` keyword. Headers can sometimes contain several additional rows of metadata, so make sure you look at your data files before loading them.

- `unpack`: by default, this function returns all of the data as a 2D numpy array in which the outer array holds each row and the inner array holds the individual values in a row. It can save you some time to unpack the data and having multiple variables on the left-hand side of the assignment operator to accept each column separately. You can do this by setting `unpack = True`.

"Of course, `genfromtxt()` has a few useful behaviors and many additional options. Most importantly, it gracefully handles missing values from files, allowing you to specify what goes in their place. More can be found using Python's `help()` function."

2.1.10 Making Appealing and Interactive Plots Using Plotly

"Now, we are ready to talk about plotting! There are an abundance of **libraries** in Python that can be used to plot data. We are going to be using the **Plotly** library. In Plotly, the instructions for data visualizations are held in a **figure object**. These Figure objects have two basic components: traces and layout. The traces contain information that includes the data points you are plotting, the type of representation that you wish to use (e.g., scatter, bar, etc.), how these representations should be displayed (e.g., color, shape, etc.), and where the representation should be drawn (e.g., if there are multiple subplots, which one to add the representation to). The layout object contains information about how the axes are formatted, the color of the plot background, any annotations you wish to include, and so on. We will explain elements of both the traces and layout components of the Figure object as we need them, but for now, it is enough to understand that the basic usage of Plotly is to create a Figure object, specify the representations you wish to include, and then format the elements of your plot."

"Finally, you can display this plot in a few different ways, but we will largely either be showing static images in the Spyder IDE or making interactive plots that open in web browsers. The interactive functionality is a particular strength of Plotly, and it provides a easy means for exploration of your data. All of these steps will be shown in the code at the end of the chapter."[9]

[9] Many details on how to understand and use Plotly can be found in the technical chapter devoted to this library (Chapter I). At this point, it is probably too early to read this chapter in detail, but it is worth keeping in mind that this resource exists.

"Okay, we are now in a position to create the code that solves our workflow problem!"

Why not use Matplotlib?
If you have any prior experience with working up data in Python, you have likely encountered Matplotlib before. If you are new to Python, you will very quickly encounter Matplotlib in examples and discussions online. It is probably the most popular plotting library in Python. However, we prefer Plotly over Matplotlib for five reasons: (1) Plotly creates prettier images, (2) Plotly creates better interactive plots by default, (3) interactive plots in Plotly are easily sharable with others, (4) Plotly provides better access to the elements of the figure, and (5) Plotly figures persist and are accessible even after showing them. These last two points means that Plotly is easier to use in the **object-oriented** approach to coding that we use in this book. Although there is nothing wrong with using Matplotlib, we honestly believe that you will be better off in the long run using Plotly.

Why not use Plotly express?
As you learn to use Plotly, you will encounter Plotly express in many examples. We do not use it here, because we are interested in presenting you with a general tool. Plotly express is good for making graphs that have single plotting areas, but as soon as you want to have multiple subplots (which we will do as soon as we start fitting data), you will find that Plotly express makes this harder than our approach. Our approach, although not as well represented on the internet, is just as simple as Plotly express, while being more flexible.

2.1.11 Outlining Your Solution Before Coding Using Comments

You sit down and open up the Spyder IDE to start coding. Your labmate points out that it is going to be very important to **document** your code with **comment**s, and so you might as well plan things out using them. They suggest that you can use the outline you produced at the very start as a basic map. Doing so, you arrive at the following initial structure:

```
'''
V0.1 A program to plot uv-vis spectra from csv
by: you
'''
# specify the name and place for data

```

```
# import data to plot as a numpy array

# construct the plot

# display the plot
```

Now, all that remains is to provide the code that carries out what your comments describe. You pass your laptop over to your labmate, who produces the following code:

2.2 CODE

```
'''
A program to plot one uv-vis spectrum from a .csv file
    Requires: a .csv file with col 1 as wavelength and col 2 as intensity
    Written by: Ben Lear and Chris Johnson (authors@codingforchemistsbook.com)
    v1.0.0 - 250131 - initial version
'''
import numpy as np # needed for genfromtxt()
from plotly.subplots import make_subplots # needed to plot
from codechembook.quickTools import quickOpenFilename

# Specify the name and place for data
data_file = quickOpenFilename()

# Import wavelength (nm) and absorbance to plot as a numpy array
x_data, y_data = np.genfromtxt(data_file,
    delimiter = ',',
    skip_header = 1,
    unpack = True)

# Construct the plot - here UVvis holds the figure object
UVvis = make_subplots() # make a figure object
UVvis.add_scatter(x = x_data, y = y_data, # make a scatter trace object
    mode = 'lines', # this ensures that we will only get lines and not markers
    showlegend = False) # this prevents a legend from being automatically created

# Format the figure
UVvis.update_yaxes(title = 'absorbance')
UVvis.update_xaxes(title = 'wavelength /nm', range = [270, 1100])
UVvis.update_layout(template = 'simple_white') # set the details for the appearance

# Display the figure
UVvis.show('browser+png') # show an interactive plot and in the spyder Plots window

# Save the spectra using the input file name but replacing .csv with the image file form
UVvis.write_image(data_file.with_suffix('.svg')) # save in the same location as the data
    file
UVvis.write_image(data_file.with_suffix('.png')) # save in the same location as the data
    file
```

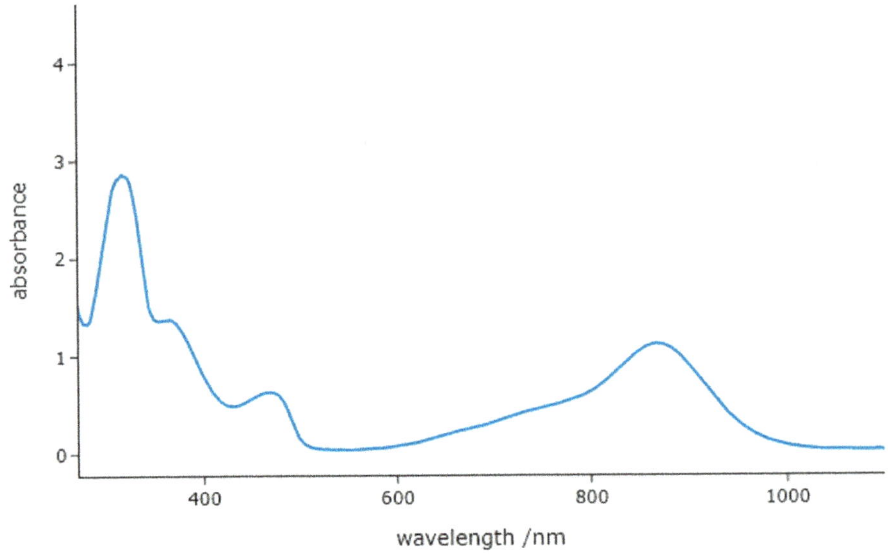

FIGURE 2.3. Plot of a UV-visible spectrum, as generated using the Plotly library and data read from the file 0.999.csv. The code in this chapter will allow you to make such plots automatically by pointing to the file of interest. The raw data and code for it can be found on https://www.codingforchemistsbook.com/.

You run the script in Spyder, and a file dialog box opens. You navigate to the folder "Research/Catalyst/Titration/UVvis" to select the file 0.999.csv, and after about a second, most of which is due to loading the Plotly library,[10] you soon see a figure that looks something like that shown in Figure 2.3.

The same basic figure opens in a browser window, is displayed in the "plots" tab of the Spyder IDE, and is saved as both 0.999.png and 0.999.svg in the same folder as the csv file.[11] The image displayed above was taken from the Spyder plot window. The formatting of the saved images—especially the number of tick marks and size of the font—might look different, but we will learn how to specify the exact format the figures as we move through the book. You can also read the technical Chapter I for more information.

[10] This delay will only happen the first time you run this code in a session. Once the library has been loaded once, it won't need to happen again unless you restart the session.

[11] If you are confused on how you might open up a .svg image, most internet browsers (i.e., Microsoft Edge, Google Chrome, etc.) will display them. Also, programs like Adobe Illustrator, Affinity Designer, or Inkscape can both open and edit them.

2.2.0 LINE-BY-LINE EXPLANATION

1–6: We start the file with a multi-line comment that states the purpose of the code, requirements for using the code, who wrote the code (and how to contact them), and then information on the versions. This is sometimes known as a "docstring" and is used to help others understand what is going on, and (perhaps more importantly) help you remember what you were doing if you ever return to the script. The version numbers and date, in particular, are very excellent things to have.

7: We import `numpy`. We will use a function from this **library** in order to read our **.csv** file. Because we don't want to type "numpy" all the time, we assign it the **alias** "np." This allows us to refer to `numpy` using simply `np`.

8: We want to use the `make_subplots()` function from `plotly.subplots` **module**, but we don't need the entire Plotly library. We write from <library>.<module> import <part> to get only what we want. The `make_subplots()` function creates an empty **instance** of a Plotly figure **object**. Once we have this object, we can add data to it and format it as we like.

9: We import a specific function in the `codechembook` library, in this case the function `quickOpenFilename()` from the `quickTools` module. This function will open up a native file explorer, in which the user can select a file of interest, and then the function will return a pathlib Path object that points to that file.

12: Here we get the path to the file of interest, using the `quickOpenFilename()` function, which returns a pathlib Path object. This path is then stored in a **variable**. It is convenient to have a Path object because it is easier to parse the components of the path, such as getting folder, filename only, extension, and so on. This is a functionality we will use at the end of the script (lines 37 and 38). There are other reasons to use `Path` **objects**, which can be read about in technical Chapter E.

15–18: These lines call the `genfromtxt() function` from `numpy` to read the csv data file and generate Numpy arrays from the data in the file. The `genfromtxt()` function of `numpy` can return the data read in a row-by-row manner or a column-by-columns (we will use it by columns). `genfromtxt()` will return a Numpy array that contains a number of other Numpy arrays equal to the number of columns we tell it to read. There are two columns in the data file, so we will get back a Numpy array containing two `numpy arrays`. We **unpack** these two arrays into two variables on the left hand side of the assignment **operator** (=).
You will also notice that our usage of `genfromtxt()` covers multiple lines of code. When we call a function, we have to place `arguments` for that function between parentheses. In Python, we can put the `arguments` on different lines, as long as they are between parentheses. When functions have lots of `arguments`, we will do this in order to achieve greater clarity and readability.
To understand this call of `genfromtxt()`, we can look at the usage of each argument.

15: A file name as a **positional argument**, stored as a `Path` **object** in the variable `data_file`. The `genfromtxt()` function will read this file and pull out the values in it.

16: We have to let genfromtxt() know what character separates the data values. This separator is called a **delimiter** and, for our .csv file, it is a comma. We supply this separator as a **string** (',') to the keyword argument delimiter.

17: The data file we are using has **headers**. In this case, the first row are labels of the columns. The genfromtxt() function has no idea that the data doesn't start until the second row, but we can use the skip_header keyword to tell it how many rows to skip before treating the rest as data. Remember that Python starts counting at zero, so you can think of this number as either the number of rows to skip or the first row in which there is data.

18: Finally, the default function of genfromtxt() is to assign each row to a numpy array. For our data, which has 911 rows, this would result in 911 arrays, each storing two values (each row has two entries). However, we want to have the columns in our file each in an array. Using the keyword unpack and setting this equal to the boolean True will give us the data organized as two columns.
We end the function **call** with a closing parentheses. This closing parentheses needs to match the opening one (line 15) and indicates to Python that we have provided all the arguments we are going to provide for this function.

21: We use the make_subplots() function to create a Figure object. There are actually several ways to make a Figure object in Plotly, but this approach will create an object that can hold many plots if we like. For this reason, it is the most general approach. Here we also see that some functions (if they have no positional arguments) can be used *without* supplying an argument. When you do this, they will just perform some default action. In this case, make_subplots returns a blank Figure object assigned to the variable UVvis that holds a single subplot.

22–24: We use the add_scatter() method of the Plotly Figure object to create another object known as a "trace" (Plotly's term for a graphical representation of data) and add the data to the Figure object. We know this is a **method** of the Figure object because it follows the form: <Figure **object**>.add_scatter() This trace is an **instance** of the scatter plot **class**.

22: The add_scatter() method requires both x and y values that we specify using the associated **keyword argument**s x and y.

23: We use the keyword argument mode to tell Python that we want to plot using only lines (as opposed to using markers as well).

24: We use the keyword argument showlegend to tell Plotly we don't want this trace included in a legend, setting it equal to the boolean False. We use a closing parentheses to indicate we have completed the function call.

We have used just a few options for the `add_scatter()` method. A more complete description of the options can be found in the technical Chapter I.

27–29: We use methods of the Figure class to control the formatting (appearance) of the Plotly figure.

27: Using `update_yaxes()`, we make changes to *all* y-axes in our figure, though in this case, we have only a single y axis. Later in the book, we will see how to specify only some of the y-axes, when we have more. This method accepts a large number of keyword arguments, but here we only use one: `title`, which accepts a `string` that will be used as the title of the y-axes. The default behavior of the axes in Plotly is to show a range slightly larger than the full range of the data. So, we don't need to specify a range to show, if we are okay with seeing the full range.

28: Similar to the y-axes, this line formats the x-axes, this time using the `update_xaxes()` method. Here we specify the title, but also add the use of another keyword argument: `range`, which is used to specify the limits of the axes. Because we don't need to show all of the ultraviolet data shown, we want to only use part of the full range of the data. We specify the start and stopping point for the axes using two numbers, between square brackets, separated by commas.

29: We define the appearance of the figure by using the `update_layout()` method of the Figure class. Again there are many possible keyword arguments, but we only use one of them: `template`. We pass a string containing the name of a pre-defined template in Plotly: `simple_white`. This tells Plotly to apply the settings of this template. This contains information regarding the tick marks, font, font size, background color, and much more. We will eventually discuss how to create your own templates, but for now, this one will work well.

32: Using the Plotly Figure method `show()`, we can display our composed, formatted figure. We provide this function a positional argument that is a string with two words, joined by a plus sign. Separately, "browser" and "png" could be used with `show()`. If we only used "browser," then the figure would be opened in your computers default internet browser (e.g., Chrome, Firefox, etc.) and you could interact with it by zooming in or hovering over data. If you only used .png then (assuming you are using the Spyder **Integrated Development Environment (IDE)**) you would get a static image in the png format, displayed in your Spyder IDE plot window. By using both, and joining them with a + sign, we are telling Plotly to do both. This approach to combining multiple options is somewhat unique to Plotly, so don't think this is a general way to do this in Python.

Running this line will open a browser with an interactive plot, and you will get a static plot in the Spyder 'Plots' pane.
You may encounter an error here, and if you do, please refer to the Plotly technical Chapter I to trouble shoot it. The "browser" option should work for sure, but the .png or .svg may not at first.

35 & 36: Both of these lines use the `write_image( )` method to create an image file on your computer. The first line will create a .png image and the second will create a .svg image. In both cases, we provide a positional argument that is the name and path to the file we want to save. We are starting with the same path we created back on line 12, but are using a method of the Path object, `with_suffix( )`, to change only the **extension** of the file (*i.e.*, it otherwise has the same name). So, while the name was initially '0.999.csv', it is now '0.999.png' (line 32) and '0.999.svg' (line 36). We need to change the extension so that `write_image( )` knows what format to save the image file as, so that the data file is not accidentally overwritten, and because operating systems use extensions to specify what program to use to open files.

And that is it! You now have a script that you can use over and over again to plot UV-vis spectra. If you need a different data file, then you just select it. If you want a different x-range, you change that. But that is all you need do. Of course, this will only work for files that have two columns of data, but as we move forward, we will learn how to handle cases where this might change.

CHAPTER 2 EXERCISES

TARGETED EXERCISES

Exercises for Section 2.1.4

0. Make a Numpy array that has all integers from 14 to 1, including both 1 and 14. Assign this to the variable `pH`. Use **methods** of Numpy arrays to accomplish the following (you may need to read the online Numpy documentation):

- Calculate the mean.
- Calculate the variance of the array.
- Get the standard deviation of the array.
- Sort the array so that the values increase monotonically.

Exercises for Section 2.1.1

1. If you only want to do some math on individual values, then it may not make sense to import the whole **Numpy** library. Write import statements to only import the following functionality from Numpy. For example, if you wanted to import the function that is needed in order to create a Numpy array, and then make an array of the integers 1–10, assigning it to the variable x, you can use:

```
from numpy import array
x = array([1,2,3,4,5,6,7,8,9,10])
```

Using this pattern, accomplish the following:

a) Import the function to compute the sine of an angle, and use it to operate on x.

b) Import the function to compute the natural logarithm and use it to operate on x.

c) Import the function to compute the exponential and use it to operate on x.

d) Import the function to compute the square root and use it to operate on x.

Exercises for Section 2.1.2

2. Pick three people you know. Using a single assignment operation assign their favorite food to a variable that is their first name.

3. Define variables for the masses of hydrogen, carbon, nitrogen, and oxygen in a single line.

4. A function that is native to Python that we have not yet seen is `divmod()`. Using the information returned by Python `help()` function, learn how to write a single line of code that will use `divmod()` to solve the following problem: You have a solvent bottle with 289 mL of solvent. You have a series of experiments to run that each require 7 mL of solvent. How many experiments can you run, and how much solvent will be left over. These should be assigned to separate variables. Again, use only a single line of code.

Exercises for Section 2.1.3

5. Which of the following are valid values of booleans?

a) `True`

b) true

c) TRUE

d) Yes

e) yes

f) YES

g) 1

h) 0

i) Maybe

Exercises for Section 2.1.5

6. Write down a **statement** to create a variable `folder_name` that contains a Path object pointing to the following folders on your computer. Use the `print(<path objet>)` to print the absolute path for each:

   a) Your Downloads folder.

   b) Your Documents folder.

   c) The folder where you are storing files for this book. This *must* be different from both the Downloads and Documents folder.

Exercises for Section 2.1.6

7. For the following combinations of folders, print out the relative paths that point from the first folder to the second folder. Use the `.relative_to()` **method** of the Path class to find the relative path between them. You may need to use the `walk_up = True` **keyword argument**. When interpreting the results, realize that two periods in a row (`..`) indicates the folder one level up from the current folder:

   a) From Downloads to Documents.

   b) From Documents to Downloads.

   c) From the folder you are using for this book to your Downloads folder.

   d) From the folder you are using for this book to your Documents folder.

Exercises for Section 2.1.7

8. Verify that `import codechembook` will run. Then use `help(codechembook)`. How many submodules are there in this library?

Exercises for Section 2.1.8

9. In this exercise, you will learn how to use `quickOpenFilename()` to generate a relative path from one file or folder to another file or folder.

   a) Use `quickOpenFilename()` to select a file in your Documents folder. This will return an absolute path to the file. Assign this absolute path to a variable called `path1`.

   b) Do the same for a file in your Downloads folder, but assign it to the variable `path2`.

   c) Use the `.relative_to()` method of the `Path` class to get the relative path from `path2` to `path1`. Note you will want to use the **keyword argument** `walk_up` to allow full flexibility for relative paths.

10. Modern file dialogs have search features that make it so that you don't have to click through your whole file structure to find a file or folder and get its path. Use the search feature of the file dialog in `quickOpenFilename()` to search for the file 0.999.csv used in this chapter and select it to get the absolute path to it. Confirm that this path exists.

Exercises for Section 2.1.9

11. Use the following files supplied on https://www.codingforchemistsbook.com/ for this exercise. Read them into Python using `genfromtxt()`.

    a) Stress-Strain1.csv

    b) Stress-Strain1.tsv

    c) Stress-Strain1.ssv

12. Repeat Exercise 11, but only for the .csv file. Assign what is returned from `genfromtxt()` to a single variable. Then, using this variable, unpack the values into variables that correspond to the columns in the original data file.

13. Use `genfromtxt()` read the same .csv file you used in Exercise 11, but extract only the second column of data.

14. As you would have noticed, the first two rows in the .csv file used in Exercise 11 contained header information. Get the information for each row, using the `genfromtxt()` function and setting `unpack = False`. You may also want to know that there are two other keyword arguments that will be useful: `max_rows` and `dtype`. The last of these specifies the type of data you are trying to read, which in this case is a string (`str`).

Exercises for Section 2.1.10

15. Go to the Plotly website and find an example plot. Explain how you might use that type of plot in your own research.

16. Run `plotlyTest()` from `codechembook` after importing it using `from codechembook.plotlyTest import plotlyTest`. You can always use `help(codechembook)` to find this function.

17. The `plotlyTest()` function you ran in Exercise 16 above should have opened a figure in your computer's browser. One nice thing about Plotly is that these figures are interactive. Using your mouse, move your mouse cursor over the data shown. What happens? Then, click and drag your mouse to try select a region of data. What happens? Play around with this interactivity, and describe what else you find you can do.

Exercises for Section 2.1.11

18. Consider the Comprehensive exercises for this chapter (see below).

a) Create a text outline for the solution, formatted as Python comments.

b) Create a flow diagram for the solutions.

c) Write a list of instructions, like you would to a human being, for how to solve them.

COMPREHENSIVE EXERCISES

19. Generate Plotly plots of the following relationships:

a) e^{-x}

b) $1/x$

c) ν_{rms} vs T using a Maxwell–Boltzman distribution for N_2, over the temperature range of 200 K to 400 K

d) K_{EQ} as a function of T, for $\Delta H = 10$ kJ·mol^{-1} and $\Delta S = -5$ J·mol^{-1}·K^{-1} over the temperature range of 300 K to 400 K.

20. Consider the decay of fluorescence intensity (I) over time (t). This typically follows the functional form $I(t) = Ae^{-kt}$. Using Plotly and the values $A = 0.1$, $k = 2$:

a) Plot I versus t over the range $t = 0$ to $t = 10$.

b) Plot ln(I) versus t.

c) Plot I versus ln(t).

d) Plot ln(I) versus ln(t).

e) Which of the above is the most useful, and why do you think that?

21. The data file CO2-IR.txt found on the book's companion website (https://www.codingforchemistsbook.com/) contains an absorption-mode Fourier transform infrared spectroscopy (FTIR) spectrum of gaseous CO_2 with the x-axis in wavenumbers. Write a script to that outputs a plot of %transmission versus wavenumber. Remember that absorbance = 2 – $\log_{10}$(transmission). Make sure you use correct axis labels! Bonus: Make the x-axis run from the largest to smallest numbers, which is the convention for plotting wavenumbers.

22. Using the same file from Exercise 21 change the x-axis from wavenumber to wavelength (nm) and plot absorption versus wavelength. When converting between wavenumber and wavelength on the x-axis, you will need apply a Jacobian correction to the absorption. You can do this by $I(\lambda) = I(\nu)1 \times 10^7/\nu^2$. Make sure you use correct axis labels!

23. One nice feature of Plotly over Matplotlib is the fact that you can change a plot after you show it and get an updated version. In Matplotlib, once you show a plot, you usually can no longer make changes and you have to start over. In the case of the final code presented in this chapter, the Plotly figure object, `fig`, is still in your computer's memory and you can update it. To prove that you can do this, run the code and adjust the x-axis range to 450–750 nm using the **`range`** keyword of the `update_xaxes()` **method** of the figure object. Then show the plot again. Then, do the same for the y-axis, choosing values for **`range`** so that the resulting spectrum fills the plot area.

24. In this chapter, we introduced a function that makes it easy to identify files on your computer: `quickOpenFilename()`. This will be nice to use when you don't know where the file is located, or when you are choosing different files each time you run your script. However, there will be times when you simply want to use the same file over and over—for instance when you are developing and testing a code. In these cases, it can be much faster to provide the path to the file in the script. Modify the final code in this chapter, replacing the `quickOpenFilename()` function with a line that creates a Path object with the absolute path to the data. Verify this works, and also comment on when you might prefer this approach over the use of `quickOpenFilename()`.

CHAPTER 3

Plotting Multiple Data Sets

- ☐ **New concepts**
  — *and their prerequisites*

- ☐ **Making ordered collections of data more flexible using list objects**
  — *Section 1.1.1: Operating on collections of data efficiently with Numpy arrays*
- ☐ **Combining lists by concatenation**
- ☐ **Adding items to lists using the `append()` method**
  — *Section 2.1.4: Exploiting built-in methods to manipulate data stored in objects*
- ☐ **Determining the number of items in a collection using the `len()` function**
  — *Section 2.1.4: Exploiting built-in methods to manipulate data stored in objects*
- ☐ **Access items in lists, arrays, or strings by indexing**
  — *Section 0.1.6: Expressing data as readable text using strings*
  — *Section 1.1.1: Operating on collections of data efficiently with Numpy arrays*
- ☐ **Access subsets of lists, arrays, or strings by slicing**
  — *Section 0.1.6: Expressing data as readable text using strings*
  — *Section 1.1.1: Operating on collections of data efficiently with Numpy arrays*
- ☐ **Ordering lists and arrays by sorting**
  — *Section 1.1.1: Operating on collections of data efficiently with Numpy arrays*
- ☐ **Obtaining a list of paths using `codechembook.quickTools.quickOpenFilenames()`**
  — *Section 2.1.7: Adding `codechembook` to your Python installation*
- ☐ **Creating collections of objects with customized organization using dictionaries**
- ☐ **Accessing information about objects using attributes**
  — *Section 2.1.4: Exploiting built-in methods to manipulate data stored in objects*
- ☐ **Performing repetitive operations using `for` loops**

3.0 PROBLEM

Now that you have plotted the spectrum for the final pH point, you realize you have 19 other spectra that you want to add to this plot. If all of the data were in a single **csv** file, you can see how you might do this. However, you have 20 different csv files—one for each spectra. You could, of course, create a common csv and copy/paste all of the data into a single csv, but you think there must be a better way. You ask your labmate, who says, "Of course! We can just go through each file one by one, and build a plot one spectrum at a time. You already know almost everything you need to do it anyway! Let's get started!"

3.1 SOLUTION

3.1.0 PLANNING

Once again, you start a rough outline and sketch of what you want to do:

```
# gather the locations of the data to plot

# for each data file...

    # read the data and put it into Numpy arrays

    # add the data to a Plotly figure as a line trace

# arrange and format the Plotly figure

# save and show the Plotly figure
```

Your labmate says this looks a lot like what you did for a single plot, except that, instead of a single file and a single line on your plot you want to identify a set of files and then *for* each of these files, you want to add it to the plot you have. Once all the data has been added to your plot, you can format it, and then display and save it.

"You might be thinking about this as if you had a list of files, and then *for* each file in this list you added a trace to your plot. Well... you would be right, because that is

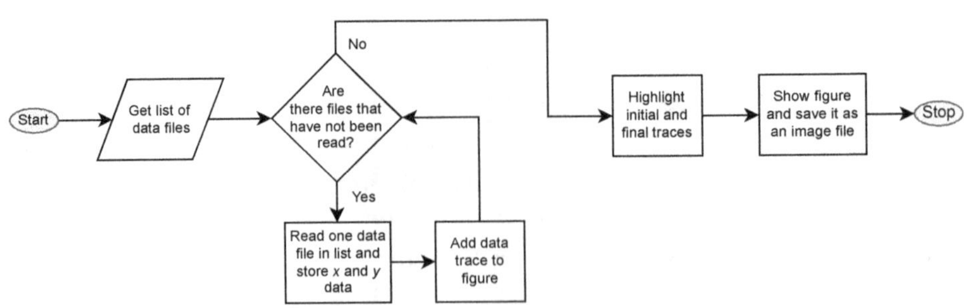

FIGURE 3.1. Flowchart of the solution to plotting multiple series of data on single plot. This uses the concept of a **`for` loop** to **iterate** through the different series.

exactly what we are going to do!" Your labmate explains that Python has a type of data type called a `list` and also a **control structure** called `for`, which can be used to iterate through these items. "We can look at these two new ideas, as well as several others, and then we will be ready to create the solution to this workflow problem." You both sketch out a plan, producing the flowchart shown in Figure 3.1.

3.1.1 Making Ordered Collections of Data More Flexible Using List Objects

"A Python **list** is a **data type** that holds several values. It is kind of like the **Numpy arrays** that we have been using, with some advantages and disadvantages. The advantage is that lists are a bit more flexible. A single list can simultaneously hold numbers, letters, and even things like Plotly **figure objects**, whereas the default behavior for a Numpy array requires all the data to be of the same data type. However, the disadvantage of the list is that it is much less convenient to do math with them than with Numpy arrays. For instance, you make a list by enclosing elements, separated by commas, between square brackets, like this:"[1]

[1] Now you know that, in past chapters, when we were making Numpy arrays, we were really making lists and then converting them to arrays using the `numpy.array( )` function.

```
In: lst = [2, 3, 5]
```

"And then you try to multiply all the elements by 2, you will get a different result than if it were a Numpy array:"

```
In: lst*2
Out: [2, 3, 5, 2, 3, 5]
```

"So you can see that the command actually just results in repeating the list elements… making a list that is twice as long."[2]

[2] When used in this way, the * is called the "repetition operator."

"This is one way to create longer lists, but it is not very flexible because we just repeat the list. We can grow lists by adding arbitrary elements one of two ways: **concatenation** and the `append( )` **method** of list objects."

3.1.2 Combining Lists by Concatenation

"Another difference between a **list** and a **Numpy array** is that we can **concatenate** lists, which is just a fancy way to say that we can add one list to the end of another. Imagine we had a second list:"

```
In: lst2 = [2.0, "yo", True]
```

"We could add this list to our old list using the + **operator**. When this operator is between two lists, it adds the second list to the end of the first one."

```
In: lst + lst2
Out: [2, 3, 5, 2.0, "yo", True]
```

"And so this is one way that we grow a list.[3] We can only do this with two lists. For instance the following results in an error:"

[3] The reason we couldn't do this with Numpy arrays is that the + operator will be interpreted as mathematical addition, rather than concatenation.

```
In: lst + 7
```

"because 7 is not a list. If we wanted to simply add the 7 to the end of the list, we could do:"

```
In: lst + [7]
```

"This will work because now both items are lists. Because we did not assign the result of this concatenation to a **variable**, the value of `lst` did not change:"

```
In: lst
Out: [2, 3, 5]
```

3.1.3 Adding Items to Lists Using the append( ) Method

"There will be times that you don't wish to convert an **object** to a **list**, in order to add it to an existing list, and for this we can use the `append()` **method** of list objects. This method accepts a single **positional argument** and then adds that object to the end of the list. For instance, if we wanted to add the number 10 to `lst2`, we could use:"

```
In: lst2.append(10)
```

"And now 10 will appear at the end of this list:"

```
In: lst2
Out: [2.0, "yo", True, 10]
```

"Look back at this example, and you will see that in the first expression, we used `append()` and did not assign the results to a variable. And, yet, when we ask for the value of the list, 10 was present! This is because `append()` modifies the list **in place**, meaning that the list we used the method on is modified directly."

"One important difference between **concatenation** and `append()` is that if the object added using `append()` is a list, the list will be added as a list element. Above, we used `[7]` for concatenate, which added 7 to the end of the list. If we do this with `append()`, we get `[7]` added to the end:"

```
In: lst2.append([7])
In: lst2
Out: [2.0, "yo", True, 10, [7]]
```

"and the last item of this list is itself a list. Thus, the original list contains another list. This highlights another reason to use append instead of concatenate, when you want to add a list as a list to another list."

3.1.4 DETERMINING THE NUMBER OF ITEMS IN A COLLECTION USING THE `len()` FUNCTION

"We can always find out how many items are in our lists, using the function `len()` **function**. This also applies to other **collections**, so you can use it to get the length of **Numpy arrays** as well:"

```
In:  len(lst2)
Out: 5
```

3.1.5 ACCESSING ITEMS IN LISTS, ARRAYS, OR STRINGS BY INDEXING

"Something that a **list**, a **string**, and a **Numpy array** all have in common is that an item in the collection can be identified and accessed by its position. The position is given by an integer number, the so-called **index**, in which the first item has the index 0, the next 1, and so on. We can get the value at any index by using the name of the variable followed by `[<index>]`. So, if we go back our list we assigned to `lst`, we can get the *second* element using:"

```
In:  lst[1]
Out: 3
```

"We get the *second* element *because index counting starts at zero in Python*."[4]

"We can also count from the back of the list forward, and in this case, the last item of the list is given the index −1, the second to last is given the index −2, and so on. So if I wanted to get the last item of our list, I can use:"

```
In:  lst[-1]
Out: 5
```

"The ability to count backwards is very convenient because it allows us to access the last element in any list without needing to know how long the list is."

"You can also modify elements of a list or Numpy array by indexing them on the left hand side of the assignment **operator** (=):"

```
In:  lst[1] = 1
In:  lst
Out: [2, 1, 5]
```

"Indexing also works for arrays:"

```
In:  array1 = np.array([3, 4, 5])
In:  array1[1]
Out: 4
```

"And because arrays can contain arrays and lists can contain lists, we can use one index after another to select elements within nested arrays. Consider this 2D array:"

```
In:  array2d = np.array([array1, array1*10])
```

[4] This is the reason the chapters, subsections, exercises, and so on in this book start counting at zero. It is a gentle reminder you should be doing the same.

"which looks like:"

```
In:  array2d
Out: array([[ 3,  4,  5],
Out:        [30, 40, 50]])
```

"And then if we wanted to select the number 50, we can use:"

```
In:  array2d[1][2]
Out: 50
```

"The first index selected the **inner array**, and the second index selected the third element from that second inner array."

"Finally, we can also index strings:"

```
In:  greeting = "Hiya there!"
In:  greeting[1]
Out: 'i'
```

"There are other types of collections where we can use indices to access objects, but these three are going to be all we need to know about for most chemistry problems. In general, you can use indices with any data type that is **sliceable**. But this begs the question: what is slicing?"

3.1.6 Access Subsets of Lists, Arrays, or Strings by Slicing

"Any data type that you can use **indices** with can be **slice**d. Slicing uses indices in order to select a range of items as well. This is done by indexing with `[<start index>:stop index >]`. This is known as **slicing**, and it will return a new list of all items in order spanning from the one at the start index (**inclusive**) up to but not including the last one (**exclusive**). To illustrate this, imagine we sliced our `lst`:"

```
In:  lst[0:2]
Out: [2, 1]
```

"You can see that we got the first item (index 0) through the second (index 1) because the first index is inclusive (we get the item at that index) and the second is exclusive (we do not get the item at that index, but the one before it)."[5]

"If you leave the first index blank, then slicing will start at the first item. If you leave the second index blank, slicing will go all the way to the end of the list. Using the slices this way, we can do:"

```
In:  greeting[:2]
Out: 'Hi'
```

"Much like indexing, you can replace a slice of a list or Numpy array by assigning the slice to an equal-length list or array:"

[5] Although it might seem strange to have the first index be inclusive and the second one exclusive, it makes it much simpler to do things like know how long the slice is because you just subtract the first index from the last.

```
In: lst[1:] = [12, 'bye']
In: lst
Out: [2, 12, 'bye']
```

"There is an important subtlety to remember: in the background, slicing lists and Numpy arrays works quite differently. When you slice a list and assign the result to a new variable, you get a new copy of the sliced data from the old list. Changes to that copy do not change the original list. However, for a Numpy array, slicing actually creates a 'view,' which is a reference to a subset of the original list. The idea is that you are only looking at part of the array. Because of this, changes to the view result in changes to the original array at the same time. This can cause problems when you're not expecting it!"

3.1.7 ORDERING LISTS AND ARRAYS BY SORTING

"There are times that we won't care what order our **lists** are in, and there are times we will. For instance, it is common to want lists of numbers in ascending order, or names in alphabetical order. We can accomplish this by using the `sorted()` **function**, and supply the list we want sorted as a **positional argument**. This function will **return** the sorted list:"

```
In: numbers = [5, 8, 1, 5]
In: sorted(numbers)
Out: [1, 5, 5, 8]
```

```
In: names = ["Krewald", "Aronsen", "Hukka", "White"]
In: sorted(names)
Out: ["Aronsen", "Hukka", "Krewald", "White"]
```

3.1.8 OBTAINING A LIST OF PATHS USING `codechembook.quickTools.quickOpenFilenames()`

"Before, we introduced the idea that we could use `codechembook.quickTools.quickOpenFilename()` to get the **path** associated with a single file. Now we will need to get the path to multiple files all at once. For this, we use `codechembook.quickTools.quickOpenFilenames()`. Note, this new function has an 's' at the end. It will allow you to select multiple files in a file browser, and then the function will return a **list** that contains absolute paths to all of the files you selected. Go ahead and try this by entering the following in a Python console:"

```
In: from codechembook.quickTools import quickOpenFilenames
In: quickOpenFilenames()
```

"If you then select multiple files, you will see an output that is the paths to these files contained within a list. Because they are in a list, we could then choose to use any functionality that we know for lists to access the path objects, such as using **indexing** and **slicing**."

3.1.9 CREATING COLLECTIONS OF OBJECTS WITH CUSTOMIZED ORGANIZATION USING DICTIONARIES

"There is one more **data type** we will need to introduce, which is a **dictionary**.[6] We will not need much of their capabilities to solve this workflow problem, so we will only introduce the concept. A dictionary is also a **collection** of elements. However, each element in the collection is actually two parts: a **key** and a **value**. Thus, when adding items to a dictionary, we need to specify a key/value pair. There are two ways to make a dictionary. The first is to use the form <key>:<value>, with each pair separated by commas, and all the pairs placed between curly brackets ({}). An example is:"

[6] The reason we are worrying about this is that we will need to use dictionaries to format certain aspects of Plotly figures. For now, the above is all you really need to know, and we will develop the idea of dictionaries more in the next chapter.

```
In: dict1 = {"first" : 1, "second": "two", "third" : 3}
```

"In this usage, the key must be specified using a string."

"The second is to use `dict()`, and then to place keys and values separated by the assignment operator.[7] An example is:"

[7] In this book, we will use the second way most of the time.

```
In: dict2 = dict(fourth = "four", fifth = 5, sixth = "six")
```

"In this approach, you do not need to use strings for the keys, but what you use must be something that can be converted to a string, which the `dict()` function will do."

3.1.10 ACCESSING INFORMATION ABOUT OBJECTS USING ATTRIBUTES

"You already know about **methods**, which we said were **functions** that could be called using an **object** that was an instance of a **class** to which the function belongs, using `<object>.<method>()`.[8] There is another construction that looks very similar, except that the name after `<object>.` is *not* a function. This is indicated by the fact that there are no parentheses used. In this case, we call this an **attribute** of the object. An attribute is essentially a **variable** that lives inside of the object—it stores data and has a name, but only after the object has been created. For instance, the **Path** object in the `pathlib` **library** has several useful attributes. To access then, we first need to import `Path` from pathlib:"

[8] You can review Section 2.1.4.

```
In: from pathlib import Path
```

"Then, we can supply a path and assign it to a variable."

```
In: myPath = Path("/<folder1>/<folder2>/<file>.csv")
```

"Now that we have the path, we have access to a lot of information about it simply by using attributes:"

- We can access the paths *up to, but not including,* the file using: `myPath.parent`, which would result in `/<folder1>/<folder2>`.

- We can access the full name, including extension using: `myPath.name` which results in <**file**>.csv.

- We can get the file name, *without the extension* using: `myPath.stem`, which would result in <**file**>.

- We can get only the extension using: `myPath.suffix`, which would result in `csv`.

"These examples used attributes of the Path object: `parent`, `name`, `stem`, and `suffix`, which returned the containing folders, the complete file name (including extension), the filename without an extension, and the extension only, respectively."

3.1.11 PERFORMING REPETITIVE OPERATIONS USING `for` LOOPS

"Earlier, we said that we are going to get a **list** of files and then go through each one, one at a time. There are many ways to do this in Python, but we will use the `for` **loop**, which is a kind of **control structure**.[9] The `for` loop allows us to run the same set of lines of code over and over a set number of times. Each time the loop code runs is called an **iteration**. To see how this will work on a list, let us just make a `for` loop."

[9] You can read more about control structures in technical Chapter F.

```
lst = [2, 3, "hi"]
for thing in lst:
    new = [thing] + ["!"]
    print(new)
print(lst)
```

"Here, `in` is a keyword that tells Python to expect the iterable collection of which we will loop through the elements. Running this code will output the following to the console:"

```
Out: [2, '!']
Out: [3, '!']
Out: ['hi', '!']
Out: [2, 3, 'hi']
```

"What happened was this: Python went through each item in our list in turn. We assigned the item to the **variable** named 'thing.' Then, we added this variable to a list, and **concatenated** it with another list, `['!']`, assigned the result of the concatenation to a new variable, and then printed the value of this new variable. Each iteration, we go to a new item in the original list and assign it to the variable named 'thing.' So, we get the results for each list item in turn. We also printed the `lst` at the end, and see that it is unaltered by its use in the `for` loop, because we did nothing to change what the elements were."

Your labmate takes a breath. "That was a lot, but there are a few important things to remember about the `for` loop:

- We start the loop with a line ending in a : character. This will be the same for other control structures in Python, like a function definition.

- We go through the entire list, one at a time. If we wanted to use only part of the list, we could **slice** it.

- The lines of code that happen in each iteration of the `for` loop are indented. So, the concatenation and the first `print()` **statement** both happen for each item. The indentation indicates they are part of the same control structure—a construction called using **white space**. We saw a similar usage when we wrote a function in Section 0.1.7.

- We end the `for` loop by ending the indentation. So the last `print()` statement, which is not indented, runs after the `for` loop completes."

Your labmate smiles at you, and says, "We are now in a place to solve your problem, so lets do it!"

3.2 CODE

```python
'''
A program to plot multiple uv-vis spectra from .csv files from a titration experiment
    Requires: .csv files with col 1 as wavelength and col 2 as intensity
    Written by: Ben Lear and Chris Johnson (authors@codingforchemistsbook.com)
    v1.0.0 - 250131 - initial version
'''
import numpy as np
from plotly.subplots import make_subplots
from codechembook.quickTools import quickOpenFilenames

# First, we need the names of the files that we want to plot
# Ask the user to select the files and then sort the resulting list
data_files = quickOpenFilenames(filetypes = 'CSV files, *.csv')
sorted_data_files = sorted(data_files)

# Next, we will process one file at a time and add it to the plot
titration_series = make_subplots() # start a blank plotly figure object

# Read the data in one file and add it as a scatter trace to the figure object
for file in sorted_data_files: # loop through the data files one at a time
    # Read the data and store it in temporary x and y variables
    x_data, y_data = np.genfromtxt(file,
        delimiter = ',', skip_header = 1, unpack = True)

    # Add data as scatter trace with formatted lines and exclude from legend
    titration_series.add_scatter(x = x_data, y = y_data,
                                 line = dict(color = 'gray', width = 1, dash = 'dot'),
                                 name = file.stem + ' eqs', showlegend = False)

# Adjust the appearance of only the first and last traces to highlight
titration_series.update_traces(selector = 0, # specify the initial trace
```

```
    line = dict(color = 'darkcyan', width = 2, dash = 'solid'),
    showlegend = True, name = 'initial')

titration_series.update_traces(selector = -1, # specify the final trace
    line = dict(color = 'darkred', width = 2, dash = 'solid'),
    showlegend = True, name = 'final')

# Move the initial trace to the end of the data, so that it is drawn on top
titration_series.data = titration_series.data[1:] + titration_series.data[:1]

# Format the plot area and then show it and then save it
titration_series.update_layout(template = 'simple_white')
titration_series.update_xaxes(title = 'wavelength /nm', range = [270, 1100])
titration_series.update_yaxes(title = 'absorbance', range = [0, 4.5])
titration_series.show('png+browser')
titration_series.write_image('titration.png', width = 3*300, height = 2*300)
```

You run this code and a file selection dialog box opens. You use it to select all of the files in the directory "Research/Catalyst/Titration/UVvis" and click Open, and the code quickly produces the output in the Spyder IDE's plot window (Figure 3.2):

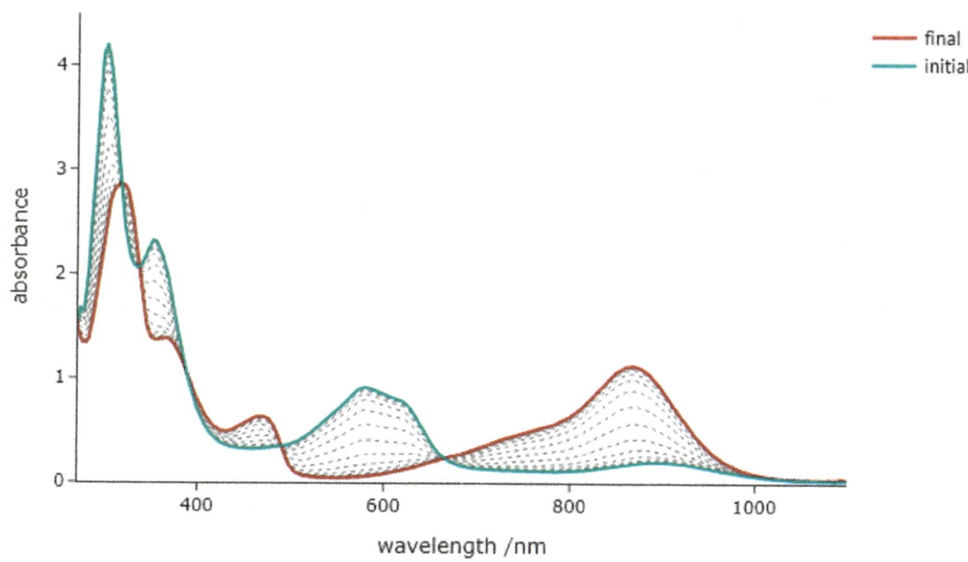

FIGURE 3.2. Multiple spectra plotted on a single plot. In this chapter you learned to do this in a single code by looping, and how to format the appearance of the plot.

Your advisor happens to walk by when you are have the "browser" version of the figure up, and sees you interacting with it. He comments how cool that is, and you offer to send him the Python script. The manner in which he cocks his eyebrows tells you this is not going to happen.

Your labmate leans over and stage whispers, "You can have Python save the figure as HTML as well. If you send him that file, he can interact with it, just like you are!"

Sure enough, you find that introducing the line `titration_series.write_image('titration.html')` produces an HTML file that can be shared and is an interactive plot of your data. So cool!

3.2.0 LINE-BY-LINE EXPLANATION

1–6: As always, we start with a multi-line comment that has information about the script.

7–9: We **import** the **libraries** we want to use. The only new import here is `quickOpenFilenames`. Last chapter we used `quickOpenFileName` (note the difference in the s at the end). The **function** we use this chapter works the same as the function from the last chapter, with the only differences that it allows the user to select multiple files and returns a **list** of the paths associated with the files.

13: Using the `quickOpenFilenames()` function, we get a list of the paths to the files holding the data we wish to plot and assign them to the variable `data_files`. Because these files are going to all be **csv** files, we also use an optional **keyword argument** in the function **call**. This keyword, `filetypes`, accepts a **string**, which should have two bits of text, separated by a comma. This text is used by the native file dialog. The first is included at text in the file explorer, as a hint for the types of files you are looking for. The second provides a pattern that the files must match. Here, we use *, which is a **wildcard**, followed by csv. This means that the native file explore will allow the user to select any files ending in csv, but also that only such files can be selected. This can be useful to use when you are going to look in folders that have lots of different kinds of files because it simplifies the choices for the user.

14: It would be nice to have the files in the order of the titration equivalents. We can't assume that the files will be in the order they appear in the file dialog; however, we have named the files such that we could sort them in order of increasing number. We can accomplish this using the `sorted()` **function**, providing the list of paths as the argument. We assign the returned sorted list to a new **variable**, `sorted_data_files`.

17: We are going to add our data to a single figure, and so we make a new Figure object using the `make_subplots()` function, assigning to the variable `titration_series`.

20–28: We start a **`for` loop**, in which we iterate through the files in our `sorted_data_files` list and add traces built from them to our figure. We use the variable **`file`** as the variable for the items returned from our `sorted_data_files` list within the `for` loop.

22 and 23: We use `np.genfromtxt()` to read the data from our files. We set the delimiter as `','`, tell it to skip the first row (which is a **header**), and set the keyword argument `unpack` to `True`, meaning that we will get data by columns. Each file is two columns of data, and so we can **unpack** the value returned by `np.genfromtxt()` to two variables.

26–28: Once we have our data we use the `add_scatter()` method of the Plotly **figure object**, `titration_series`, to create a trace and add it to the figure. We start by specifying the x and y data, before other **keywords**, as explained next.

27: Because there are so many closely spaced points, Plotly will default to producing a line (rather than using markers). Here, we specify the formatting of this line, using the keyword argument `line` and supplying it within a **dictionary**. In this dictionary, we specify the color and width of the line, using these words as **key**s, and providing a named color (gray), a line width (thickness) in pixels, and a dash style (dot).

28: We use the `name` keyword argument to specify the text that will be used to label the data traces. If a legend is shown, this will be included in the legend. However, even if there is no legend, these names will appear when hovering over the traces in an interactive plot (like when rendered in the browser). The name we use is the stem **attribute** of the Path object associated with the file we are plotting data from. This is a **string** that is the name of the file, without the file **extension**. Because of the naming we used for the files, this is the same as the equivalents we added and for this reason, we **concatenate** the stem with the string "eqs."
Setting the keyword `showlegend` to `False`, we tell Plotly that we should not include this trace (and its name) in the figure legend. It will still appear when hovering over traces in interactive mode.
The closing parentheses of this line ends the `add_scatter()` function call, and this is the last command in the `for` loop.

31–33: We take the first trace in our plot (`selector = 0`), and then format only that trace, using the `update_trace()` method of the Plotly Figure object. Because our filenames were in a sorted order, the first trace is the initial spectrum. Using the `line` dictionary, we set the color to `darkcyan` (a named color in Plotly) and make the line two pixels wide, and change the dash style to be a solid line (`solid`). We also change the keyword `showlegend` to `True`, which means this trace will appear in the legend, and we use the keyword `name` to specify what the trace will be called in the legend.

35–37: We repeat the above process for the last trace (`selector = -1`) using the same sort of indexing we would use for a list. This means that we are selecting the last trace we added. We also specify a different color and provide a different name for the trace.

40: When Plotly draws traces, it will draw them in order of index, meaning that the first trace in the list is drawn, then the second, and so on. The traces can overlap, and make it difficult to see the traces under them. For this reason, it would be nice to draw the first and last traces last, while all the intermediate traces are drawn first. That way, we can easily see the first and last traces. We can accomplish this by changing the order of the traces in the collection that holds them. In this line, we accomplish this reordering. The traces to be drawn, in the order they will be drawn is held in a list in the .data **attribute** of the Figure object. In this line of code, we access this attribute, take a **slice** of the list from the second trace through the end, and then concatenate this with a slice of the same list that is only the first trace.[10] We then assign this concatenated list back to the data list. This has the effect of moving what was the first plot in the list to the last position. It will now be plotted last. This will ensure it is easy to see in the final figure.

[10] We could not use `titrationSeries.data = titrationSeries.data [1:] + titrationSeries [0]`. The reason being that `titrationSeries[0]` returns the *item* from the list, but concatenation happens between two *lists*. So we need a slice for both things we concatenate.

43: Using the `update_layout()` method of the Plotly Figure object, we specify the template we wish to use.

44 and 45: Using the `update_xaxes()` and `update_yaxes()` methods of the Figure object, we can format the axes. In this case, we simply add titles to the axes.

46: We show the final figure using "`png+browser`," which opens up a interactive plot in a web browser and outputs to the Spyder plot window (if you are using the Spyder IDE).

47: We save the Plotly figure to an image file, providing Path and a size to save it as. We also use two keyword arguments for the `write_image` method: `width` and `height`, which are used to specify the width and height of the figure. The units are in pixels, and so the numbers must be able to be written as whole numbers. Here, when we scale 3.3 and 4.3 by 100 to give 33 and 43, respectively. If you then produced this figure, and scaled it to 3.3 and 4.3 inches, it would be drawn at a resolution of 300 pixels per inch. This is a fine resolution for print and web, although some print applications might desire higher resolution, such as 600 pixels per inch.
When looking at how the image is saved, we take one of the `Path` objects from our list of sorted file names, and then use the `parent` attribute of the object to get the folder that the file was in. We then supply a filename. The end result is that we save the figure output to the same folder as our data came from. Now, even though there is a .png file in this folder.

CHAPTER 3 EXERCISES

TARGETED EXERCISES

Exercises for Section 3.1.1

0. Make a list that contains no items.

1. Make a list that contains, in order, the Numpy array([1, 2, 3]), the string "frog," and the number 3.14.

2. Make a Numpy array with 20 elements from 0 to 10 and convert it to a list.

3. Make a list that has 12 items with 1, 2, 3 repeating four times without typing in 12 numbers (i.e., use the arithmetic properties of lists).

4. Make a list that has four items, where each item is one of the lists that you just made in Exercises 0–3.

5. Lists have a property known as **mutability**, which means they can be changed after assignment (for more information, see Chapter B). In many programming languages, if you want to make a copy of list or arrays, you simply assign it to a new variable. You might make a list and assign it to the variable `A`, then use `B = A` to copy it to a new variable. Then you could change `A` and `B` independently. What happens when you do this for a list in Python?

6. Lists have a property known as **mutability**, which means they can be changed after assignment (for more information, see Chapter B). One consequence of this is that a function can change the list, and this can result in changes that are accessible outside the function. Write a function that will accept a list as an argument, and then the function replaces the first element in the list with the string `"this element was changed inside a function"`. Make a list, and print this list. Pass the list to the function, and then print the list again.

Exercises for Section 3.1.2

7. How many items would be in the list that was made by concatenating all five lists you made in Exercises 0–4?

8. Repeat Exercise 3 using concatenation.

9. Add the number 5 to the end of the list you made in Exercise 1.

Exercises for Section 3.1.3

10. How many items would be in the list that was made by appending the lists you made in Exercises 1–4 onto the list made in Exercise 0?

11. Repeat Exercises 7–9 using `.append()` instead of concatenation.

12. Sometimes it would be better to use a list, and sometimes a Numpy array. In the following cases, which do you think would be better, and why?

    a) You want two variables to hold the x and y points, respectively, of an X-ray diffraction pattern.

    b) You are doing a titration experiment that involves saving UV-vis spectra for each titration point, and you want to keep a record of the files that you saved, but you don't know how many files you will need to save.

    c) You want a single variable to store the history of responses of a user to a code that periodically asks them to record the lab temperature.

    d) You finished the experiments being discussed in this chapter and now you want to write a code that plots only the intensity of the spectrum at 450 nm (a single value) versus the titration equivalent for each spectrum. You want to store the intensity points and titration equivalents as two separate variables for plotting.

Exercises for Section 3.1.4

13. Prove that your answer to Exercise 7 is correct, using `len()`.

14. Starting from an empty list, create a list that goes from 0 to 4 using only `append()` and `len()`.

Exercises for Section 3.1.5

15. Create the following list and store it in a variable named `data`: $[5, 1, 9, 4, 2, 8, 2, 7]$

    a) What is the index of the item containing 4?

    b) What does the **expression** `data[4]` yield?

    c) Write an expression that yields the item containing the value 8.

    d) Write an expression that yields the third-to-last item in the list, and will do so, even if the length of the list changes.

    e) Replace the number 4 with the number 6.

Exercises for Section 3.1.6

16. Create a list containing `['yes', True, None, 3, 5.5, 'no']`:

    a) Write an expression that yields a subarray containing the first five items of the array.

    b) What does the **expression** `data[2:5]` yield?

c) Write an expression to create new list containing every other item in this list, starting with the second item.

d) Write an expression to save the third through sixth items of the list as a new list stored in a variable named `data_reduced`.

e) Write an expression that yields the last four items in the list and will work even if the length of the list changes (but continues to have at least four items).

Exercises for Section 3.1.7

17. Make the following list: `[12, 6, 0, 2, 31, 4, 7]`

a) Sort this list to run in increasing order.

b) Using resources outside of this book, find and use the **keyword argument** that lets you sort the same list in the reverse order.

Exercises for Section 3.1.8

18. Using `quickOpenFilenames()`, print the paths for all the files in your downloads folder.

19. Using the `filetypes` keyword argument in `quickOpenFilenames()`, print paths for *only* the pdf files in your downloads folder. If you don't have any pdf files in your downloads folder, then download a scientific paper as a pdf, add it to your downloads and try this exercise again.

Exercises for Section 3.1.9

20. Create a dictionary in which the keys are the atomic numbers of the atoms on the second row of the periodic table, and the values are the atomic symbols.

a) Write an expression to get the atomic symbol of fluorine from this dictionary.

b) Write an expression to produce a string containing the formula for lithium fluoride using only this dictionary.

c) Add hydrogen and helium to this dictionary.

21. Make dictionaries of solvent properties for Benzene. Each entry of the dictionary should be a different property, and you should choose keys that have descriptive names to make it easy to remember which key goes with which property. The properties you need are:

a) IUPAC name

b) Molecular weight

c) Boiling point

d) Density

22. Repeat Exercise 20, but for the following solvents:

    a) Benzene
    b) Cyclohexane
    c) n-Hexane
    d) Toluene
    e) Methanol
    f) Acetonitrile

23. Create a dictionary where the keys are the names of the solvents and the values are the dictionaries that you made in Exercises 21 and 22. This means you are creating a dictionary that holds other dictionaries—a behavior known as nesting.

Exercises for Section 3.1.10

24. Starting with the list of files in your Downloads folder that you generated in Exercise 18, sort this list and then, use the `stem` attribute of Path objects to print just the filename (without the extension) of the first file.

25. Numpy arrays also have attributes. Create a Numpy array from 0 to 5 with 0.2 spacing. Using the `size` attribute, print the number of elements of this array.

Exercises for Section 3.1.11

26. Using a `for` loop, print each item in the list that you made in Exercise 4 on a separate line.

27. In Exercise 19, we found all the pdf files in your downloads folder. The output would be unwieldy. Using a `for` loop, place each on its own line.

28. Suppose you have the list `[3, 6, 2, 7, 5]`. Using two nested `for` loops, print out the products of all pairs of items in the list. In other words, you will want a series of text output like: `6 x 2 = 12`. Note that you should have 25 total output statements.

COMPREHENSIVE EXERCISES

29. Make a single Plotly plot that shows the Maxwell-Boltzmann distribution for N_2 at 11 different temperatures equally spaced between 0°F and 100°F. The atmosphere is mostly nitrogen, and so these curves represent how fast the majority of molecules in the atmosphere are moving under various "typical" conditions.

30. Make a single Plotly plot that shows the Maxwell-Boltzmann distribution for N_2, O_2, Ar, CO_2, Ne, He, CH_4, Kr, and H_2 at room temperature. These are the major components of the atmosphere.

31. Modify the final code in this chapter so that it produces a plot of each spectrum as a separate image and saves each one as a different file, all using a single `for` loop.

32. You now see that there are multiple properties for lines in Plotly. Let's practice using them. You can use the online Plotly documentation or the Plotly technical chapter (Chapter I) to learn about other formatting. Modify the final code of this chapter to:

    a) Change the colors of the lines to ones that we have not yet used.
    b) Change the line width to a value that we have not used yet.
    c) Make the lines dashed, rather than solid or dotted.
    d) Change the font for the entire figure to "Courier New."
    e) Change the font size so that it appears roughly double the default font size.
    f) Change the x-axis range to span from 270 nm to 1000 nm.

CHAPTER 4

Generating Molecular Structures and Annotating Plots

- ☐ **New concepts**
  — *and their prerequisites*

- ☐ **Accessing items from dictionaries using keys**
  — *Section 3.1.9: Creating collections of objects with customized organization using dictionaries*
- ☐ **Describing chemical structures using SMILES**
  — *Section 0.1.6: Expressing data as readable text using strings*
- ☐ **Drawing molecules with RDKit**
- ☐ **Using tuples to create unchangeable lists**
  — *Section 3.1.1: Exploiting built-in methods to manipulate data stored in objects*
- ☐ **Specifying colors using the rgb standard**
- ☐ **Decoding data from binary files**
- ☐ **Adding images to Plotly figures**
  — *Section 2.1.10: Making appealing and interactive plots using Plotly*
- ☐ **Adding annotations to Plotly figures**
  — *Section 2.1.10: Making appealing and interactive plots using Plotly*
- ☐ **Formatting text in Plotly using HTML tags**
  — *Section 0.1.6: Expressing data as readable text using strings*
  — *Section 2.1.10: Making appealing and interactive plots using Plotly*
- ☐ **Producing HTML-formatted text quickly using `codechembook.symbols`**
  — *Section 0.1.6: Expressing data as readable text using strings*
  — *Section 2.1.7: Adding `codechembook` to your Python installation*
- ☐ **Generating sequences of numbers using `range()`**
  — *Section 0.1.2: Accessing capabilities beyond basic math using functions*

- ☐ **Synchronizing iteration through multiple collections using `zip()`**
  - — *Section 0.1.2: Accessing capabilities beyond basic math using functions*
  - — *Section 1.1.1: Operating on collections of data efficiently with Numpy arrays*
  - — *Section 3.1.1: Exploiting built-in methods to manipulate data stored in objects*
  - — *Section 3.1.11: Performing repetitive operations using `for` loops*
- ☐ **Specifying the result of function evaluation using the `return` keyword**
  - — *Section 0.1.7: Accessing capabilities beyond basic math using functions*
- ☐ **Describing how to use your functions in docstrings**
  - — *Section 0.1.7: Accessing capabilities beyond basic math using functions*

4.0 PROBLEM

You now have a plot of your titration that you could share with someone else. The only problem is that you don't like how it is labeled. For one thing, it has a legend, which you don't like because it brings attention away from the data. For another, there is no title. Although you could add one, you are also a proud chemist, who easily thinks in the symbolic language of chemistry: chemical structures. Rather than writing a title that is something like "Titration of $Ni[pz]_2$ with an Acid," you would rather draw out the chemical reaction and add this to the plot as an annotation. And if you match the color of your structures to the lines in your plot, you don't need the legend anymore. If only you had a way to draw chemical structures...

You bring this up to your labmate, who says, "Oh yeah! It turns out that there is a way to draw chemical structures in Python. If you want them to look beautiful, it will require a bit of work, but if you simply what a drawing, then it is pretty straightforward. Also, learning this will help illustrate a few other concepts in Python you should know about. So why don't we go ahead and add the chemical reaction to your plot!"

4.1 SOLUTION

4.1.0 PLANNING

The rough outline for this workflow problem is rather straightforward:

1. Use the same plot from Chapter 3.
2. Create drawings for the reactant and product, matching the colors to the structures to the colors you used in plotting the data.
3. Add the molecules to your plot, connected by a reaction arrow that indicates this is a titration (i.e., H^+ is above the arrow).

Your labmate looks over this and agrees that this is a good basic plan. And they note that you really are just trying to add functionality to code you already have, so you can just return to the plotting script you already have (Section 3.2) an add in new code (remembering to advance the version number as discussed in Section 1.1.8). But, before you can implement it, you will need to understand a few more concepts. We can also sketch out the workflow as a flow chart (Figure 4.1).

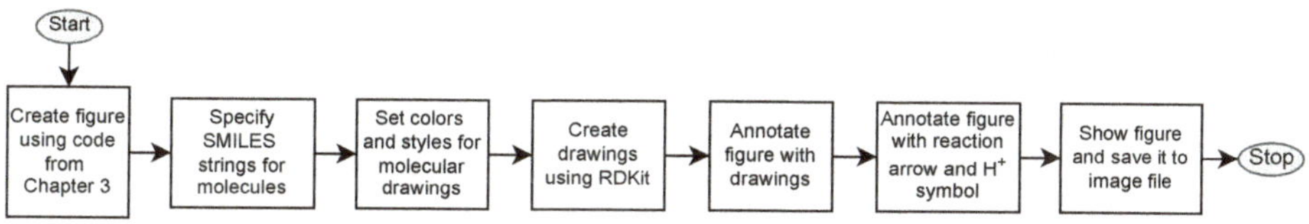

FIGURE 4.1. Flowchart for adding a skeletal formula to our Plotly figure of UV-vis spectra.

4.1.1 ACCESSING ITEMS FROM DICTIONARIES USING KEYS

"Last time, we saw how to *create* **dictionaries** using a set of **key/value** pairs, which could be created like so:"

```
In: numbers = dict(Ben = 11, Chris = 3.14)
```

"Well, we can access the values from a dictionary by using the dictionary and supplying the key between square brackets."

```
In: numbers["Ben"]
Out: 11
```

"If we try to iterate through a dictionary on its own (for instance using a **`for` loop**), the keys will be passed to the loop variable one at a time:"[1]

```
for k in numbers:
    print(k)
Out: Ben
Out: Chris
```

"We could use this to access the values in a within a `for` loop. Go ahead, change the code above to read `print(numbers[k])`, and you will then print the values, rather than the keys."

[1] You might read online that dictionaries are **unordered**, and so we cannot be guaranteed that we can iterate through them in the same order every time. However, since Python 3.7, dictionaries are ordered, and so we will iterate through the items in the same order each time. At the time of this writing, the most recent version of Python was 3.13.2.

4.1.2 DESCRIBING CHEMICAL STRUCTURES USING SMILES

"One reason that skeletal formulas are so useful is that people are visual creatures. We can look at a molecule and quickly identify if it is benzene, toluene, or a xylene. Computers are not (yet) visual creatures. So we need to have a way to communicate to a computer how atoms are connected in a way that it can understand. One way to do this is using **SMILES** (Simplified Molecular Input Line Entry System).[2] SMILES takes the form of a

[2] There are MANY different formats that computers use, and this is just one. Other familiar ones are .pdb, .cdxl, .mol, and .xyz, among others.

string of letters, numbers, and other symbols that specify what atoms are present, how they are connected to one another, and what charges they bear."

"If you want to try to either write your own SMILES or read a SMILES, then you can start learning by reading the Wikipedia page on them. However, there are also free online utilities that can help."[3] Using them, you will find that toluene is Cc1ccccc1, acetic acid is CC(=O)O, and acetate is CC(=O)[O-].

[3] As a free resource, we recommend: https://www.cheminfo.org/flavor/malaria/Utilities/SMILES_generator___checker/index.html. However, if you have access to commercial software, such as ChemDraw, they can also generate SMILES from structures, and vice versa.

4.1.3 DRAWING MOLECULES WITH RDKIT

"If you have the **SMILES** for a molecule of interest, you are now ready to have Python draw it for you, using the RDKit **library**.[4] Let's assume we want to draw acetate. To do so, we can run the following code:"

[4] Actually, you can also use things like Chemical Abstracts Service (CAS) number, etc. But we are using SMILES so that you know how to draw things that may not yet be known.

```
from rdkit import Chem
from rdkit.Chem import Draw

# Define the SMILES for acetate
acetate_smiles = "CC(=O)[O-]"

# Convert the SMILES to an RDKit molecule object
acetate_mol = Chem.MolFromSmiles(acetate_smiles)

# Display the structure
Draw.ShowMol(acetate_mol)
```

You run this code in the Spyder IDE, and it outputs the following image to the Python console[5]:

"What happened here is we made an RDKit `mol` **object** (the main way that molecules are described in RDKit) from a SMILES string. Then, we use the `ShowMol()` **method** of the `Draw` **class** to generate the image (Figure 4.2).[6] It is so simple!"

[5] For some people, the image will not output to the console. You can open the image in a new window by using `Draw.ShowMol(acetate_mol)` instead.

[6] If you would like to save a file, instead, see Problem 1 in Chapter 14.

"This is a fine drawing, but RDKit, by default, will color atoms and bonds associated with them. This can look a bit busy. If you want to control the colors of the bonds, or change other aspects of the molecule (bond lengths, widths, other colors, and so on) things become even more complex. A more complete example will be shown in the final code

> **Why RDKit?**
> A quick web search will turn up many different libraries for cheminformatics, and many of these can draw molecules. However, we felt that RDKit was the library that provided the quickest access to this capability. If you become interested in this area, check out the ACS InFocus book *Python for Chemists*, which focuses on cheminformatics.

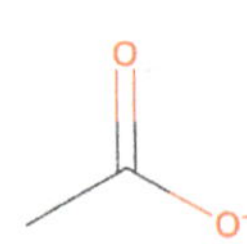

FIGURE 4.2. Skeletal formula for acetate. This was drawn programmatically using the RDKit **library** in Python. Using this library, one can specify the **SMILES string** (among many options) for the molecule, which is `CC(=O)[O-]` for acetate.

for this chapter, but to understand that, we have to understand tuples and color specification in Python. Let's talk about that."

4.1.4 USING TUPLES TO CREATE UNCHANGEABLE LISTS

"So far we have seen a few types of **collections**: **Numpy arrays**, **dictionaries**, **strings**, and **lists**. Now we will encounter another: the **tuple**. A tuple is created by placing elements between parentheses, separated by comments."

```
In: tup = (2, "me", False)
```

"You can **slice** a tuple, which means you can also access elements by **index**:"

```
In: tup[1]
Out: 'me'
```

"You can see that tuples can hold another other **objects**, and so it behaves a bit like a list. However, there is one important difference: tuples cannot be altered, meaning that you cannot change their elements, add elements, or sort them. This is a property known as **immutability**. Trying to alter a tuple will generate an error:"

```
In: tup[1] = "you"
Out: Traceback (most recent call last):
Out:   File "<pyshell#59>", line 1, in <module>
Out:     tup[1] = "you"
Out: TypeError: 'tuple' object does not support item assignment
```

"It might seem strange to create a data object you cannot change, but there are few reasons to do so. For one, it makes the data a bit more secure—you cannot accidentally change the values in a tuple. For another, they can be more **memory** efficient and faster to use. However, I do not find these advantages to be compelling reasons to use tuples over lists most of the time. Unless I'm forced to use tuples, I will just use lists. Specifying colors within RDKit is one example that requires tuples, which is why we are supplying them here."

4.1.5 SPECIFYING COLORS USING THE RGB STANDARD

"Now, we need to learn some nuts and bolts of color so that you can format the structures and their positions in the figure the way you want. So far, we have only needed to use color in our Plotly figures, and we have seen that Plotly has some named colors it knows about (e.g., 'darkcyan' and 'darkred').[7] But there are times that you will want to use a color that is not named. Or you may need to change the color of an object that doesn't use named colors. For instance, RDKit does not have named colors. In these cases, you need to specify colors in a way that a computer can recognize."

[7] For a complete list, see the Plotly documentation online.

One common way to do this is with so-called rgb colors (red-green-blue), which specify the intensity of three primary colors (red, green, and blue) we want. This is done using a **tuple** of these values. However, there are two common approaches to specifying the intensity we want for each primary color—and we will see both of them in our code below. In both specifications, the minimum amount of light you can have is 0. In one specification, the maximum amount of light is 1, and we specify color changes as any number between 0 and 1. In the other, the maximum value it is 255, and we specify color changes using any integer between 0 and 255.[8]

[8] The reason for 255 is color choices are digital, not continuous, so the 0–255 range in integers actually better reflects how a computer works (2^8 values).

"In the latter system, pure red is (255, 0, 0), pure blue is (0, 0, 255), magenta is (255, 0, 255), and so on. We can convert to the other system by dividing each number in the list by 255, which is how we will proceed in our code below."[9]

[9] For more information on color, see Chapter I.

"But there is even one more color specification worth knowing about here: the so-called rgba system. This system provides four intensities, where the first three are the amount of red, green, and blue light, and the fourth value is known as the alpha-channel, and controls how transparent the color is. If this is given a value of zero, then the color is completely transparent, whereas a maximum value (either 1 or 255), means the color is completely opaque. When working with Plotly, there is a small wrinkle: the `rgb` values are all between 0 and 255, but the value is between 0 and 1. This is something worth remembering."

4.1.6 Decoding Data From Binary Files

"Text and images are not the native language of a computer. Instead, zeros and ones are. Thus, from the computer's perspective, data that is represented by 0s and 1s (i.e., in **binary format**) is more efficient to store, faster to process, and more accurate. For this reason, it is often desirable to move between text-based representations of data to so-called binary formats. The base64 library accomplishes this for one particular kind of binary file, using so-called **base 64 encoding**. We are going to use this to embed an image in a Plotly figure. The Plotly library can use **scalable vector graphic**s, but the graphics must come in as binary information, which we will accomplish using the base64 library."

4.1.7 Adding Images to Plotly Figures

"Okay, so we know how to represent images, but how do we add them to figures? The easiest way is to use the `add_layout_image()` **method** of a Plotly Figure object. To do so, we need to specify the image source, which can be either a bitmap file on your computer, a **uri binary stream** (for instance, for an .svg image), or a URL (web address). In addition, we can specify other information, like the position, size, and so on. This is all explained in detail in the online Plotly documentation and partially explored in Chapter I."

4.1.8 Adding Annotations to Plotly Figures

"Annotations can be a great way to add clarifying elements to your plots. We can add annotations, like arrows, lines, shapes, and text to a Plotly Figure object using the `add_annotation()` **method**. The simplest use case is to provide the text label you want using the `text` **keyword argument** for the function, and to specify the location on the plot using the `x` and `y` keyword arguments. The default behavior is to put this text on the figure and draw an arrow at the location provided, although we can remove the arrowhead (i.e., draw a line), change alignment of text, change how positioning is handled, and much more."[10]

[10] The code provided at the end of the chapter demonstrates what we feel will be the most common changes to annotations. Again, this is explained in detail in the online Plotly documentation and partially explored in Chapter I.

4.1.9 Formatting Text in Plotly Using HTML Tags

As we learn to do more and more analysis automatically, we are going to want to automatically generate annotations in Plotly that look like a final product. That way, we won't have to further edit the plots to make them look nice for presentations or publications. Plotly originated as the plotting library for the Javascript language, which is the backbone of many of the interesting behaviors on webpages. Thus, the "native" way in which we format text in Plotly is through HTML tags. **HTML** stands for hypertext markup language, and it is the standard for coding webpages. It is built around tags that surround textual elements and specify how to display them. For instance, if you wanted to create a superscript, we place the text to be superscripted between `<sup>` and `</sup>` tags—these are called the opening and closing tags. Similarly, subscripts are placed between `<sub>` and `</sub>` tags.

> **Learning about HTML**
> HTML is a versatile language for displaying information in web browsers and is used heavily for a wide array of Python tools. A complete discussion of HTML is beyond the scope of this book, but many resources exist online to teach yourself to use HTML. One good place to start is http://www.w3schools.com.

"There are also instances when you don't need both open and closing tags. For instance, if you want to simply introduce a line break (to start a new line) you can use `<br>` and no closing tag is needed."[11]

[11] To enter a character that is not on your keyboard, see Section 5.1.6.

4.1.10 Producing HTML-Formatted Text Quickly Using `codechembook.symbols`

Although it is true that one can always directly use the **HTML** formatting, as introduced above, this can be a bit frustrating. In keeping with the theme to try to reduce frustration through programming, the `codechembook` **library** contains **functions** that will format things in HTML code for you. For instance, rather than using `<sup>text<\sup>` to place text in superscript, you can import `codechembook.symbols.typesettingHTML.textsup()`, providing the text you want to have in superscript as an argument to this function. This function returns the properly encoded HTML, and we will see it used in the code at the end of this chapter. Right now, this probably looks longer to type out, but if you use `from codechembook.symbols.typesettingHTML import textsup`, you can have direct access to this function, and this will save you time when you have multiple superscripts to use. Of course, you can always decide directly use the HTML.

4.1.11 Generating Sequences of Numbers Using `range()`

"There are going to be many times that you want to make a series of numbers in numerical order. They can easily be generated using the `range` **data type**. Range **object**s are created using the `range()` **function**, by providing between one and three positional arguments. The first **argument** provides the starting number (**inclusive**), the second argument provides the ending number (**exclusive**), and the third number provides the step size. If you leave out the last argument, the default step size is one. Thus, if you wanted numbers between 1 and 10, you could use:"

```
In: one_to_ten = range(1, 11)
```

"This doesn't produce a **list**, because `range` is its *own* data type. You can see this by printing the **variable**."

```
In: print(one_to_ten)
Out: range(1, 11)
```

"Instead of producing a list of numbers, the `range` object is a **generator**, which means it is an **object** that contains instructions for generating these numbers,[12] and we can access these instructions... for instance in a `for` **loop**:"

[12] Perhaps when generating the numbers from 1 to 10, the advantages of this approach is not clear. But imagine wanting to generate the numbers from 1 to a trillion. Clearly the instructions for doing so will take less memory than the actual list.

```
one_to_eleven = range(1, 11)
for num in one_to_ten:
    print(num)
```

Running this code prints out the numbers from 1 to 10, and you can see how this could be a convenient way to get sequences of numbers without having to type them out explicitly.

4.1.12 Synchronizing Iteration Through Multiple Collections Using `zip()`

"There will be many times that you have multiple **collection**s of **object**s (e.g., **list**s) and you wish to iterate through all of them in a single **`for` loop**. For instance, imagine you had a list of numbers = [1, 2, 3] and a list of names = ["one", "two", "three"]. You might like to loop through both of these lists so that you can make a statement that <number> is <name>. There are a few ways to accomplish this. In this chapter we will show one, the **function** `zip()`."[13]

[13] In the next chapter, we will show a different approach: using `enumerate()` and **indices**. Both ways have value, which is why we will teach both.

```
numbers = [1, 2, 3]
names = ["one", "two", "three"]

for number, name in zip(numbers, names):
    print(number, " is ", name)
```

"This produces the following output:"

```
Out:    1 is one
Out:    2 is two
Out:    3 is three
```

"Basically, `zip()` will accept take any number of **iterable objects** (e.g., **lists** or **Numpy arrays**) as **arguments**. These objects must be exactly the same length or you will get an error. When using `zip()`, the `for` loop iterates through the indices and returns a tuple comprised of values taken from the corresponding index from each list. This tuple can be unpacked to the variables given before `in`. The end result is we have a variable for each list we gave, which will get a value equal to one of the items in each list as we go through the lists. This is a straightforward way to ensure we synchronize going through several lists at the same time."

4.1.13 SPECIFYING THE RESULT OF FUNCTION EVALUATION USING THE `return` KEYWORD

"During our first lunch, we saw that it can be very helpful to write your own **functions**.[14] In that example, we had the function directly produce our final desired result. However, we often want to take the results of a function and use it in other places in our code. To do this, we can use the **`return` keyword**, placing any values you want passed back from the function on the same line after this **keyword**."

[14] Section 0.1.7.

"Let's make a simple example: imagine you dropped a ball, and wanted to know its velocity at any time. We know the equation for this is simple: $v = 9.8\ t^2$. If we define a function that we could **call** to return the velocity of the ball, we would do the following:"

```
1  def velocity_at_time(time):
2      '''
3      A function to find the velocity of a dropped object at a given time.
4  
5      Required Params:
6      time (float): the time that has elapsed since the drop
7  
8      Returns:
9      v (float): the velocity at that time
10     '''
11     v = 9.8 * time**2
12     return v
13 
14 v1 = velocity_at_time(2)
15 print(v1)
16 print(velocity_at_time(3))
```

Running this code gives two outputs on different lines: 39.2 and 88.2. Your labmate explains, "The returned values were generated *inside* the function and then passed back to code outside the function for later use. You can also see that we can either assign the returned value to a variable for later use (line 14), or use the returned value directly (line 16). When using the return keyword, the definition of the function still ends when you end the indentation **block**, not necessarily where the return statement shows up."

"One last thing," your labmate says. "You may have noticed a large **comment** directly after the line with the `def` keyword and the rest of the function. This is a multi-line comment explains the workings of the function and is known as a **docstring**, and it is very powerful. Let's talk about that."

4.1.14 Describing How to Use Your Functions in Docstrings

"**Docstrings** are **string literals** that are used to document **functions**, **methods**, **classes**, and **modules**. They are enclosed in triple quotes (`"""` or `'''`) and are placed immediately after the function or method definition. When a function's docstring is defined, it can be accessed via the `__doc__` **attribute** of the function. Some IDEs (like Spyder), will give you a pop-up that displays the docstring. Finally, Python provides a built-in `help()` function. When you provide a function as the argument, this will return the docstring.[15] A well-written docstring enhances code readability and maintainability by clearly explaining the purpose and usage of the function."

[15] See Section 0.1.2.

"Docstrings should help a programmer to understand what the function does, what arguments it accepts, and what it returns. There are many different standard formats for docstrings, but typically they include the following:

1. A one-line description of the purpose of the function.

2. The types of **arguments** (also called **parameters**),[16] the **data types** that can be passed to them, and a one-line description of their purpose.

[16] For disambiguation on the difference between arguments and parameters, refer to the technical chapter on functions.

3. Value(s) **return**ed, their **data types**, and a one-line description of what is contained in them. If you are not going to use a return **statement**, then it is a good idea to explicitly state this in the docstring.

It is good practice to always make a docstring for a function. You may think that it slows you down, but you will save time in the long run as you reuse this function again and again."

Formatting docstrings
The formats that you will see for docstrings are only a few examples of how people structure them. The primary thing to focus on is clear and easy to read documentation. We roughly follow the Google style guide for docstrings because this is the style that is automatically generated in Spyder, if you type `'''` after a function definition line. But you can also consider using the Numpy style guide for docstrings because it is also widely used.

Your labmate smiles and says, "We know enough now to solve your problem. We just need to tack on a bit more code to the end of the plotting script you just made (Section 3.2), so let's just focus on the new code. We could place all this new code just *before* `titration_series.show("png+browser")`. So lets do it!"

4.2 CODE

Because you are just going to modify some code you already have, you also update the **docstring** at the top of the script to track the changes and the version numbers:

```
1 '''
2 A program to plot multiple uv-vis spectra from .csv files from a titration experiment
3     Requires: .csv files with col 1 as wavelength and col 2 as intensity
4     Written by: Ben Lear and Chris Johnson (authors@codingforchemistsbook.com)
5     v1.1.0 - 250201 - added structure drawings showing protonation reaction
6     v1.0.0 - 250131 - initial version
7 '''
```

You then go down to the line just after `titration_series.update_yaxes(title = "absorbance", range = [0, 4.5])` in the script (what used to be line 45, but is now line 46 because of the new additional line about the new version in the docstring). You enter a blank line, and then start typing new code on line 48:

```
48 from rdkit import Chem
49 from rdkit.Chem import Draw
50 from rdkit.Chem.Draw.MolDrawing import DrawingOptions
51 import base64
52 from codechembook.symbols.typesettingHTML import textsup
53 
54 def make_mol_uri(smiles, color, bonds_wide, bonds_tall):
55     '''
56     Function to generate an svg of molecular drawing using rdkit and then return it in a
       format that is appropriate for inclusion in a Plotly figure object.
57 
58     REQUIRED PARAMETERS
59     smiles (string):  Valid chemical smiles string.
60     color (tuple):    Color designated in the rgb format.
61     bonds_wide (int): Specifies how many bonds wide the structure is
62     bonds_tall (int): Specifies how many bonds tall the structure is
63 
64     RETURNS
65     (bytes): Uri of the image for use in Plotly.
66     '''
67     # Create an RDKit molecule object from a SMILES string
68     mol = Chem.MolFromSmiles(smiles)
69 
70     # Create a dictionary that defines the color of all atoms (here, the same)
```

```
    for key in range(1, 119):  # 119 is exclusive, so it goes up to 118
        DrawingOptions.elemDict[key] = color

    # Set options for drawing
    drawer = Draw.MolDraw2DSVG(50*bonds_wide, 50*bonds_tall) # create a canvas to draw on
    drawer.drawOptions().updateAtomPalette(DrawingOptions.elemDict) # the colors of atoms
    drawer.drawOptions().setBackgroundColour((0, 0, 0, 0)) # the canvas color

    # Now that the options are set, we can process the molecule drawing
    drawer.DrawMolecule(mol) # create the drawing instructions
    drawer.FinishDrawing() # use the drawing instructions to make the drawing

    # Generate the SVG instructions then convert to a Base64-encoded data URI with a Plotly-required preamble
    svg = drawer.GetDrawingText().replace('svg:', '')
    svg_data_uri = f"data:image/svg+xml;base64,{base64.b64encode(svg.encode('utf-8')).decode('utf-8')}"

    return svg_data_uri # return the binary code for the image

# Define the SMILES string for the base and acid species
base_smiles = 'C1(N=CC=N2)=C2O[Ni]3(O1)OC4=NC=CN=C4O3'
acid_smiles = '[H][N+]1=C2O[Ni]3(OC2=NC=C1)OC4=NC=CN=C4O3'

# Define colors for the base and acid species
base_color = (0, 139/255, 139/255) # dark cyan
acid_color = (139/255, 0, 0) # dark red

# Loop through both molecules and create their respective images
for species, color, position in zip(
        [base_smiles, acid_smiles], # the SMILES strings
        [base_color, acid_color], # the colors
        [580, 870]): # the positions of the images on the plot area

    # Add image to plot
    titration_series.add_layout_image(
        dict(source=make_mol_uri(species, color, bonds_wide = 8, bonds_tall = 3),
            xref='x', yref='y', # references for coordinate origin
            x=position, y=2.45, # x- and y-coordinate position of the image
            xanchor='center', yanchor='top',# alignment of image wrt x,y coords
            sizex=200, sizey=1)) # width and height of the image

# Add an arrow to denote the reaction
titration_series.add_annotation(
    dict(ax = 680, ay = 2, # arrow start coordinates
    x = 770, y = 2, # arrow end coordinates
    axref = 'x', ayref = 'y', xref = 'x', yref = 'y', # references for coordinates
    showarrow = True, arrowhead = 1, arrowwidth = 1.5, arrowcolor = 'grey')) # arrow format

# Add H+ label
titration_series.add_annotation(
    dict(text = 'H' + textsup('+'), font = dict(color = 'grey'),
    xref = 'x', yref = 'y', # references for coordinate origin
    x = (770+680)/2, y = 2, # x- and y-coordinate position of the image
```

```
123     xanchor = 'center', yanchor = 'bottom', # location of text box wrt position
124     showarrow = False))       # we want no arrow associated with this annotation
125
126 # eliminate the legend entries for all traces
127 titration_series.update_traces(showlegend = False)
128
129 # Now output the plot
130 titration_series.show('png+browser')
131 titration_series.write_image('titration.png', width = 6*300, height = 4*300)
```

Running the new script the same way as for your previous plotting script,[17] you get the output shown in Figure 4.3.

[17] See Chapter 3.

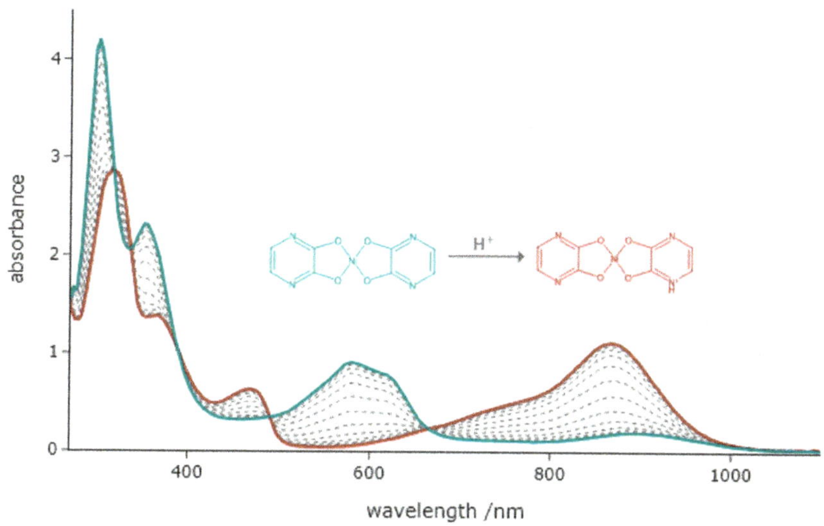

FIGURE 4.3. A plot of a series of UV-visible spectra from a titration experiment, annotated with skeletal formulas diagramming the chemical change and labeling the spectra. This was enabled by being able to represent molecules using RDKit and learning how to add annotations in Plotly.

4.2.0 LINE-BY-LINE EXPLANATION

1–5: Import libraries we will want to use. The `rdkit` imports are needed to produce the desired drawing of our molecule, and `base64` is needed to encode the drawing for use in Plotly. We also import the `textsup()` function from the codechembook library, which helps us give Plotly the HTML formatting it needs for superscripts.

Even though we are going to be working with a Plotly Figure object, we do not need to import Plotly because we are simply going to add this code to the code from last chapter, which already imported the parts of Plotly we need. If you wanted to, you could still include the import of `make_subplots()`—doing so is not a problem (it will not be loaded twice), it is just not needed.

7–47: This is the code from the previous chapter.

48–52: Import the libraries needed for drawing the molecular structure, handling the image data, and typesetting HTML.

54–87: We will define a function that will produce the desired molecular drawing from `rdkit`, formatted correctly for inclusion within a Plotly Figure object. Though we could just include this code directly in the `for` loop defined later (line 51), the use of a function helps us separate out the generation of the figure. This kind of compartmentalization of code can help with readability, as it separates out the code by task.

54: Start with the definition of the function, with the **parameters** that are expected. We are going to need a SMILES string, a color for the molecule, how wide to make the drawing, and how tall to make the drawing. These are captured by the arguments `smiles`, `color`, `bonds_wide`, and `bonds_tall`, respectively. They will all be required.

55–65: The docstring for the function. This explains the purpose of the function, what the parameters are, and what the function will return. The parameters are defined by giving the name of the parameter, followed by the **data type** that is expected. Then, there is a terse description. The return does not have a name because we don't know what it will be called outside the function, but does have the data type and a terse description.

68: Using the `Chem.MolFromSmiles()` function, we produce a RDKit molecule **object**, which we assign to a variable `mol`. This creates the information that RDKit will need to produce the bonds and atom labels that form the drawing. We specify the molecule is using a SMILES string held within the variable `smiles`.

71 and 72: On these two lines we add entries to a **dictionary** that RDKit will use to specify colors of atoms and bonds. Essentially, this dictionary contains **key**:**value** pairs that are <atomic number>:<color>. This dictionary is an attribute of the DrawingOptions object (see below) in RDKit, and when we tell RDKit to use this dictionary, for each atom, it will look at this dictionary to see what color that atom should be—identifying the atom by its atomic number.
We are going to draw a molecule so that all bonds and atoms are the same color. To ensure we have the same color for all atoms, we will simply create entries for all atomic numbers (from 1 to 118). We use a `for` loop and iterate through a `range` object. For each entry, we use the atomic number as a key and assign the value as the appropriate color, using the variable "color." Thus, we have the color designated for any atom we could possibly be using.

75: We create an RDKit `Drawer` object that will actually generate the data that specifies the scalable vector graphic (svg) and assign this object to the variable "`drawer`." When doing this, we can tell it how wide and how tall we want the drawing to be. This is done in units of pixels. There is a default size for font, so we want to make sure the drawing is large enough for the heteroatom labels and the bonds to show when the labels use this font size. We have found that thinking in terms of the number of bonds moving horizontally or vertically and then scaling that number by a factor of 50× produces reasonable results. You can change this scaling if you like, and see if you prefer the output when the figure is scaled differently.

76: We tell the drawer object to use the dictionary we created of all atoms colors.

77: We set the background color for the drawing. Here, there are four numbers provided. These are the **rgb**a values, where a stands for alpha and controls how opaque the color is. By setting it to zero, the background will be completely transparent, which just makes it a bit easier to use the image as an annotation because you don't need to worry about the image overlapping with any other aspects of the plot.

80 and 81: Start drawing the molecule, and then finish the drawing. At this point, we don't need to worry about anything else other than both of these are needed to get the final molecule.

84: The drawer object contains *instructions* for drawing in text form. Here, we get this text and then remove the label "svg" from it, by replacing this bit with an empty string (''). This is needed before we encode the instructions to base64, simply because of how the base64 library works.

85: Encode the svg text into base64 uri because this is needed to add to a Plotly Figure object. We assign this to a variable, and this variable can be used with Plotly to add a figure to a plot. Don't worry too much about the details of what this line is doing—this is just boilerplate that we have to do.

87: Using the `return` keyword, we tell the function to **return** the value of the base64 uri.
This completes the function. We will move on to the rest of the script, where we will call the function from inside a `for` loop.

90 & 91: Provide the SMILES strings for the base and acid forms of the compound that was titrated. Save them as a variable for later use.

94 & 95: We wish to match the colors of drawing we will make to the traces on our plot, so we need to define what they are. Here we have colors that match dark cyan and dark red from Plotly, specified in the rgb format, with values that range from zero to one. These will ultimately be used with Rdkit to set the colors. If you ever need to find out what the rgb values are for named colors, you can look in the Plotly documentation, or use a color-picker tool to determine the rgb of the color.

98–109: We are now ready to add the drawing of both the base and acid forms of our molecule (the SMILES of which were given in Lines 43 and 44). We will need to perform similar operations for both, but color them differently and place them in different places on the final plot, so we begin by using `zip()` function so that we can iterate through the SMILES strings, colors, and positions together. We provide these lists on lines 52–54, where we end with the the : character that specifies that we are moving into the `for` loop. The SMILES strings are assigned to the variable `species`, the colors to the variable `color`, and the positions to the variable `positions`.

104–109: We call the `add_layout_image()` method of the Plotly Figure object to add an image to the plot. We specify how this is added, using a **dictionary**. Here we will ultimately add two images, one for each iteration of the `for` loop.

105: The value for the `source` needs to be the image to be embedded, and here we will provide the base64 encoded uri for the image. Since this is also what the function we wrote above returns, we can simply call the function, specifying the species, color, and width and height for the image. The returned object will then be assigned as the value associated with the `source` keyword.

106: The values for the `xref` and `yref` keys specify how the x and y values are referenced. That is, do they use the values on the axes, or relative to the "paper" (plot area), etc. We will use the actual x and y coordinates.

107: The values for the `x` and `y` keys provide the coordinates at which to place the image. We use a common y-value, but the position variable from the for loop for the x. This will result in both images being placed at the same vertical position, but separated horizontally (and placed above the position of the band we are interested).

108: The values for the `xanchor` and `yanchor` keys specify the alignment for the image. We will center this horizontally and align at the top, vertically.

109 The values for the `sizex` and `sizey` keys specify the maximum width and height of the image being drawn, in the same units as the axes. You

will note these are different from those that were passed to our function because those were used to determine the size of the image generated by Rdkit, whereas this is the size of the images when added to Plotly. The image is sized according to the x and y coordinates (because we specified `xref="x", yref="y"` on line 159), so this will be 200 nm wide and 1 absorbance unit tall on the plot. So the image returned is resized when added to Plotly. Because we used an svg, the image can be resized without a loss in resolution and this will produce a nice looking image. We also provide the closing parentheses needed. The first closes the dictionary, while the second closes the `add_layout_image()` method.

112–116: Using the `add_annotation()` method of the Plotly Figure object, we add an annotation to the plot. We also specify the details using a dictionary. Here, we are going to draw an arrow between the two images, to symbolize the chemical reaction.

113: The values for the `ax` and `ay` keys provide the coordinate of the *beginning* of the arrow.

114: The values for the `x` and `y` keys provide the coordinate for the *end* of the arrow.

115: The values for the `axref`, `ayref`, `xref`, and `yref` keys provide the type of reference for the coordinates of the beginning and end of the arrow, respectively. Again, we will use the coordinates of the x and y axes.

116 We want an arrow, so we tell Plotly to show it (strictly speaking, this is not needed because the default option is to show the arrow, but we write this out in order to make it very clear we have an arrow. Sometimes it can be nice to be a bit more verbose in order to make code more readable). We also specify the size of the head and the width of the line, both in pixels, and we specify the color of the arrow. Because this method is in the Plotly library, we can use a named color, in this case, `grey`. Finally, we close both the dictionary and the `add_annotation()` method.

119–124: We repeat the above annotation process, now *without* an arrow, but with text that specifies we have H^+. This is nearly identical to above; however, there are a few differences:

120: We provide text that will be used as the annotation and also provide a color and size for this font. Note that we need to use HTML code to specify a superscript for the + symbol, and we use the `textsup()` function from `codechembook` to do this for us.

121: We specify the reference point for the coordinates specifying the annotation position. By setting these references to `x` and `y`, the positions will be with respect to the axes scales.

122: We next set the position of the annotation by specifying the x and y values at which we want the annotation to appear.

123: We set the alignment of the text with respect to the position of the annotation using the `xanchor` and `yanchor` keyword arguments. In this case, we have it horizontally aligned to the center and vertically aligned at the bottom.

124: By default, Plotly will include an arrow (actually, just a line), and so we have to set `showarrow` to `False` in order to avoid this.

127: Because we are going to have color-coded structures, we don't need a legend anymore. So, we update the Plotly traces to not show a legend using the `update_traces()` method of the Plotly figure. This applies the specified changes to *all* traces in the figure, unless we specify otherwise. We could have done this by only updating the last two traces (which were the only two left in the legend) by calling this function twice, once for each trace. But since *none* of the traces need a legend anymore, this usage requires fewer lines of code.

130 & 131: These are the final two lines of the code from the last chapter to output the plot—now containing our molecular structures annotated on it.

Your labmate then says, "I know this was a lot, but you don't always need to go through all of this. If you are okay with the default settings, RDKit is actually pretty straightforward to use. Now you know how to use more of its functionality. It is a good thing to be a chemist and know how to draw chemical structures."

CHAPTER 4 EXERCISES

TARGETED EXERCISES

Exercises for Section 4.1.1

0. Starting with the dictionary that you made in Exercise 23 of Chapter 3, print the following properties for the given solvent:

a) Boiling point of benzene.

b) Density of cyclohexane.

c) Density of toluene.

d) IUPAC name of *n*-hexane

e) Density of a mixture of 75% acetonitrile and 25% *n*-hexane (assume ideal behavior).

1. It can be useful to get a collection of the keys and values of a dictionary. If you have a dictionary, my_dict, you can get a list of the keys using list(my_dict.keys()) and a list of the values using list(my_dict.values()). Using this approach, print out all the keys and values for the dictionary you used in Exercise 0.

2. You can iterate over dictionary keys, using for key in my_dict: Using this approach, and the dictionary from exercise 0, develop a pair of nested for loops that will print out the string "the value <property value> is associated with the <property> of <solvent>", where <property value> is the dictionary value of each solvent dictionary, `<property>` is the key associated with the value, and `<solvent>` is the solvent under consideration. The last two will be keys of the dictionary.

Exercises for Section 4.1.2

3. Update the dictionary from Exercise 0 to include the SMILES strings of each solvent.

Exercises for Section 4.1.3

4. Using rdkit, create an image of the structure of each molecule in the dictionary from Exercise 0. If you use the updated version (Exercise 3), then you can simply access the SMILES from the dictionary.

Exercises for Section 4.1.4

5. Using two different methods, make a tuple from the list [3, 'frog', 9.73, False, 6.02e23].

Exercises for Section 4.1.5

6. Make a dictionary of the rgb color designations for the following colors and assign it to a variable. Use the convention that the maximum for any given channel is 255:

   a) Pure red

   b) Pure blue

   c) Pure green

   d) Pure cyan

   e) Pure purple

f) Pure orange

g) Black

h) 50% gray

7. Find the rgb color designations for the colors of your current school or a past school. Continue using the convention that the maximum value is 255.

Exercises for Section 4.1.6

8. There are no Exercises for this section.

Exercises for Section 4.1.7

9. Find a picture of the logo of your current/past school, lab, or employer and add it to a blank Plotly plot. If you find a bitmap image file (e.g., .png, .jpg, .jpeg, or so on), you can use `source = <path to image>`, though you can also read the online documentation for Plotly to find out how to do so. Additionally, if you wish to hide the axes, you can supply the keyword/value `visible=False` in the `update_xaxes()` and `update_yaxes()` methods.

Exercises for Section 4.1.8

10. Starting with the final code from Chapter 2, add an arrow pointing to the highest peak in the spectrum. The arrow line should have a width of two pixels and be red. The annotation text should also be red and say "`Measure here (<wavelenth> nm)`."

Exercises for Section 4.1.9

11. On a blank Plotly canvas, produce an annotation that is the properly formatted chemical formula for your five favorite molecules or complexes. This means using subscripts and superscripts.

Exercises for Section 4.1.10

12. Using `codechembook.symbols.typesettingHTML`, write out the following, using using proper formatting:

a) $\mathrm{p}K_a$

b) ^2H

c) $y = a \cdot x^2 + b \cdot x + c$

13. Repeat Exercise 11 using the functions in `codechembook.symbols.typesettingHTML`. If you find formatting the chemical formulas frustrating, don't worry, in the next chapter you will learn a better way to do this.

Exercises for Section 4.1.11

14. Create a list (not a generator) of integers from 0 to 10, including 10 using an expression that includes `range()`.

15. Create a list (not a generator) of atomic numbers for the third row atoms using an expression that includes `range()`.

16. Create a list (not a generator) of integers from 0 to −9, including −9 using an expression that includes `range()`.

17. Create a tuple(not a generator) of integers between 0 and 100, including 100, spaced by 10, using an expression that includes `range()`.

Exercises for Section 4.1.12

18. You have the following two lists: one that holds the masses in grams of solvent in your sample (`[0.567, 0.523, 0.263, 0.983, 0.634]`), and one that holds the corresponding solvent name ((`['n-Hexane', 'Cyclohexane', 'Acetonitrile', 'Toluene', 'Benzene']`). Write a single loop that prints out the volume of each solvent you used, including its name. Use the dictionary you made in problems 21 and 22 of Chapter 3 to get the densities.

Exercises for Section 4.1.13

19. Make a function that computes the length of a list and returns it as an integer. Do not use `len()`.

20. Make a function that accepts a pair of x & y values and computes the slope and y-intercept associated with the line that would pass through both points and then and returns the value for the slope and the intercept.

21. Modify the function from Exercise 19 to instead return a dictionary with two keys, "`slope`" and "`intercept`."

Exercises for Section 4.1.14

22. Make docstrings for the functions you created in Exercises 18–20. Include a description of the function's purpose, the parameters it uses (with type and description), and the return value(s) (with type and description).

COMPREHENSIVE EXERCISES

23. Starting with the dictionary from Exercise 3 of this chapter, write a code that will draw each chemical structure on a blank Plotly canvas. Arrange them in a grid. If you wish to hide the axes, you can supply the keyword/value `visible=False` in the

`update_xaxes()` and `update_yaxes()` methods. In this chapter, we demonstrated how to display a svg image, if you would like use a png instead, you can use the function `rdkit.Draw.MolToImage()`, supplying the mol object as the first **positional argument**.

24. You decide you want to include the formulas of the molecules in the final figure we created this chapter. Add them using proper superscripting and subscripting for the formula and its charge. Make the formula text color also match the colors of the molecules they label using the `font` parameter of the `add_annotation()` method. If you need help, you can refer to the technical chapter on Plotly (Chapter I).

25. Change the colors used in the final code of this chapter to ones of your choice. You should change the colors of the initial and final data, as well as the molecules being drawn. Make sure the colors for the molecules match those for the data.

26. Use the Spyder variable explorer or `print()` to inspect the Plotly figure object created by the code in this chapter. You will find that it is structured like a dictionary. Find the keys for the following elements:

a) x-data.

b) x-axis title.

c) Color of the line representing the last data series.

PART TWO

MODELING AND FITTING DATA

CHAPTER 5

FITTING EQUATIONS TO DATA

- ☐ **New concepts**
  — *and their prerequisites*

- ☐ **Prompting the user for information using `input()`**
  — *Section 0.1.2: Accessing capabilities beyond basic math using functions*
  — *Section 0.1.6: Expressing data as readable text using strings*
- ☐ **Turning numeral characters into numbers**
  — *Section 0.1.6: Expressing data as readable text using strings*
- ☐ **Inserting calculation results into strings using f-strings**
  — *Section 0.1.6: Expressing data as readable text using strings*
- ☐ **Formatting the representation of numbers in strings**
  — *Section 0.1.6: Expressing data as readable text using strings*
- ☐ **Generating symbols that aren't on your keyboard using `codechembook.symbols`**
  — *Section 0.1.6: Expressing data as readable text using strings*
  — *Section 2.1.7: Adding `codechembook` to your Python installation*
- ☐ **Producing formatted chemical formulas using `codechembook.quickHTMLFormula()`**
  — *Section 0.1.6: Expressing data as readable text using strings*
  — *Section 2.1.7: Adding `codechembook` to your Python installation*
- ☐ **Finding paths to files in folders using `glob()`**
  — *Section 2.1.5: Describing the location of files or folders on your computer using Pathlib*
- ☐ **Dividing a string into multiple substrings with `split()`**
  — *Section 0.1.2: Accessing capabilities beyond basic math using functions*
  — *Section 0.1.6: Expressing data as readable text using strings*
- ☐ **Finding the element closest to a given number in a Numpy array**
  — *Section 0.1.2: Accessing capabilities beyond basic math using functions*

- *Section 1.1.1: Operating on collections of data efficiently with Numpy arrays*
- *Section 3.1.5: Access items in lists, arrays, or strings by indexing*

☐ **Fitting models to data with lmfit**
- *Section 1.1.0: Giving your computer instructions using Python commands*

☐ **Analyzing fit results in lmfit**
- *Section 2.1.4: Exploiting built-in methods to manipulate data stored in objects*

☐ **Building figures with multiple subplots in Plotly**
- *Section 2.1.10: Making appealing and interactive plots using Plotly*

5.0 PROBLEM

The next thing your advisor wants from the UV-vis data is the extinction coefficients for the metal-to-ligand charge-transfer (MLCT) bands of the two endpoints (the deprotonated and protonated states), because they will need to be reported in a manuscript. You do a simple experiment in which you measure the absorption spectrum for a series of concentrations, and you know you need to fit this to a line where the slope allows you to calculate the extinction coefficient. Basically, this comes from Beer's law:

$$A = \epsilon c l$$

where A is the observed absorption at some specific wavelength, ϵ is the extinction coefficient, c is the concentration, and l is the path length.

You work with an undergraduate you recently started mentoring and teach them to perform the data collection on a series of concentrations of your compound. Later, they send you this data as a series of individual csv files (one for each concentration) and also include an image of the spectrum, taken from the spectrometer. You could do this analysis and fitting in Excel, but if you could do it in Python, it would save time over the Excel workflow that requires opening each file manually, making a chart, adding a trendline, adding the trendline to the display, and using a special function to determine the uncertainty of the fit parameters. Additionally, there are aspects of fitting in Excel that you don't like. For instance, although Excel will give you the slope and intercept of the best-fit line, it doesn't provide estimates of uncertainties in these parameters, and it doesn't show you the residuals of the fit.

You talk to your labmate about this, who says that you absolutely can fit a line to data in Python. They say that there are several ways you can do this, but they think the best is to use a **library** called "lmfit." "This library is powerful, and is going to allow you to fit things that are much more complex than a line. It also helps keep track of things like parameter values, uncertainties, residuals, and so on. So, it is a good library to know."

Your labmate says that all you need is the data you already have, and they can even teach you how to quickly select only the csv files and not the images from a folder. Combined with just a few new ideas about coding, this knowledge can be used to produce a nice result.

> **Why lmfit?**
> A very common practice in experimental science is to try to fit a quantitative model to your data using least-squares regression. The Python library SciPy provides functions for this, but they can be a bit tedious to use. Additionally, analyzing the results (such as uncertainty in estimated parameters) can take some work. The library lmfit streamlines this workflow, and so provides a much nicer interface to work with. Although we think lmfit is the best, it might be worth looking at the library "symfit" as well.

5.1 SOLUTION

5.1.0 PLANNING

Because there are only two parameters you need to fit (slope and the intercept), you could simply report them (and their uncertainties) as text. But it would be nice to produce a figure that shows your data, the fit to the data, the residuals of the fit. You like the idea of showing the residual as a second plot under the first.

"This is going to require that we understand how `lmfit` works and how to include multiple subplots in Plotly. Also, `lmfit` has a lot of capabilities, and we are going to be unable to explain everything about it the first time you encounter it, but we will use more of them soon. Finally, we can introduce a new way top introduce values to code and to identify multiple files to work with." You can diagram the process as shown in Figure 5.1.

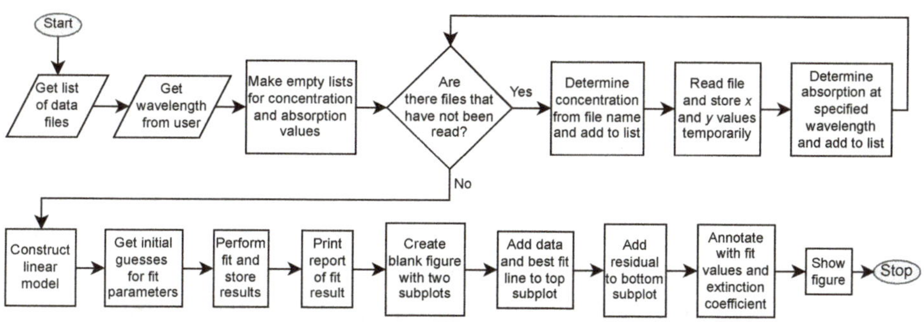

FIGURE 5.1. Flowchart for the code to determine the extinction coefficient of your catalyst. It reads the csv files in the Extinction data folder containing UV-vis spectra at a range of concentrations, extracts the absorbance value at a specified wavelength, and fits this data to a linear model.

5.1.1 PROMPTING THE USER FOR INFORMATION USING `input()`

"There are going to be many times that you want to run the same basic process, but change only a few specifics within in it. For instance, we are going to be using absorbance values at the peak of a specific absorption. But what if you wanted the extinction coefficient for a different spectral feature? You would do the exact same process, but only change the wavelength of interest. Up until now, the only way you could do this is to make a **variable** and assign it the value of interest in the code. If you wanted to change the value, you would need to change the code. This is a solution known as **hard coding**."

"Hard-coding makes it inefficient to change your code dynamically. A better solution might be to write a program that doesn't know the value of interest ahead of time, but instead asks the user to supply this value only when the code is run. There are several ways you might do this, but one simple way is to use the `input()` **function**. The default behavior of `input()` is to expect the user to supply an input to the console using their keyboard. Once the user presses return, the input from the user is **return**ed as a string by the `input()` function, which can then be assigned to a variable."

"It is also common to provide `input()` a string (as a **positional argument**) that prompts the user. For instance, if we wanted to ask what wavelength is going to be of interest for our program we could use:"

```
In: l_max = input("What is the position of the feature of interest? ")
```

"If we run this code, we obtain the following output:"

```
Out: What is the position of the feature of interest?
```

You notice there is a flashing cursor after the question mark. "Now, Python is waiting for your input," your labmate says. "If you want to specify a value, you might supply a number like 580. If you do so, and press return, then that value will be assigned to the variable `l_max`."

You type 580, then hit return. "Also, notice that we left a space at the end of the message so that your input would not start immediately after it. We can verify that it worked by checking the value of `l_max` now:"

```
In: l_max
Out: '580'
```

"One *extremely* important thing to note here is that the value of `l_max` is a *string*—as indicated by the fact that the number 580 is placed between single equation marks. You may find this odd because you input a number. But remember this is accepting input from a keyboard, and you could just as easily typed in 'house,' instead of '580.' Everything that `input()` returns is going to be a string. In our case, it is would be much more convenient if we have a number, and so we are going to need to convert our string '580' into a number. But to do that correctly, we need to understand that, in Python, there are many different kinds of numbers."

5.1.2 TURNING NUMERAL CHARACTERS INTO NUMBERS

"So far, we have treated all numbers as the same. However, Python actually distinguishes between three different kinds of numbers: `int`, `float`, and `complex`. The `int` types are integers (numbers without decimal places), the `float` types are numbers with decimal places, and `complex` types are complex numbers.[1] Complex numbers are indicated by using the construction <digits> + <digits>j,[2] where the first set of digits are the real portion and the second are the complex portion. We have already been using `int` and `float` this whole time, although we have not had to worry about it until now."

"For now, we simply need to know two things. First, we can create an **instance** of each type using the function with the same name: `int( )` will create integers, `float( )` will create floats, and `complex( )` will create complex numbers. You've seen that normally, we don't have to specify the type, and Python just figures it out. However, when converting between data types (say, from a **string** to a number type), then we have to worry about it. Second, we need to know that, from the perspective of Python, these numbers are different. When we want to do something like convert a string to a number, we need to specify which type of number we want. For instance, if we had the string "2.6" and wanted to convert it, we will get different results for each number type:"

```
In: textnumber = "2.6" # this is a string
In: int(textnumber)
Out: 2
```

```
In: float(textnumber)
Out: 2.6
```

```
In: complex(textnumber)
Out: (2.6+0j)
```

"If we really want to have numbers with decimals, we need to ensure we are using `float( )` to create them."

[1] If you want to know more about these numeric data types, you can read about them in Chapter B

[2] Many readers may be more familiar with $i = \sqrt{-1}$. Python uses the notation that $j = \sqrt{-1}$ instead. The reason is that i typically is used for other purposes in programming, such as keeping track of index values.

5.1.3 FINDING THE ELEMENT CLOSEST TO A GIVEN NUMBER IN A NUMPY ARRAY

The next thing you start to worry about is what happens if the user specifies a wavelength that isn't exactly one of the data points in the spectra. You mention this to your labmate, who says, "There are going to be many times that we want to find the index of the value that is the closest to a target value. A simple way to do this for **Numpy arrays** is to subtract the target value from the whole array, take the absolute value of the differences, and then find the minimum of that. The number that was closest to the target value will result in the smallest number."

"Because we know we can use this approach, the Numpy **function** `argmin()` becomes very useful. What this does is examine a Numpy array and return the **index** for the *first* number that is the smallest in the array. We can use this index to then get the number of interest."

```
import numpy as np

arr = np.array([0, 3, 1, 2.3, 3.1])
target = 2.7
```

```
ind = np.argmin(abs(arr - target))
print("closest to", target, "is", arr[ind], "at index", ind)
```

```
Out: closest to 2.7 is 3.0 at index 1
```

"Here, we get the index for the number closest to 2.7, and then get that number using the index. One important thing to realize about `argmin()` is that it does return only the first instance of this smallest value. So, if we do:"

```
arr = np.array([0, 3, 1, 3, 3.1])
target = 2.7
ind = np.argmin(abs(arr - target))
print("closest to", target, "is", arr[ind], "at index", ind)
```

"we will obtain the exact same result as before, even though there is also a value of '3' at index 3."

```
Out: closest to 2.7 is 3.0 at index 1
```

"This is worth keeping in mind as you write your codes."

"If there is an `np.argmin()`, it should not be surprising that there is a `np.argmax()` as well. This does the same thing, but returns the first occurance of the index for the maximum value in the array."

5.1.4 INSERTING CALCULATION RESULTS INTO STRINGS USING F-STRINGS

"It is going to be convenient to write out the results of our fit using strings. But because we also have uncertainty in our fit, we would like to include that. Ideally, would be nice to write a sentence like '<value> ± <uncertainty>.' In our program, <center> and <uncertainty> will be held in **variables**. We do have a way to print this, using a construction like print(<center value>, " +/- ", <uncertainty>), but there are two problems with this solution. First, it is a bit cumbersome to write and read. Second, this doesn't actually produce a **string**, so this construction provides no way to pass the sentence we want to any other function, like Plotly's `add_annotation()`."

"There are several ways to solve the problem of getting a string object with values from variables," your labmate says. "But I think the best option is using **f-strings**. These are strings, but we place an `f` prior to the first set of quotation marks. We can use this to produce a 'normal' string:"

```
In:  f"Hi"
Out: 'Hi'
```

"But we can also place variable names between curly brackets ({}) and the variables will be evaluated."

```
In: pi = 3.14159
```

```
In: f"A nice approximation of pi is {pi}."
Out: 'A nice approximation of pi is 3.14159.'
```

"Actually, f-strings do more than this because really what is happening is that any Python **expression** we place between the curly brackets is evaluated and the value included in the string."

```
In: f'The square of pi is {pi**2}'
Out: 'The square of pi is 9.869587728099999'
```

"These evaluations could include function calls, or really any valid Python statement. As you can imagine, f-strings can be useful in a huge range of contexts, even when you're not outputting text."

5.1.5 FORMATTING THE REPRESENTATION OF NUMBERS IN STRINGS

"If we are going to be placing numeric values inside of **strings**, then it is going to quickly become important to understand how to format these values. For instance, imagine you had 10 g of a chemical you wanted to equally divide up between seven vials. You could use f-strings to provide a nice statement."

```
In: f"Place {10/7} grams in each vial."
Out: 'Place 1.4285714285714286 grams in each vial.'
```

"This is not a great output, because it is hard to read and you likely do not have a balance that reads with attogram precision. Even if you did, there is no need for such precision. Instead, weighing to 1 mg might be sufficient, so we only need to report this to three decimal places. We can format the string by placing a colon (:) after the value and then formatting instructions before the end of the curly brackets."

```
In: f"Place {10/7:.3f} grams in each vial."
Out: 'Place 1.429 grams in each vial.'
```

You note that the number has even been rounded, so that you don't have to do it yourself as a separate step.

"In this usage, the 0.3 means that you want three decimal places, and the f indicates that this should be formatted as a standard float. There are other options where we can ask for scientific notation, **e notation**, and so on."[3]

[3] More information can be found in technical Chapter C.

5.1.6 GENERATING SYMBOLS THAT AREN'T ON YOUR KEYBOARD USING `codechembook.symbols`

"As we do more sophisticated things in Python, we are going to want to produce sophisticated *looking* text. For instance, rather than writing +/− to indicate plus-minus values, we would want to use the character: ±."

You note that your keyboard does not have a ± key, so you wonder how you could type this character. Your labmate says that what you need is the designation for this character that your computer uses. Because a computer views everything as a number, characters are also just numbers. The exact number depends on the type of **binary encoding** being used, but a common one (and one that Plotly can use) is called **unicode**.[4] In unicode, all characters are represented by a string of **hexadecimal** numbers that follow `U+`. For instance, the character `t` is `U+0074`. Thus, when using unicode encoding, the keys on your keyboards are just shortcuts to these designations. If you want to render a character that is *not* on your keyboard, you could look up what the corresponding unicode is, and then enter it directly. For instance, the unicode for ± is `U+00B1`.

[4] If you want a more complete discussion of this, you can refer to Chapter C.

"This approach will work, but can be cumbersome, so the **codechembook library** has a module with shortcuts that use variable names as aliases for the unicdoe strings of common glyphs that you might need. For instance, the plus-minus glyph can be accessed using:"

```
In: codechembook.symbols.math.plusminus
```

"and this will output ±."

"There are useful math glyphs in the `codechembook.symbols.math`, such as `proportional_to` (∝) and `equivalent_to` (≡)."

> **Finding and Using Unicode Characters**
> While `codechembook.symbols` provide a number of useful symbols, they are only a small fraction of all possible symbols (including an ever-growing number of emojis). To find other symbols, you can use lookup websites like https://unicodelookup.com/, which will give you the **hexadecimal** code that specifies the symbol. Unicode characters can be used in strings by including the hexadecimal code (minus the beginning `0x` after `\u`: the superscript plus, for instance, can be inserted into a string by placing `\u207A` where the superscript plus should be.

"The `codechembook.symbols` module has many different symbols, which are organized into several sub-modules.[5] These are `math`, `greek`, `script`, `chem`, and `typography`. We have already seen examples from `math`. The `greek` sub-module contains Greek letters, such as `Gamma` (Γ) and `gamma` (γ). The `script` sub-module contains script versions of Latin letters, such as `l` (ℓ) and `H` (ℋ). The `chem` sub-module contains symbols that might be of use to chemistry, such as `degree` (°), `angstrom` (), and `wavenumber` (cm^{-1}). Finally, the typography sub-module contains glyphs that do not fall into the above categories, such as `copyright` (©) and `ellipsis` (...)."

[5] You can find a description on https://www.codingforchemistsbook.com/, and they are also discoverable when using them in an IDE, meaning that auto-complete hints should pop up with them.

5.1.7 PRODUCING FORMATTED CHEMICAL FORMULAS USING `codechembook.quickHTMLFormula()`

"Since we are chemists, we often want to include chemical formulas on plots that we make—for instance, we can add a title to a Plotly plot using the `title` **keyword argument** inside the `layout()` **method** of the figure object. We know that, for annotations and labels in Plotly, we need an HTML formatted string. We already know how to produce a properly formatted chemical formula using the HTML superscript and subscript tags `<sup></sup>` and `<sub></sub>` or the quicker functionality of `codechembook.symbols.typsettingHTML`. However, doing it this way is tedious because you have to format each **string** element separately. Fortunately, there are tools out there that can parse a chemical formula and produce a useful string for you to include in a plot or anything else. One approach is included in the `codechembook.quickTools` module."

"We can use `codechembook.quickTools.quickHMTLFormula()` to produce these HTML-formatted chemical formulas. To use this function, we simply pass the formula you want as a string for the first **positional argument**. For instance, if we wanted the condensed formula for n-octane, we could type:"

```
In:  from codechembook.quickTools import quickHTMLFormula as qf
In:  qf("CH3(CH)6CH3")
Out: 'CH<sub>3</sub>(CH)<sub>6</sub>CH<sub>3</sub>'
```

"You can see that this function returns a string that automatically contains HTML code specifying the subscripts needed. If you want to specify charge, you don't do it in the chemical formula, but use the `charge` keyword argument."

```
In:  qf("SO4", charge = -2)
Out: 'SO<sub>4</sub><sup>2-</sup>'
```

"If you want strings formatted for other applications, there are other functions. For instance, you can get a unicode-formatted string using `quickUnicodeFormula()` if you want to output, say to the command line instead of Plotly. You can get a LaTeX-formatted string using `quickLatexFormula()`. For instance:"

```
In:  from codechembook.quickTools import quickUnicodeFormula as quf
In:  quf("SO4", charge = -2)
```

which will produce: '$SO_4{}^{2-}$'

These functions from `codechembook` are implementations of the `chemformula` **library**,[6] but we have included them in our library, because `chemformula` was not available on Conda (at the time of this writing). The `ChemFormula` object that the `chemformula` library produces also contains a number of other useful attributes and methods. You can also import the `codechembook.chemformula` module, which will give you this functionality, and read the documentation for the `chemformula` library on how to use it.

[6] https://github.com/molshape/ChemFormula

5.1.8 FINDING PATHS TO FILES IN FOLDERS USING `glob()`

"Sometimes, you will want your code to select files for you, rather than specifying them yourself using code or the file dialog. This is especially true if you have a folder with many different types of files in it. For this, the `glob()` **method** of the **Path object** is very useful. It accepts an argument that specifies a pattern, and then `glob()` looks through the Path object and finds all items that match this pattern."

"However, we would rather not have to know or specify the entire path for every file, since your files will have unique names. Instead, we often specify a pattern to match file names to that uses what is called a '**wildcard**.' Essentially, any file name that contains the text specified in the pattern, *and* that contains text that can be replaced by the wildcard (*) and still match the pattern, will be valid. For instance, imagine we wanted to find all of the Python script files (files with **extension** .py) in a folder, named <folder>. If we make a Path object to this folder, then we can use: `<folder>.glob('*.py')`."

"In this usage, `glob()` will look through the <folder> object and find all files that have '<anything>.py.' Because the .py is specified, the files must contain it, and because there is a wildcard (*) before that, we will find all files that end in .py. Because Python files end in .py, `glob()` will find all the Python files in <folder>."

"One technical aspect to be aware of is that `glob()` produces an object called a **generator**. We encountered this for the first time when we introduced the `range()` function.[7] A generator is an object that doesn't have the objects you will use, until you ask it for them (i.e., it generates them on demand). One way to get these objects all at once is to ask for them as a **list** using: `list(<glob generator object>)`. Here, the generator will generate **strings** that contain all the filenames and their paths, and put them in a list."

[7] See Section 4.1.11.

"Finally, we can use the `codechembook.quickTools` module function `quickSelectFolder()` to open a dialog box that allows us to get the path to a folder that we want to glob, or any other scenario where we might want to have the path to a folder instead of file(s)."

5.1.9 DIVIDING A STRING INTO MULTIPLE SUBSTRINGS WITH `split()`

Now you start to think about getting the values for the concentration of each spectrum. You remember that they are recorded in the file names of your spectra. You could make a new file with these separately and read that, or you could hard code them. "Or!" your labmate says, "this is an example of a time where it is useful to break apart text **strings** to get their contents. As a simple example, think about what we would do with a csv file, where values are separated by commas. We would break apart the text where these commas are and put the pieces into a list. We can accomplish this using the `split()` **method** of a string **object**. This **function** accepts a **positional argument**, which is the pattern we will split on."

```
In: data = "1325, 32512, 232314, 12311,"
In: split_data = data.split(",")
In: print(split_data)
Out: ["1325", " 32512", " 232314", " 12311", ""]
```

Here you can see that `split()` **return**ed a **list** of strings, where the strings are the bits that were separated by commas in the original string. You also notice that the list items are still strings, so they are not numbers (from a **data type** perspective). Finally, it is clear that the text split *exactly* on the supplied argument. Because you provided "," with no spaces, it has split the string only on the comma, and preserved the spaces after the commas. At first, it is confusing that list ends with an empty string. You realize that the original string ended with a comma, so there was nothing that followed it, and the split function returns the "nothing" that followed it, but as a string—in other words the empty string.

"In your case, all you would do is split your file name using whatever character is in front of the concentration!"

5.1.10 FITTING MODELS TO DATA WITH LMFIT

You recall that the process of fitting **model**s to data includes the following steps: identify a model that produces our expected behavior, determine the fitting parameters that control the behavior of our model, and optimize the values for these parameters until the predicted behavior is "close enough"[8] to the experimental behavior. At this point the fitting is finished and you use the final parameter values as the best-fitting values for those parameters. In an ideal world, you can also analyze the process of the fitting in order to estimate uncertainties in these final parameter values, although you remember this being mathematically complex.

[8] There are multiple ways that this is defined. The default for lmfit is that if the difference in the sum of residuals squared between two parameter guesses is below some threshold. If you wish to have more information, consult the online documentation for lmfit.

Your labmate agrees. "Fitting can be quite involved, but fortunately, there is a **library** that can simplify this for us: lmfit. If you really want to understand how lmfit works, then you can read the online documentation. However, the basics of its use are as follows:"

1. Create the model we wish to use, and the parameters needed for that model. In our case, we want to use a linear response model, and so the parameters are the slope and intercept of a line.

2. Define the parameters and make an initial guess at their values.

3. Fit the model object to our experimental data using these initial parameter value guesses. This will generate a 'ModelResult' **object** that contains a significant amount of information about the fit. We can use **attribute**s and **method**s of this object to access this information.

"Again, more information can be found in the online documentation for lmfit. However, this list provides enough structure to follow how we will use it below."

5.1.11 ANALYZING FIT RESULTS IN LMFIT

"One nice thing about `lmfit` is that the result of the fit—a ModelResult **object**—has a **method** for generating a report of the fit: `fit_report()`. Using this method without any

arguments will produce a formatted report that gives important information regarding the fit. Specifically, it will provide the model that was fit to the data, the method of fitting, several goodness-of-fit parameters (including reduced-χ^2 and R^2), the final value of the parameters and their uncertainties, and any correlation between the parameters. Thus, simply by invoking this function, we can see most of the information we would be concerned about."

"Although metrics like reduced-χ^2 and R^2 are commonly used to indicate goodness of fit, there are other considerations that are just as important. For instance: does the resulting fit line make sense? Is there a pattern in the residuals (the difference between the fit line and the data) that needs to be explained? For this reason, we recommend, *at a minimum*, always plotting the data, the fit result, and the residual. Fortunately, `lmfit` also provides ready means to do so. The fit result object has many **attribute**s, but the ones we will care most about are `best_fit` and `residual`. They provide the y-values for the best fit and the residuals of the fit, respectively."

5.1.12 BUILDING FIGURES WITH MULTIPLE SUBPLOTS IN PLOTLY

"We are already using the `make_subplots()` function. So far, we have not provided it any arguments, so it produces the default of a single subplot. However, we can tell it how many subplots we want to use, which is done using the `rows` and `cols` **keyword arguments** (notice the `s` at the end of these keywords). Each of these keywords accepts an integer that specifies how many rows and columns of subplots you want. Then, when using other functions that modify a plot (e.g., `add_trace()`, `update_xaxes()`, etc.) we use the `row` and `col` keyword arguments (notice the *lack* of an `s` at the end of these keywords) to specify which subplot to apply them to. Counting for the subplots starts at one, row counting starts at the top, and column counting starts on the left. Thus, `row=1` and `col=1` always selects the top left subplot."[9]

[9] Much more information on this can be found in Chapter I.

5.2 CODE

You are now ready to produce code that can ask the user for the wavelength they wish to find the extinction coefficient for, read the data from csv files and extract the targeted data, and then fit this to a linear model, outputting the results and a plot of the results.

```
'''
A script to fit a calibration experiment to a line, starting wtih the UV/vis
spectra
    Requires: .csv files with col 1 as wavelength and col 2 as intensity
              filenames should contain the concentration after an '_'
    Written by: Ben Lear and Chris Johnson (authors@codingforchemistsbook.com)
    v1.0.0 - 250204 - initial version
'''
import numpy as np
from lmfit.models import LinearModel
from plotly.subplots import make_subplots
from codechembook.quickTools import quickSelectFolder, quickHTMLFormula
```

```
import codechembook.symbols as cs

# identify the files you wish to plot and the place you want to save them at
print('Select the folder with the calibration UVvis files.')
filenames = sorted(quickSelectFolder().glob('*csv'))

# Ask the user what wavelength to use for calibration
l_max = float(input('What is the position of the feature of interest? '))

# Loop through the file names, read the files, and extract the data into lists
conc, absorb = [], [] # empty lists that will hold concentrations and absorbances
for f in filenames:
    conc.append(float(f.stem.split('_')[1])) # get concentration from file name and add to list
    temp_x, temp_y = np.genfromtxt(f, unpack = True, delimiter=',', skip_header = 1) # read file
    l_max_index = np.argmin(abs(temp_x - l_max)) # Find index of x data point closest to l_max
    absorb.append(temp_y[l_max_index]) # get the closest absorbance value and add to list

# Set up and perform a linear fit to the calibration data
lin_mod = LinearModel() # create an instance of the linear model object
pars = lin_mod.guess(absorb, x=conc) # have lmfit guess at initial values
result = lin_mod.fit(absorb, pars, x=conc) # fit using these initial values
print(result.fit_report()) # print out the results of the fit

# Print out the molar absorptivity
print(f'The molar absorptivity is {result.params["slope"].value / 1:5.2f} {cs.math.plusminus} {result.params["slope"].stderr:4.2f} M{cs.typography.sup_minus}{cs.typography.sup_1}cm{cs.typography.sup_minus}{cs.typography.sup_1}')

# Construct a plot with two subplots.  Pane 1 contains the best fit and data, pane 2 the residual
fig = make_subplots(rows = 2, cols = 1) # make a blank figure object that has two subplots

# Add trace objects for the best fit, the data, and the residual to the plot
fig.add_scatter(x = result.userkws['x'], y = result.best_fit, mode = 'lines', showlegend=False, row = 1, col = 1)
fig.add_scatter(x = result.userkws['x'], y = result.data, mode = 'markers', showlegend=False, row =1, col = 1)
fig.add_scatter(x = result.userkws['x'], y = -1*result.residual, showlegend=False, row = 2, col = 1)

# Create annotation for the slope and intercept values and uncertainties as a string
annotation_string = f'''
slope = {result.params["slope"].value:.2e} {cs.math.plusminus} {result.params["slope"].stderr:.2e}<br>
intercept = {result.params["intercept"].value:.2e} {cs.math.plusminus} {result.params["intercept"].stderr:.2e}<br>
R{cs.typography.sup_2} = {result.rsquared:.3f}'''

fig.add_annotation(text = annotation_string,
                   x = np.min(result.userkws['x']), y = result.data[-1],
                   xanchor = 'left', yanchor = 'top', align = 'left',
                   showarrow = False, )
```

```
57
58 # Create annotation for the extinction coefficient and uncertainty
59 fig.add_annotation(text = f'{cs.greek.epsilon} = {result.params["slope"].value:5.2f} {cs.
       math.plusminus} {result.params["slope"].stderr:4.2f} M<sup>-1</sup>cm<sup>-1</sup>',
60                    x = np.max(result.userkws['x']), y = result.data[0],
61                    xanchor = 'right', yanchor = 'top', align = 'right',
62                    showarrow = False, )
63
64 # Format the axes and the plot, then show it
65 fig.update_xaxes(title = 'concentration /M')
66 fig.update_yaxes(title = f'absorbance @ {l_max} nm', row = 1, col = 1)
67 fig.update_yaxes(title = 'residual absorbance', row = 2, col = 1)
68 fig.update_layout(template = 'simple_white', title = f'calibration for {quickHTMLFormula
       ("(C4H2N2S2)2Ni")}')
69 fig.show('png')
```

You run this code, navigate to the folder "Titration," which is a subfolder of "Catalyst," and select it. You enter 581 when the code asks you for the wavelength at which to calibrate, and it quickly produces the fit report output and figure, both shown in Figure 5.2.

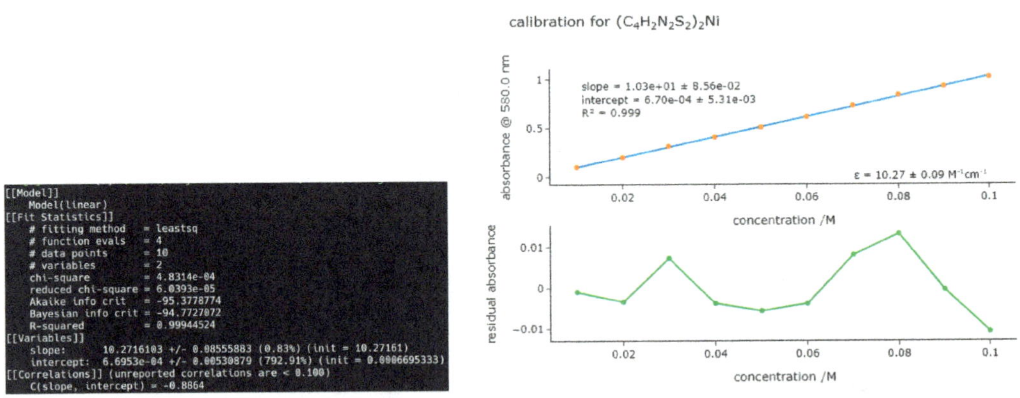

FIGURE 5.2. Results obtained using the lmfit **library** to fit a linear model to data of absorbance versus concentration. The left shows the output from the `.fit_report()` method of the lmfit result **object**, as output into a **console**. The right shows a plot of the data, fit, and residuals of the fit, plotted in two subplots using Plotly.

5.2.0 LINE-BY-LINE EXPLANATION

9–13: Import the libraries we need. One new library is lmfit, from which we import the `LinearModel` from the `models` class. This will provide the behavior we want to fit to our data. We also introduce the `quickSelectFolder()` function from the `codechembook` library. This function allows the user to select a folder using the native file explorer, and returns the path to this folder as a Pathlib Path object. We further import the `quickHTMLFormula()` function to handle formatting of our chemical formula for the title of the plot. Finally, we import the `symbols` module of `codechembook`, which we are going to use to get **unicode string** for characters that are commonly used in science.

16: We are going to have the user identify a folder. Because we have been using the console to communicate with the user, we can print a message here that tells them what we want them to be doing. This can be helpful when people other than the coder run the code, or if you come back to your code after a long time. Moving forward, we will explore other means to prompt the user, but for now, this is a nice solution.

17: We use the `quickSelectFolder()` function to have the user identify a folder where the data is. This will return a Path object, so we can use the `glob()` method of this Path object. This method will find all files that match the pattern supplied. In this case, the pattern is a wildcard (`*`) followed by `.csv`, so `glob()` will match to all files ending in ".csv". Glob returns a **generator** for the specified Path objects in this folder, and we pass this result to `sorted` function, so that this list is sorted, and then we assign the result to the variable `filenames`. It is not strictly necessary that the files be in order, but it does make things a bit nicer down the line and costs very little time or typing.

20: We are **nesting** functions here. By this we mean that we are taking one function (`input()`) and then using the value that is returned as the argument for another function (`float()`). Nested functions evaluate from the innermost set of parentheses out. In this case, the effect is to obtain input from a user, which we expect to be a number, and then convert this input from a string (the **data type** `input()` returns) to a **float**. If the user supplies a data type that cannot be converted to a float, then the code will stop and automatically generate an **error** message.
When running this code on your own, we think a good wavelength to use here is 581 nm, but feel free to play around with other values!

23: To form a calibration curve, we need two sets of data: the concentrations of our species and the absorption at these concentrations. We make two empty lists to which we will add these values as we pull them from files and assign them to two variables on the same line using **unpack**ing.

24–29: We start a **`for` loop**, in which we go through each item in our list of filenames one by one. This means we are iterating through the Path objects that are the paths to our data files.

25: We need to know the concentration of the species. This is contained in our filenames, so we can extract it from the filenames. Because the `f` variable points to a Path object, we can get just the name (without the extension) using the `stem` attribute of the Path object. Once we have the filename, we can use the `split()` method of the string to split it into a list where each item in the list is a part of the filename before or after a "_". We then get the *second* item in the list (index 1) returned from the `split()` method, which is the concentration. However, this is returned as a string, and we want it to be a float, so all the prior commands (which return the concentration as a string) are contained in the `float()` function to convert the string to a floating point number. Finally, all of this is contained in the `append()` method of our `conc` list. The end result of this line is that we take the file we are currently on in the `for` loop, and we extract the concentration as a float and add it to our list of concentration values. It is worth noting that this only works because of the file naming convention that we have used—so in your own work, it might be worth carefully considering how you name your files.

26: We use the `genfromtxt()` function, just as we have in past chapters, to extract the data of interest from the file, unpacking the returned arrays into two variables. These hold the x and y data.

27: We use the `argmin()` function of Numpy to find the index corresponding to the x-value that is closest to the value of `l_max`. The way this works is that we subtract the value of `l_max` from our Numpy array of x-values. We then take the absolute value of this result, using the `abs()` function. In the resulting Numpy array, the item with the smallest value will be the item at the same index as the item in the original array that was closest in value to `l_max`. We then use `argmin()` to find the index of this smallest value, and assign this index to a new variable.

28: Using the index we obtained in the line above, we obtain the value held in the Numpy array of y-values at that same index. This results in pulling the y-value for the index that was closest to our target `l_max`. We add this to the `absorb` list using the `append()` method of the list.

31: We create a lmfit Model object, using the `LinearModel()` function that we imported on line 10. This Model includes instructions to create a line and manage the parameters *slope* and *intercept*. We assign this to a variable for use later on.

32: We invoke the `guess()` method of the `LinearModel` object, passing it the y-values as a positional argument and x-values as a keyword argument. This method then determines initial guesses for the model parameters. Because this is a linear model, the parameters are the slope and intercept of the line. We assign these initial guesses object to a variable.

33: We use the `fit()` method of the `LinearModel` object to fit the model to our data. So we provide the y-values and the initial values of the parameters as positional arguments, while we also provide the x-values as a keyword argument. This will return a fit object, which we assign to a variable.

34: Using the `fit_report()` method of the result object, we get the fit report, and then print it to the console. This produces a report from which we can read the goodness of fit ($R^2 > 0.99$ in this case), as well as the final values and uncertainties for the parameters.

37: Print out the molar absorptivity to the user in the console. Here we make use of f-strings and the `chemSymbols` feature of the `codechembook` library to give us the $\pm$ symbol and the superscripts we will use on the units.

40: We make a Plotly Figure object that will hold two subplots. We want two rows and one column, which we specify using keyword arguments. This will produce a vertical arrangement of the two subplots.

43–45: We add traces to the Figure object. The first two are added to the top plot (row = 1, col = 1), and the third is added to the lower plot (row = 2, col = 1). All three use the same x-values, which we can access using the unintuitive `result.userkws` attribute. This attribute is a dictionary in which the independent variable array is stored using the key that is specified for the fit model, in this case, simply `'x'`. The first line adds the results of our fit (using the `best_fit` **attribute** of the result object), the second adds the experimental y-values, and the last adds the residuals (using the `residuals` attribute of the result object). We also choose to show the experimental data using markers only, the fit result using lines only, and the residuals using markers and lines. Because the last of these is the default behavior, we do not need to specify a mode. Finally, we include none of these traces in the legend.

48–51: We are going to annotate our plot with three pieces of information: the final slope value (and uncertainty), the final intercept value (and uncertainty), and the R^2 value of the fit. There are many ways to do this, but the simplest is going to be to use a multi-line f-string. Multi-line strings are placed between “ ” ” characters, and this allows us to specify strings that span multiple lines, which will help us visualize the string we are writing. Normally, the line breaks in the multi-line strings also translate to breaks in the string when printed or otherwise represented. However, because we are going to put this string as an annotation in a Plotly figure, “normal” line breaks do not produce line breaks. Instead, we have to the use HMTL tags for breaks: <br>.
We will create a line for three pieces of information from the fit: the slope (and uncertainty), the intercept (and uncertainty), and the R^2 value. These are passed

to the f-string, put between curly brackets ({}). We can use the `.value` attribute of the parameters accessed via a dictionary in the fit result object to access the values, and the `.stderr` attribute to access the uncertainty in these parameter values.
When we supply these values, we can also format them, providing the information after a colon (:) within the curly brackets, after the value we are formatting. In this case, we specify we want **e notation** with two decimals of precision using .2e.
One final note: we also use the chemical symbols function from the `codechembook` library to provide quick access to the ± symbol in our plot.
We assign this string to a variable, so that we can use it below.

53–56: We add the above string to the Plotly figure using the `add_annotation()` method of the Figure object. We provide the string we just created above, and specify that we do not want an arrow and where to place this annotation on the plot, using the x- and y-axis coordinates.

59–62: Also using the `add_annotation()` method, we add the molar absorptivity to the plot. This time, we construct the string we want inside the function call.

65–67 We add a title to the x- and y-axes. However, the title for the y-axis is different for the top (experimental data) and bottom (residuals), so we specify a title for each y-axis individually, using the `row` and `col` keyword arguments. Note that, when assigning the title to the x-axis, we do not specify a row or column, and so the title is applied to *all* of the x-axes in the plot.

68: We apply the `simple_white` template to our plot. We also add a title using the `title` keyword argument and supply a properly formatted HTML string, returned by the `quickHTMLFormula()` function.

69: We show the result of the plot.

CHAPTER 5 EXERCISES

TARGETED EXERCISES

Exercises for Section 5.1.1

0. Using the dictionary of solvent properties that you made in Exercise 3 from Chapter 4, write a code that asks the user for a solvent, then asks the user for a property, and then prints a sentence that tells the user the value of that property for the solvent they chose.

Exercises for Section 5.1.2

1. Modify the script from the end of Chapter 0 so that it asks the user for the concentration of HCl, ionic strength, and concentration of the catalyst, then performs the calculation and outputs results in the same way as before.

Exercises for Section 5.1.4

2. Using f-strings and the `print()` function, output a grammatically correct sentence that expresses the following (perform the math, if necessary, and write out the result to the user in plain English including units). For each item, you can use a single line of code to accomplish this:

   a) Number of moles of water in 14.7 g.

   b) Density of toluene in g/mL.

   c) Number of mL in one gallon.

   d) 72.5°F in Celsius.

Exercises for Section 5.1.5

3. In two lines of code, define the variable specified and print the following statement using f-strings to display the value of the variable, with units. Remember that Python uses banker's rounding:

   a) Given `normal_temp = 293.15`, print "Normal temperature is 293.1 K."

   b) Given `vol = 1.672834e-3`, print "The volume is 1.673 mL."

   c) Given `conc_NaCl = 0.0043782`, print "[NaCl] = 4.38 mM."

   d) Given `ethane_CC_bond_length = 153.5`, print "Ethane's C-C bond is 0.153 nm."

Exercises for Section 5.1.6

4. Create the f-strings that produce the following text, using the `codechembook.symbols` module to generate special characters:

   a) λ_{max}

   b) 3.27 μL.

   c) $[CH_3CH_2COO-]$ = 5.14 mM.

   d) $N_2 + 3H_2 \rightarrow 2NH_3$.

   e) The C=C bond length of ethylene is 1.33 Å.

   f) For p orbitals, $\ell = 1$.

Exercises for Section 5.1.7

5. Repeat Exercise 13 from Chapter 4, using `codechembook.quickTools.quickHTMLFormula()`.

Exercises for Section 5.1.8

6. Using the `.glob()` method of pathlib Path objects:

   a) Get a list of all the files in your Downloads folder.

   b) Sort the list of files you just made.

   c) Using `len()`, print the number of .pdf files you have in your Downloads folder.

Exercises for Section 5.1.9

7. You have an irritating instrument that outputs data in text files interspersed in sentences. For instance, it will produce a string that goes like this:

   "at 1.58 seconds, conc = 40.2 mM;at 3.24 seconds, conc = 78.2 mM;at 4.98 seconds, conc = 119.4 mM;at 6.02 seconds, conc = 154.7 mM"

   You want to plot concentration versus time, but you first need to process this string. Using the `.split()` method for strings, write a script that:

   a) Breaks apart the above string to extract the times and concentrations.

   b) Processes these times and concentrations them into Numpy arrays.

   c) Plots these arrays with proper axis labels.

   d) Makes no assumptions about how many data points there are.

Exercises for Section 5.1.3

8. If you used the approach we demonstrated in this chapter to find a number closest to a target number, what happens if there are two numbers that are equally close to the target number?

9. Write code that will take a Numpy array and find a number in it, closest to some target number. Write this code such that, if there are two (or more) numbers that are equally close to the target number, you identify the smallest number.

Exercises for Section 5.1.10

10. You have measured the equilibrium constant for a chemical reaction at multiple temperatures. You wish to use a Van't Hoff plot to determine ΔH and ΔS for the reaction. The equation for this is $\ln(K_{eq}) = -\frac{\Delta H}{RT} + \frac{\Delta S}{R}$. Thinking about this proceedure, you realize that this form of the equation was historically used because it produces a linear relationship between $\ln(K_{eq})$ and $\frac{1}{T}$ and this was simpler to fit

in the time before computers. However, both of these transformations can significantly change the values of the parameters that are estimated by the fit. So, what you should actually do is plot K_{eq} versus T and fit that plot. Write the equation you would use as the model for this fitting and identify the independent variable, the dependent variable, the parameters you would adjust in a fit, and any constants that would not change in the fit.

Exercises for Section 5.1.11

11. Use the Spyder IDE's variable explorer or `print(vars(result))` to inspect the `fitResult` object created by the final code in this chapter. You will find that it is presented as if it had some dictionary structure. The 'keys' are the attributes of the result object and are used to access the associated 'values.' For instance, if you wanted to find the R^2 value of the fit, you could use `result.rsquared`, which should return something like `0.9994452445068337`. Find the attributes for the following elements:

    a) The best-fitting values.

    b) Their uncertainties.

    c) Whether the fit was aborted.

    d) The model used.

    e) χ^2.

Exercises for Section 5.1.12

12. Make and display a Plotly figure with two subplots, one on top and one on bottom, in which the top plot is the first spectrum from our titration series in Chapter 3 and the bottom plot is the last one.

13. Make and display a Plotly figure with four subplots, in two rows and two columns, in which the top left plot is the first spectrum from our titration series in Chapter 3, the bottom right is the last spectrum, the top right is the first spectrum minus the last spectrum, and the bottom left is the last spectrum minus the first spectrum.

COMPREHENSIVE EXERCISES

14. In this chapter, we introduced a function that makes it easy to identify multiple files on your computer: `quickSelectFolder( )`. This will be nice to use when you don't know where the file is located, or when you are choosing different files each time you run your script. However, there will be times when you simply want to use the same file over and over—for instance when you are developing and testing a code. In these cases, it can be much faster to provide the path to the file in the script. Modify the final code in this chapter, replacing the `quickSelectFolder( )` function with a line that creates a Path object with the absolute path to the folder and on

which you use the `glob()` method. Verify this works, and describe when you might prefer this approach over the use of `quickSelectFolder()`.

15. You remember from analytical chemistry that there is a "linear regime" for absorbance values. You wonder about the behavior in transmittance. Convert the values used in the final code of this chapter to transmittance, and fit that to a line, displaying the data, fit, and residual of the fit. If you use R^2 of 0.99 as a threshold for "good" linearity, then how is this fit?

16. It is common to fit models that depend exponentially on some value, like kinetics, by taking the natural log of the dependent variable and performing a linear fit instead of fitting the raw data to an exponential function:

$$[A](t) = [A]_0 \exp -kt \rightarrow \ln \frac{[A](t)}{[A]_0} = -kt$$

In the absence of noise, these two approaches to fitting are equivalent. However, if your data is somewhat noisy, you can get different results from the two different approaches:

a) Download the file expfit.csv and save it to your class folder. This file has three columns: time (sec), exact concentration (M), and noisy concentration (M). Load the data and create new arrays containing the natural log of the concentrations. Then, fit both forms of the raw data to exponentials, and both forms of the natural log data to lines. Compare the rates that you get from the linear model to the exponential model for the exact and noisy cases. What do you notice? Make a table here of your results and include your code when you turn in the assignment.

b) The above problem can be mostly remedied. Because LMFIT is basically trying to minimize χ^2, which depends on the uncertainty of each point. In the data you've been given, the uncertainty of each point is the same ($\sigma = 0.0001$). When you take the natural log of the y-values, you also have to compute new uncertainties in the logarithmic space, which are not all the same! Thus, not all points have equal weight when calculating χ^2. LMFIT allows you to give each point a different weight using the weight keyword in `Model.fit()` (check it out in the documentation). The weights should be $\frac{1}{\sigma_i^2}$ for each point. When you take the log, the new uncertainties ${\sigma'}_i^2$ in the logarithmic y-axis can be computed by ${\sigma'}_i = \frac{\sigma_i}{y_i}$. Using this relationship, revise your code to do the linear fit with appropriate weights, and confirm that it makes the results more consistent.

17. Using the solvent dictionary from Exercise 3 in Chapter 4, make a Plotly figure that has a grid of subplots equal to or greater than the number of solvents in that

dictionary. Using RDKit, add a drawing of one solvent molecule to each subplot. It okay if you end up with blank subplots. In order to specify which subplot you are adding a figure to, you can use the `xref=f"x{i}"` and `yref=f"y{i}"` within the `add_layout_image()` method of the Plotly figure object. Here, `i` indicates the subplot number. Plotly counts subplots left to right, top to bottom (like you read English), and starts at the number 1.

CHAPTER 6

FITTING DATA WITH MULTIPLE COMPONENTS

☐ **New concepts**
— *and their prerequisites*

☐ **Getting yes/no answers to questions posed by comparison statements**
— *Section 2.1.3: Representing answers to yes/no questions using booleans*

☐ **Slicing arrays by values instead of indices using comparison expressions**
— *Section 1.1.1: Operating on collections of data efficiently with Numpy arrays*
— *Section 6.1.2: Having Python do basic math by evaluating mathematical expressions*
— *Section 2.1.3: Representing answers to yes/no questions using booleans*

☐ **Slicing based on multiple conditions using `and` and `or`**
— *Section 1.1.1: Operating on collections of data efficiently with Numpy arrays*
— *Section 6.1.2: Having Python do basic math by evaluating mathematical expressions*
— *Section 2.1.3: Representing answers to yes/no questions using booleans*

☐ **Visualizing fit results using `codechembook.quickPlots.plotFit()`**
— *Section 2.1.7: Adding `codechembook` to your Python installation*
— *Section 5.1.10: Fitting models to data with lmfit*

☐ **Learning to use libraries by reading documentation**

☐ **Combining lmfit models**
— *Section 5.1.10: Fitting models to data with lmfit*

☐ **Accessing the number of times a loop has iterated using `enumerate()`**
— *Section 0.1.2: Accessing capabilities beyond basic math using functions*
— *Section 3.1.11: Performing repetitive operations using `for` loops*

6.0 PROBLEM

Looking at the spectra for your initial and final titration points, you notice that there is some structure in the MLCT band. Specifically, for the band around 600 nm in the spectrum at the initial point, you wonder how many features are present and what their central positions, widths, and intensities are. The band clearly has two partially resolved features, but there also appears to be a shoulder on the high-energy side that indicates that the band is really three underlying features. Seeing that you can now fit equations to data, you want to fit this band to try to identify the underlying components. Because this is a solution phase UV-vis spectrum, you reason that any features are heterogeneously broadened and should be treated as Gaussian lineshapes. Thus, you would really like to fit the band to three Gaussian bands, plus a background. If you treat the background as linear, you already know how to use the linear model in lmfit.[1] So, you wonder if there might be a similar model for a Gaussian and a way to combine multiple instances of it.

[1] See Chapter 5.

6.1 SOLUTION

The solution in your mind is pretty simple: figure out how to use the lmfit **library** to fit the overall shape of your MLCT band to the sum of three Gaussians and a linear background. Then you can make a plot showing the three individual Gaussian components and annotate it with their positions.

6.1.0 PLANNING

Your labmate says that basically what you want to do is:

1. Load in the spectrum you want to fit.
2. Isolate the region of the spectrum you want to fit.
3. Make a composite model that is a linear background + 3 Gaussians.
4. Fit the model to the desired region of the spectrum and extract parameters for the Gaussians.
5. Plot the results. You start by drawing up a flowchart for the process you want to code, producing Figure 6.1.

They point out that you already know how to do #1, #4, and #5. They say they can teach you how to do #2 quickly, but that they is leaving on a backpacking trip with friends and so you might have to figure out #4 by reading the online documentation. They say you should have enough knowledge to read the documentation, and so you should give

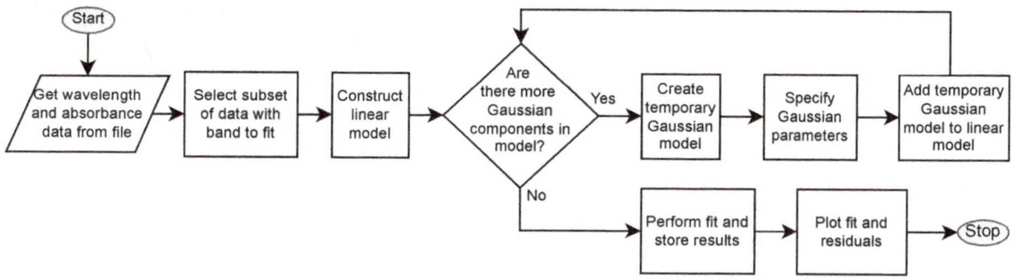

FIGURE 6.1. Flowchart for the code that extracts spectral data from a csv file, builds up a model of this spectrum using multiple Gaussian line shapes and a linear background, and then fits this model to the data, displaying the results.

it a try. But first, they launch into a quick tutorial on how to isolate the spectral region of interest.

"This relies on combining logical operations with array slicing, so let's see each in turn."

6.1.1 GETTING YES/NO ANSWERS TO QUESTIONS POSED BY COMPARISON STATEMENTS

When looking for a region of a spectrum, you really mean you are looking for wavelengths that are between two values. In other words, you want to find values that are greater than a specific value *and* also less than another value. "We can use logical operations to test for this," your labmate says. "These operations evaluate to one of the two **boolean** values: `True` or `False`,[2] so you can use them to decide which data points are in the range you want."

[2] This is the same **data type** introduced in Section 2.1.3.

```
In: 1 > 0
Out: True
```

```
In: 0 > 0
Out: False
```

"You can also test that something is less than (<), equal to (==), and so on.[3] Note that the test for equal to uses two equal signs. This done to distinguish it from the assignment operator (=)."

[3] A complete list of such operators is found in Chapter B.

"We have encountered the **keyword** `in` when building `for` **loop**. This was used to access the elements in a **collection** (e.g., **list**, **array**, and so on). We can also use this in a logical test, to check if something exists in a collection."

```
In: mylist = [1, "one", 2, 5]
In: 3 in mylist
Out: False
```

"You can see that `in` checks if the item on the left is in the collection on the right. If it is, it will return `True` and if it is not it will return `False`. You could try to change the entries in `mylist` to make the `in` statement `True`."

6.1.2 SELECTING A SUBSET OF A NUMPY ARRAY USING COMPARISON EXPRESSIONS

"The logical tests we just saw can be applied to a Numpy array. When you do so, an array of the same length will be returned, but this array will only contain the values `True` and `False`. For instance, suppose you had a **Numpy array** of numbers from 1 to 20:"

```
In: import numpy as np
In: nums = np.linespace(1, 21, 1)
```

"You can see which elements are greater than five using:"

```
In: nums > 5
Out: array([False, False, False, False, False, True, True, True, True, True, ↩
     ↪True, True, True, True, True, True, True, True, True, True])
```

You see that this returns an array of the same length as `nums`, but where each entry specifies if that number was greater than five.

"Something else that is interesting is that one can use an array of booleans to specify which elements of a Numpy array to select. This is used similarly to specifying indices—that is you place this between two square brackets, after the array. If we assign the array of booleans to a variable, then we can use this variable to select a subset of the array."

```
In: mask = nums > 5
In: nums[mask]
Out: array([ 6,  7,  8,  9, 10, 11, 12, 13, 14, 15, 16, 17, 18, 19, 20])
```

Using an array of booleans to select a subset of an array is known as **mask**ing and it creates a *copy* (not a **view**) of the array.

You think that maybe you could skip the intermediate step of assigning the result of a logical operation to a mask, so you try another version of the code that perform the logical test directly within the square brackets.

```
In: nums[nums > 5]
Out: array([ 6,  7,  8,  9, 10, 11, 12, 13, 14, 15, 16, 17, 18, 19, 20])
```

It worked, and looks very clean! So, masking is a powerful way to select specific elements in an array.

Then, you think this could be useful for **list**s, too. So you try the same thing:

```
In: num_list = list(range(1, 21))
In: num_list > 5
Out: Traceback (most recent call last):
Out:
Out:   Cell In[17], line 1
Out:     num_list>5
Out:
Out: TypeError: '>' not supported between instances of 'list' and 'int'
```

Your labmate explains, "Masking only works because of the **broadcasting** behavior of Numpy arrays. If you tried this on something like a **list**, you will get an error."

6.1.3 SELECTING ARRAY SUBSETS BASED ON MULTIPLE CONDITIONS USING `and` AND `or`

You see that **mask**ing is great, but it still doesn't do what you want. You can test whether a point is greater than the minimum wavelength that you want to keep, or if it is less than the maximum, but not both. Your labmate says, "Sometimes we want to apply multiple tests to an object. For instance, we want to know if a point x is greater than 7 *and* less than 10. Mathematically, we would just write $10 > x > 7$, but we would understand that this is really two distinct logical operations: $10 > x$ *and* $x > 7$. For the expression to evaluate to true, x must satisfy the first *and* the second."

"In Python, the same behavior is accomplished using the `and` operator,[4] but Python allows the same shortcuts: `10 > x > 7` is a shorthand for `10 > x and x > 7`. More complicated tests can be created using multiple operations (e.g., `and` and `or` statements), like `10 > x and x > 7 or x == 0`. For these statements, there is an order of operations that is followed, but it is usually better to use parentheses to make it very clear what order you intend the conditionals to be evaluated. For example: `(10 > x and x > 7) or x == 0` may evaluate to a different result than `10 > x and (x > 7 or x == 0)`."

[4] This is true of any boolean operator, like or or xor.

"So, you **masked Numpy arrays** using a single **comparison operation**, but in this case, you want to take a subset of the arrays that corresponds to a range of values in the middle of the array. A quick test reveals the problem:"

```
In: wavenumber = np.arange(400, 4000, 1)
In: sliced_wavenumber = wavenumber[1900 > wavenumber > 800]
Out: Traceback (most recent call last):
Out:   File "<pyshell#11>", line 1, in <module>
Out:     sliced_wavenumber = wavenumber[1900 > wavenumber > 800]
Out: ValueError: The truth value of an array with more than one element is ↩
       ↪ambiguous. Use a.any() or a.all()
```

Trying

```
In: sliced_wavenumber = wavenumber[1900 > wavenumber and wavenumber > 800]
```

yields the same result.

Your labmate explains, "You gave Python a mathematically reasonable statement, but since `wavenumber` is a Numpy array, Python evaluates one conditional and returns an array of equal size but containing only `True` and `False` values at each **index**. When you link two conditionals like this, Python interprets it as you looking for values that satisfy the first condition *and* the second one simultaneously and implicitly inserts the `and` **operator** for you. Unfortunately, the `and` operator is only designed to evaluate to a single `True` or `False` value, and cannot handle the **vectorization** that operations between Numpy arrays typically rely on. Basically, it is asking "are these arrays both true?" rather than "are the ith elements in each of these arrays both true?" This obviously is a problem."

"What you really want to do is an "elementwise" logical operation, performing a logical operation on two corresponding elements in two different arrays to form a new array holding the result of that logical operation. For this, use the 'elementwise and' operator (&):"

```
In: sliced_wavenumber = wavenumber[(wavenumber > 800) & (wavenumber < 1900)]
```

"Basically, you use & rather than the Python `and` command and everything else works as expected. Similarly, `or` can be replaced by `|`.[5] Note that *the parentheses here are important to set the order of operations correctly*. If you leave them out, the line of code will fail because it evaluates the operators in order from left to right and the expression `wavenumber > 80 & wavenumber < 1900` will not work. This is because the statement is evaluated left to write and this is the equivalent of `((wavenumber > 80) & wavenumber) < 1900`, which will reduce to comparing the `True`/`False` values in an array produced by the first logical test to the numerical data contained in the `wavenumber` array. This will either fail, or produce results that you do not expect."

[5] On a U.S. keyboard, you get this character by holding shift and pressing the button above the Enter key

6.1.4 ACCESSING THE NUMBER OF TIMES A LOOP HAS ITERATED USING `enumerate()`

"We saw that we could **iterate** through multiple iterable **collection**s together using `zip`.[6] Sometimes you need to know what number iteration you are on (corresponding to an index of the collection). Or, sometimes the arrays or lists that you want to iterate over are not the same length. In either of these cases, there is a better solution: the `enumerate()` function."

[6] See Chapter 4.

"The enumerate **function** accepts an **iterable object** as a **positional argument** and **returns** a **tuple** of two values for each element in the object. The first value is the iteration number, and the second is the element at the index corresponding to the iteration number. These values can be **unpacked** and assigned to two **variable**s. In some sense then `enumerate(mylist)` is equivalent to `zip(range(0, len(mylist)), mylist)`. The behavior can be seen in the following:"

```
for i, name in enumerate(["one", "two", "three"]):
    print(f"the index is {i} and the value at that index is {name}")
```

```
Out: the index is 0 and the value at that index is one
Out: the index is 1 and the value at that index is two
Out: the index is 2 and the value at that index is three
```

"This can be a very powerful way to keep track of where you are in a list."

6.1.5 VISUALIZING FIT RESULTS USING `codechembook.quickPlots.plotFit()`

You had previously created a figure with two subplots that plotted data and a fit to this data in the top plot, and a plot of the residuals (data–fit) in the bottom plot. This allowed you to easily visualize the fit and how it deviates from the data, but it was a fair amount of coding.

This approach is generally useful enough that the **`codechembook` library** includes a **function** to accomplish this quickly: `fitPlot()`. To use this function, you simple provide a lmfit fit **result object** as the only **positional argument**, and this function will generate a plot with data and fit on top and residual on the bottom. There are also a few useful **keyword argument**s for this function:

- `residual`. The default value for this is `False`, and the default behavior is to plot only the top potion (data and fit). Passing a value of `True` results in showing the residual below the data. If `True` is used, then the data and fit subplot will be four times taller than the residual subplot—a behavior that makes it very easy to see the residual, even when it is small. However, passing `residual = 'scaled'` to the function will produce a plot in which the scale of the y-axis on the top and the bottom are the same, meaning that one can directly compare the residuals to the data.

- `components`. The default value for this is `False` and the default behavior is to plot only the final fit line. If this keyword is set to `True`, then the plot will also include individual components of the model. For instance, if you had a model that was the sum of a line and a Gaussian profile, this will show the sum, as well as just the line and just the Gaussian. This can be useful for understanding the behavior of your model.

There are a few other keywords, and you can consult the codechembook library to find them, but for now, this is sufficient to solve your current problem.

6.1.6 LEARNING TO USE LIBRARIES BY READING DOCUMENTATION

"Alright, I have to run!" Your labmate grabs their messenger bag. "Remember, you know what you need to in order to read the documentation for **lmfit**. Here is a hint, start by Googling what you already know: 'lmfit linear model.'"

Following your labmate's advice, you find a page titled "Built-in fitting models of the models module." Just a short time ago, this would have sounded like nonsense to you, but now you know that a **module** is just part of a **library**, and so the page is describing how that part of the lmfit library works. Scrolling through this page, you see the linear model you have been using. And you also see a Gaussian model! The documentation as of the writing of this book is reproduced in Figure 6.2.

GaussianModel

class **GaussianModel**(*independent_vars=['x'], prefix='', nan_policy='raise', **kwargs*)

A model based on a Gaussian or normal distribution lineshape.

The model has three Parameters: *amplitude*, *center*, and *sigma*. In addition, parameters *fwhm* and *height* are included as constraints to report full width at half maximum and maximum peak height, respectively.

$$f(x; A, \mu, \sigma) = \frac{A}{\sigma\sqrt{2\pi}} e^{[-(x-\mu)^2/2\sigma^2]}$$

where the parameter *amplitude* corresponds to A, *center* to μ, and *sigma* to σ. The full width at half maximum is $2\sigma\sqrt{2\ln 2}$, approximately 2.3548σ.

For more information, see: https://en.wikipedia.org/wiki/Normal_distribution

Parameters:
- **independent_vars** (list of str, optional) – Arguments to the model function that are independent variables default is ['x']).
- **prefix** (*str, optional*) – String to prepend to parameter names, needed to add two Models that have parameter names in common.
- **nan_policy** (*{'raise', 'propagate', 'omit'}, optional*) – How to handle NaN and missing values in data. See Notes below.
- ****kwargs** (*optional*) – Keyword arguments to pass to **Model**.

Notes

1. *nan_policy* sets what to do when a NaN or missing value is seen in the data. Should be one of:
   - *'raise'* : raise a *ValueError* (default)
   - *'propagate'* : do nothing
   - *'omit'* : drop missing data

FIGURE 6.2. Screenshot of the online documentation for the GaussianModel **class** within the lmfit **library.**

Reading this documentation, you find that the model is actually a **class**. You also find a discussion of the **parameter**s used by the Gaussian model function. You find that they are all optional, but decide that you will want to supply the `independent_vars` (the x-values), just as you did when using the linear model. You also see other optional parameters, but aren't totally sure what they are used for now. What you don't see is anything about how to pass the parameters that you want to fit (which the model says are amplitude, center, and sigma). So, next you need to find this, as well as how to combine models.

6.1.7 COMBINING LMFIT MODELS

Scrolling more on the same page with the documentation for the GaussianModel, you find a section called "Example 3: Fitting Multiple Peaks—and using Prefixes" (Figure 6.3). The first part sounds exactly like what you are wanting to do, and the second part references "prefixes," which you recall is one of the optional **parameters** of the Gaussian Model. Looking at the other built-in models described on the page, you see that "prefix" is an optional parameter for all of them. Looking over the text of the example, you find that you can take existing models and add them together to make new models! Specifically, on line 24, you see: `mod = gauss1 + gauss2 + exp_mod`, and you note that `gauss1`, `gauss2`, and `exp_mod` are all variables that were given to calls of the built-in

Example 3: Fitting Multiple Peaks – and using Prefixes

As shown above, many of the models have similar parameter names. For composite models, this could lead to a problem of having parameters for different parts of the model having the same name. To overcome this, each Model can have a prefix attribute (normally set to a blank string) that will be put at the beginning of each parameter name. To illustrate, we fit one of the classic datasets from the NIST StRD suite involving a decaying exponential and two Gaussians.

```
# <examples/doc_builtinmodels_nistgauss.py>
import matplotlib.pyplot as plt
import numpy as np

from lmfit.models import ExponentialModel, GaussianModel

dat = np.loadtxt('NIST_Gauss2.dat')
x = dat[:, 1]
y = dat[:, 0]

exp_mod = ExponentialModel(prefix='exp_')
pars = exp_mod.guess(y, x=x)

gauss1 = GaussianModel(prefix='g1_')
pars.update(gauss1.make_params(center=dict(value=105, min=75, max=125),
                               sigma=dict(value=15, min=0),
                               amplitude=dict(value=2000, min=0)))

gauss2 = GaussianModel(prefix='g2_')
pars.update(gauss2.make_params(center=dict(value=155, min=125, max=175),
                               sigma=dict(value=15, min=0),
                               amplitude=dict(value=2000, min=0)))

mod = gauss1 + gauss2 + exp_mod

init = mod.eval(pars, x=x)
out = mod.fit(y, pars, x=x)

print(out.fit_report(correl_mode='table'))
```

FIGURE 6.3. Screenshot of the online documentation for the lmfit **library**, showing an example of how different model **objects** can be combined.

models. So they appear to be adding together models to make a combined model that has two Gaussians and an exponential background.

You also notice that although the guess( ) **method** of the ExponentialModel was used to get initial **parameter** values for that model (and store them in a variable), this was not done for the Gaussian models. Instead, it seems these were added directly to the same variable using what looks to be a **method**: update( ). You think this is a method because it follows the **syntax** for using methods **<object>**.method( ). Given that this is holding parameters to be fit, you suspect this must be something like a "parameter object."

You notice on the top of the web page that there is a link to a page called "Parameters," and so you click on that and find out that there is, indeed, a Parameters **class** (Figure 6.4 shows part of this documentation), which means there are objects associated with it. Thus, the guess( ) method created a parameter object, that update( ) added to.

Looking at Example 3 more (Figure 6.3), you find that parameter values were supplied using the make_params( ) method of the Gaussian model object. That is, first, a Gaussian model object is created (and assigned to either the variable gauss1 or gauss2),

The **Parameter** class

class **Parameter**(*name, value=None, vary=True, min=-inf, max=inf, expr=None, brute_step=None, user_data=None*)

A Parameter is an object that can be varied in a fit.

It is a central component of lmfit, and all minimization and modeling methods use Parameter objects.

A Parameter has a *name* attribute, and a scalar floating point *value*. It also has a *vary* attribute that describes whether the value should be varied during the minimization. Finite bounds can be placed on the Parameter's value by setting its *min* and/or *max* attributes. A Parameter can also have its value determined by a mathematical expression of other Parameter values held in the *expr* attribute. Additional attributes include *brute_step* used as the step size in a brute-force minimization, and *user_data* reserved exclusively for user's need.

After a minimization, a Parameter may also gain other attributes, including *stderr* holding the estimated standard error in the Parameter's value, and *correl*, a dictionary of correlation values with other Parameters used in the minimization.

Parameters:

- **name** (*str*) – Name of the Parameter.
- **value** (*float, optional*) – Numerical Parameter value.
- **vary** (*bool, optional*) – Whether the Parameter is varied during a fit (default is True).
- **min** (*float, optional*) – Lower bound for value (default is -numpy.inf, no lower bound).
- **max** (*float, optional*) – Upper bound for value (default is numpy.inf, no upper bound).
- **expr** (*str, optional*) – Mathematical expression used to constrain the value during the fit (default is None).
- **brute_step** (*float, optional*) – Step size for grid points in the *brute* method (default is None).
- **user_data** (*optional*) – User-definable extra attribute used for a Parameter (default is None).

FIGURE 6.4. Screenshot of the online documentation of the Parameter **class** for the lmfit **library**.

and then make_params() was used on this to set values of the parameters to be fit (center, sigma, and amplitude).

Comparing this to the description of the GaussianModel (Figure 6.2), you see that Gaussian model has three parameters: amplitude, center, and sigma, and that these are all used as the name of keyword arguments in the make_params() method in the example (Figure 6.3). Moreover, each of these keyword arguments accepted a dictionary, where value was always specified, and sometimes **min** and **max**.

Going back to the parameter page (Figure 6.4) you decide the **dictionary** was specifying the starting value (value), the lower bound for the fit range (**min**), and the maximum value for the fit range (**max**). You can see there are other arguments that could be supplied, such as if you want the parameter to vary at all (vary), but for now, you think this is probably all you need.

Reading this, you also realize that what make_params() must be doing is taking the dictionary for each parameter, and then passing this to the pararmeter() function to define each parameter.

The last thing from the example (Figure 6.3) that you notice is that you can directly use the fit() method on the composite model (assigned to the variable mod). So, you now think you now understand how to set up and carry out the fit.

There remains one last problem: you want to get the values of the parameters of the final fit—specifically the center positions and their uncertainties. You could go through the same process as above, but in the interest of space, we present the following information, assuming a usage like: result = <model>.fit(), so that the final result is stored in a variable called "result":

- The final parameter values are an **attribute** of the fit result object. So we can get all parameter information using `result.params`.
- You can treat the above attribute like a dictionary, so if you want information on a single parameter, you can use `result.params["<parameter name>"]`.
- The above contains all the accessible information about that parameter, and you can access the value of the parameter and its uncertainties using the attributes `value` and `stderr`, respectively.

With that information, you can sit down and both fit and extract the information from that fit.

6.2 CODE

```
'''
Fit data to multiple gaussian components and a linear background
    Requires: a .csv file with col 1 as wavelength and col 2 as intensity
    Written by: Ben Lear and Chris Johnson (authors@codingforchemistsbook.com)
    v1.0.0 - 250207 - initial version
'''
import numpy as np
from lmfit.models import LinearModel, GaussianModel
from codechembook.quickPlots import plotFit
import codechembook.symbols as cs
from codechembook.quickTools import quickOpenFilename, quickPopupMessage

# Ask the user for the filename containing the data to analyze
quickPopupMessage(message = 'Select the file 0.001.csv')
file = quickOpenFilename(filetypes = 'CSV Files, *.csv')

# Read the file and unpack into arrays
wavelength, absorbance = np.genfromtxt(file, skip_header=1, unpack = True, delimiter=',')

# Set the upper and lower wavelength limits of the region of the spectrum to analyze
lowerlim, upperlim = 450, 750

# Slice data to only include the region of interest
trimmed_wavelength = wavelength[(wavelength >= lowerlim) & (wavelength < upperlim)]
trimmed_absorbance = absorbance[(wavelength >= lowerlim) & (wavelength < upperlim)]

# Construct a composite model and include initial guesses
final_mod = LinearModel(prefix='lin_') # start with a linear model and add more later
pars = final_mod.guess(trimmed_absorbance, x=trimmed_wavelength) # get guesses for linear coefficients
c_guesses = [532, 580, 625] # initial guesses for centers
s_guess = 10 # initial guess for widths
```

```
a_guess = 20 # initial guess for amplitudes
for i, c in enumerate(c_guesses): # loop through each peak to add corresponding gaussian component
    gauss = GaussianModel(prefix=f'g{i+1}_') # create temporary gaussiam model
    pars.update(gauss.make_params(center=dict(value=c), # set initial guesses for parameters
                                  amplitude=dict(value=a_guess, min = 0),
                                  sigma=dict(value=s_guess, min = 0, max = 25)))
    final_mod = final_mod + gauss # add each peak to the overall model

# Fit the model to the data and store the results
result = final_mod.fit(trimmed_absorbance, pars, x=trimmed_wavelength)

# Create a plot of the fit results but don't show it yet
plot = plotFit(result, residual = True, components = True, xlabel = 'wavelength /nm',
    ylabel = 'intensity', output = None)

# Add best fitting value for the center of each gaussian component as annotations
for i in range(1, len(c_guesses)+1): # loop through components and add annotations with centers
    plot.add_annotation(text = f'{result.params[f"g{i}_center"].value:.1f} {cs.math.plusminus} {result.params[f"g{i}_center"].stderr:.1f}',
                        x = result.params[f'g{i}_center'].value,
                        y = i*.04 + result.params[f'g{i}_amplitude'].value / (result.params[f'g{i}_sigma'].value * np.sqrt(2*np.pi)),
                        showarrow = False)

plot.show('png') # show the final plot
```

You run this code, select the file 0.001.csv from the folder Titration/UVvis in your data directory, and wait for about a second. It produces the output shown in Figure 6.5.

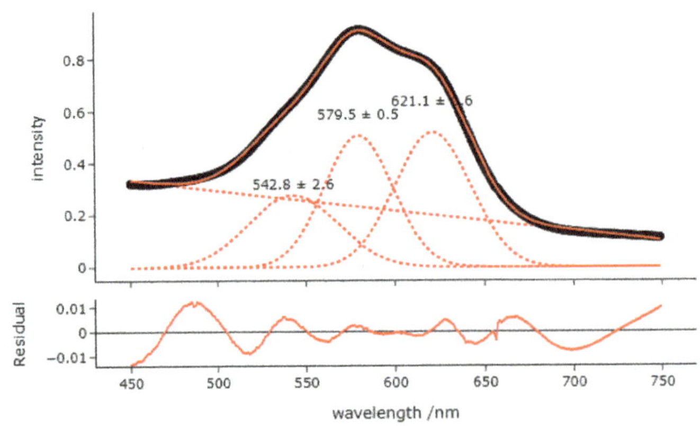

FIGURE 6.5. A plot of the area of the spectrum saved as 0.001.csv, including indications of the background and the three Gaussian components of the fit and their centers. The key new concepts demonstrated here are the construction of a composite fit model and the use of methods and attributes of the `lmfit` `ModelResult` object.

6.2.0 LINE-BY-LINE EXPLANATION

8–11: We import the libraries (or parts of libraries) we will be using. New to this chapter are: `GaussianModel()` (used to fit Gaussian lineshapes to data), `plotFit()` (which we will use to plot the final result of our fitting), and `quickPopupMessage()` (used to generate messages to remind the user of what they need to do next).

14: We use the function `quickPopupMessage()` to tell the user what we want them to do. This will bring up a native operating system message box that shows the message. Including such helpful hints can be nice for users who might wait a long time between using the code, and so might forget what they are supposed to be doing. In this case, we want them to select the 0.001.csv file because that is the data that this fit is going to be set up to handle. If you want, you can try a different file, and see what happens. By the time you get to the final titration point, this code will fail because the initial guesses and bounds provided for the parameters below will not work for that data.

15 and 16: We identify the Path to the data we wish to fit, using the `quickOpenFilename()` function to specify that we can only open up csv files. As a reminder, this will return a pathlib Path **object**.

18: Using Numpy's `genfromtxt()` function, we read the data file, and return to **Numpy arrays** that have our x (wavelength) and y (absorbance) values of interest.

21: We define the lower and upper limits for our fitting. This is the range of wavelengths we want to consider in our fitting.

24: We mask the wavelength array such that we only get values that are greater than or equal to our lower limit and less than our upper limit, and assign the resulting sliced Numpy array to a new variable.

25: We can use the same mask expression and apply it to the absorbance values. Thus, we use the logic test on the wavelength arrays (which is the only place it makes sense, given the limits we specify). This logic test tells us which indices to include. But because we want the same indices kept in the absorbance array, we can just apply the same mask to it. The end result is that we have a trimmed absorbance that matches the trimmed wavelength Numpy array.

28: Our final model is going to be the sum of three Gaussians and a linear background. We will see below that we can simply add together model objects. So, to start with, we create a linear model object, just as we did in the last chapter. There is one difference: we now have used the `prefix` keyword argument. This will append the string that is provided to the front of all parameters. Although not truly needed quite yet, we will find this behavior useful as we build up the multiple Gaussian contributions below.

29: Also, just as we did in the last chapter, we have Python provide guesses for the parameters of the linear model.

30: We are going to have three Gaussian components. It is possible to ask Python to guess the starting parameters, just as we did for the linear model. However, when we have three closely overlapping Gaussians the guesses may not be good, and it is useful to learn how to provide guesses for the parameters manually. Here we create a list that contains the guesses we will use for the center positions of the three Guassians. This will allow us to iterate through them.

31: We will use a single guess for all three widths, which we assign to a variable.

32: This variable holds a number we will use as a guess for the amplitude for all three Gaussians. Although we don't anticipate all three Gaussians to be the same amplitude, starting with the same amplitude will be a good enough guess to start the fitting.

33–38 This is the **`for` loop** that will allow us to iterate though our guesses for the center of the Gaussians. We also use the `enumerate()` function in order keep track of the iteration number, which also happens to correspond to the index for each element we have retrieved within the loop. Basically, for each Gaussian, we'll make a temporary model object with only one Gaussian model, set its parameters, and then add it to the final composite model.

34: We produce a Gaussian model object. We again use the `prefix` keyword argument to supply a string that is prepended on all parameter names. We use an f-string to provide a prefix that starts with a g (for Gaussian) and then has a number associated with it, based on the iteration we are on within the loop. We add 1 to this, so that we are going to number our Gaussians g1, g2, and g3.

35–37: We explicitly add parameters to our Gaussian model. We do this by nesting the `make_params()` method of the model object inside the `update()` method of the parameter object we created in line 29. Within the `make_params()` method, we specify the name of the parameter, and then several other items within a dictionary. We have to at least provide a value for the starting guess, which we do using the `value` key.

35: We assign the values for the center of the Gaussian distribution, using the value we took from the `c_guesses` list.

36: We set up the `amplitude` parameter, providing a starting value, but also another keyword: **`min`**. This allows us to specify a value below which the parameter is not allowed to go, when fitting. The starting guess must, of course, be higher than this value. Sometimes it is quite useful to set such limits because the fitting procedure will vary the parameter value until it fits the data best, without regard to if it makes sense. Setting the lower limit of the amplitude allows us to avoid negative amplitudes, which we do not expect in our UV-vis experiment.

37: We assign the value for the `sigma` parameter (the width of the distribution) using the same approach. However, now we provide **`min`** and **`max`**, which specify the lower and upper limits for this parameter, respectively. We are telling the fitting program that it can vary the value of `sigma`, but it cannot ever be smaller than 0 or greater than 25. Although there might be a good physical reason to not allow the width of the feature to be less than zero, there is not really an excellent physical justification for this upper limit. Instead, we are setting this because our chemical intuition tells us that we might expect all three bands to have similar widths, and this prevents one band from having very different widths from the other two. When restricting parameters to ranges, it is important to understand this may result in a non-optimal fit, so use such constraints with caution. Good practice would be to remove the upper limit after the first fit, re-run the fit with the best-fitting values as the initial values to see if you get different results, and assess how different they are.

It is important to realize that each of the parameters we create above will be prepended with the prefix specified in line 34, as it is added to the parameter object. This will be important later, when we extract information from the fit result object.

38: We now add the Gaussian model object we created to the final model object and then assign this to the final model object. In this way, we are ending each **`for`** loop by adding the newly created Gaussian model to our existing model. By the time all iterations are completed, our final model will consist of four components: one linear and three Gaussian.

41: We use the `fit()` method of the final model object to fit data. As we did last chapter, we supply the y-values, parameters, and x-values to this function, and store the result of the fit to a variable.

44: We use the `plotFit()` function to plot the results of our fit. We supply the result as a **positional argument** and then specify several keyword arguments that (in turn) tell the function that we want the residuals to be plotted, that we want to show all components of the model (in this case, linear and three Gaussians), the title of the x-axis, the title of the y-axis, and we also tell the function that it doesn't need to show the results. The default behavior would have output the results, but we are going to annotate the figure a bit before we show it. The `plotFit()` function also returns the Plotly Figure object it creates, so we can assign this to a variable so that we can continue to modify it.

47–51: We are going to include annotations on the plot, specifically the final center positions of the Gaussians and their uncertainties. To do so, we need access to the values of the individual components of the model, which are instances of Parameter objects that are entries in the dictionary stored in `params` object. Because the **keys** in this dictionary are numbered according to which Gaussian we are working with, it is convenient to get numbers for them. We can generate them using the `range()` function. The Gaussian components start at one; we start at one. Because the range function is exclusive for the final number, we add one to the length of the `c_guesses`, which corresponds to the number of Gaussian components we have. Iterating through the prefixes defined with the `range()` function will provide the numbers we need to access the correct dictionary entries. Once we have the corresponding Parameter object, we can use its attributes to get its value (`value`) and uncertainty (`stderr`).

48: We access the values and standard errors (uncertainties) of our parameters in turn, using the numbers that we generated in the `range()` function above. We include these in an f-string that is passed to the text keyword argument for use in the annotation.

49 & 50: Add the annotation to the Plotly figure to specify the x- and y-positions of the annotation.

51: Finally, we state that we don't want an arrow drawn for these annotations.

53: We show the final assembled plot.

CHAPTER 6 EXERCISES

TARGETED EXERCISES

Exercises for Section 6.1.1

0. What do the following conditional expressions evaluate to (`True` or `False`)? First, write down what you think it should be before running the code, then run it to see if you are right. If you were wrong, explain what you did wrong. Suppose the following variables are already defined:

```
trial_number = 4
pH = 2.53
acid = 'HCl'
conc_stock = 1.0 # M
```

a) `pH >= 2`

b) `acid != 'HF'`

c) `trial_number > 2`

d) `pH > 1`

e) `conc_stock / 1000 < 0.001`

f) `acid == 'HF'`

g) `'H' in acid`

1. Write a conditional expression to evaluate the following tests. Use the variable name given in parentheses for your answer. Suppose the following variables are already defined:

```
pKa = 7.6
atomic_number = 19
pH = 4.4
atom_symbol = 'K'
solvent_name = 'Isopropanol'
formula = '(CH3)2CHOH'
solvent_polarity = 3.9
```

example: Is the pK_a (pKa) less than 5.0?: `pKa < 5.0`

a) Is the atomic number equal to 11?

b) Is the pH at least 4 but less than 6?

c) Is the atom symbol K?

d) Is the solvent Methanol or Acetonitrile? (Note: do this with a single comparison.)

e) Is the string OH in formula?

2. For the dictionary of solvent properties created in Exercise 3 of Chapter 4, write a code that determines if an alcohol is in the dictionary by testing to see if `'ol'` is in any of the keys. You can do this by getting a list of the keys using `list(<dictionaryname>.keys())` and then iterating through that list, or by using `for k in <dictionaryname>:`

Exercises for Section 6.1.2

3. Create a sine wave with period of one with an x-axis that goes from 0 to 20 with 100 elements. Make a plot of this sine wave where positive values are green and negative values are blue using slicing to produce two different y-arrays and two corresponding x-arrays that you can plot separately. If you set `mode = "markers"` when you add scatter traces, then you will not get a line at zero.

Exercises for Section 6.1.3

4. What do the following conditional expressions evaluate to (`True` or `False`)? First, write down what you think it should be before running the code, then run it to see if you are right. If you were wrong, explain what you did wrong. Suppose the following variables are already defined:

```
trial_number = 4
pH = 2.53
acid = 'HCl'
conc_stock = 1.0 # M
```

a) `pH >= 2 and pH < 7`

b) `acid != 'HF' and acid != 'HBr'`

c) `trial_number > 2 and acid != 'HCl'`

d) `pH > 1 and pH < 7`

e) `acid == 'HCl' or acid == 'HF'`

f) `(pH > 2 and acid == 'HF') or (pH > 3 and acid == 'HCl')`

5. Write a conditional expression to evaluate the following tests. Use the variable name given in parentheses for your answer. Suppose the following variables are already defined:

```
pKa = 7.6
atomic_number = 19
pH = 4.4
atom_symbol = 'K'
solvent_name = 'Isopropanol'
formula = '(CH3)2CHOH'
solvent_polarity = 3.9
```

a) Is the solvent anything other than toluene or n-hexane and is the polarity less than 5?

b) Is the pH greater than 4 and the pK_a greater than 4?

c) Does the formula contain a methyl group and an alcohol group?

d) Is the atom potassium and the pH less than 2?

Exercises for Section 6.1.5

6. Modify the final code from Chapter 5 to use the `plotFit()` function from `codechembook` and confirm that it works.

Exercises for Section 6.1.6

7. Read the documentation for the `plotFit()` function and make plots for the code from the end of the chapter but with the following changes:

a) Confidence intervals of 95%.

b) Hide the components of the fit.

c) Show the residual but scale it to the same scale as the main plot.

8. Sometimes you may want to have a string representation of a whole array. Read the documentation for `numpy.array2string()` and use it to create a string that represents the sin wave from Exercise 3 rounded to the nearest tenth. Make sure that the output does not contain any newline (`\n`) characters using keywords for the function.

Exercises for Section 6.1.7

9. There are times where the behavior of a system is best described by the sum of the equations for two lines. For instance, you could have a reaction mechanism with distinct rate-determining steps, where one dominates at early times and another takes over later. You might also encounter this in thermal expansion of multi-phase materials. Although the model needed to describe such behavior is simple, it would be somewhat challenging to use it in a fit within Excel and extracted parameter estimates and their uncertainties. Write the code to produce a lmfit model that is made up of two linear models.

Exercises for Section 6.1.4

10. If you had two lists `[1, 2, 3]` and `[4, 5, 6, 7]`, use `enumerate()` to add the first list to the first three elements of the second list.

11. For the list `[5, 1, 9, -3]`, construct a new list using a nested `for` loop. This new list must hold only the unique six pairwise products.

COMPREHENSIVE EXERCISES

12. It is a good idea to check how robust the fitting results are—that is, how different can your initial guesses be, and still get the same final fit values. The fit for the Gaussians used in the final code of this chapter has three kinds of initial guesses: those for the position of the Gaussian, the intensity of the Gaussian, and the width of the Gaussian. For each of these parameter groups:

    - Change the initial guesses in a way that you still get the same final fit (within the errors of the parameters).
    - Change the initial guesses in a way that you do not get the same final fit (within the errors of the parameters).
    - Comment on how robust this fitting seems to you.

13. Using the final code for the chapter, modify this so that there is no upper bound on the Gaussian width parameter. When you run this version, do you get a different result? If so, discuss why you might, or might not use this upper bound, and if you could justify using it.

14. You think that there may be a small fourth Gaussian contribution in the data that was fit in the final code of this chapter, and it looks like it would be around 500 nm. Change the code to allow for fitting of this fourth peak. Discuss the results of this new fit with respect to the one that was presented at the end of the chapter.

15. Modify the code we produced at the end of this chapter to add annotations that label the peaks of the actual signal (instead of the Gaussian components) with the best-fitting peak centers. The annotation should include an arrow that points to this exact position in the data, which can be accomplished directly in the `add_annotation()` method of the Plotly figure.

16. It is technically not a good idea to fit bands to Gaussian functions when your x-axis is in wavelength. You should really be working in an energy unit because the effects that give rise to peak shapes depend on energy rather than wavelength, and intensities in the two are related by a Jacobian transformation (see Exercise 22 of Chapter 2 for details). Rewrite the code at the end of this chapter to first convert the x-axis to electron volts (eV), then perform the fit in those units. Note: you will need to convert the guesses as well to get a good start. How do the results compare between these two approaches?

17. In Chapter 3, we plotted many of the same spectra that we fit here. Generate a script that automatically fits each of these spectra and produces a single Plotly plot for each spectrum that was fit, each in its own subplot. Do not include the residuals. You may find that this code eventually "breaks." This is likely be because the guesses for parameters are no longer good places to start for later titration points.

CHAPTER 7

Fitting Simulations to Data

- ☐ **New concepts**
  - — *and their prerequisites*

- ☐ **Automatically running different code under different conditions using `if-then-else` statements**
  - — *Section 6.1.1: Getting yes/no answers to questions posed by comparison statements*
- ☐ **Setting the working directory using `os.chdir()`**
  - — *Section 2.1.5: Describing the location of files or folders on your computer using Pathlib*
- ☐ **Plotting in a single line of code using `codechembook.quickPlots.quickScatter()`**
  - — *Section 2.1.7: Adding `codechembook` to your Python installation*
  - — *Section 3.1.1: Exploiting built-in methods to manipulate data stored in objects*
  - — *Section 1.1.1: Operating on collections of data efficiently with Numpy arrays*
- ☐ **Importing any Python code from anywhere on your computer**
  - — *Section 1.1.0: Giving your computer instructions using Python commands*
- ☐ **Handling errors without crashing using `try-except-finally`**

7.0 PROBLEM

The UV-vis spectra you have taken during the titration of your compound with HCl demonstrate how clean the titration is, but you also want to know the pK_a of your compound. Because you also recorded the pH values after each acid addition, you can generate the titration curve shown in Figure 7.1.

Normally, you would simply look at the titration curve, find the equivalence point, and then determine the pK_a by finding the half-equivalence point, just as you had been taught in your undergraduate laboratory class. However, you have three issues with this approach. First, it is somewhat slow... You have to manually go in and analyze the data yourself and you are starting to get a taste for having Python do things for you. Second, it always feels a bit like guessing. Looking at the data you collected, you wonder how accurately you can really find the half-equivalence point using the procedure you learned as an undergraduate. This is especially problematic because there is some noise in the data, making knowing the "true" value difficult. Finally, you are getting used to knowing the uncertainties in parameters using the lmfit **library**. You think it would be great to have the uncertainties in your compound's pK_a value, but there is no built-in model for titration behavior in `lmfit`.

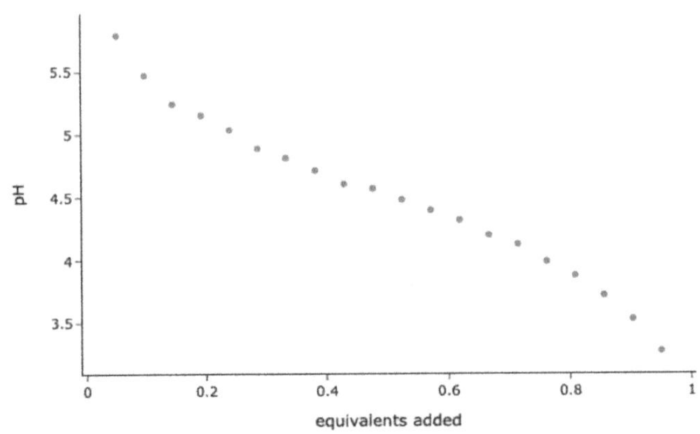

FIGURE 7.1. You would like to fit this titration curve so that you can extract the pK_a associated with the complex you titrated.

Looking at the lmfit documentation, this is doable because lmfit allows you to define custom models that you can fit to data. So you simply have to code a model that will reproduce the titration response you wish to fit.

You know that your titration curve is determined by the Henderson-Hasselbalch equation:

$$pH = pK_a + log\left(\frac{[A^-]}{[HA]}\right) \tag{7.1}$$

where $[A^-]$ and $[HA]$ are the concentration of the deprotonated and protonated forms of your compound, respectively.

Although you don't know the relative concentrations of the deprotonated and protonated for of your compound at every point in the titration, you know the initial concentration of the compound and that it was in the deprotonated form. In other words, you know an initial concentration, $[A^-]_i$. You also know the amount of a strong acid (i.e., HCl) added at each point, so you also know the concentration of H^+ you added. In other words, you know an initial concentration, $[H^+]_i$. Thus, for any given point on the titration curve, you can solve for the concentration of protonated and deprotonated forms present using an ICE (Initial, Change, Equilibrium) table and the value of the equilibrium constant ($K_a = 10^{-pK_a}$). You don't yet know the pK_a, but you think you can use it as a fit parameter and fit your custom model to your titration data.

You work out the familiar ICE table in your lab notebook:

	HA	$\rightleftharpoons$	A^-	+	H^+
initial	0		$[A^-]_i$		$[H^+]_i$
change	x		-x		-x
equilibrium	x		$[A^-]_i$ - x		$[H^+]_i$ - x

Just as you did when you took general chemistry, you solve for the value of x:

$$10^{-pK_a} = \frac{([A^-]_i - x)([H^+]_i - x)}{x}$$
$$10^{-pK_a} \cdot x = [A^-]_i \cdot [H^+]_i - [A^-]_i \cdot x - [H^+]_i \cdot x + x^2$$
$$0 = x^2 - ([A^-]_i + [H^+]_i + 10^{-pK_a}) \cdot x + [A^-]_i \cdot [H^+]_i$$

using the quadratic equation to solve for x...

$$x = \frac{-b \pm \sqrt{b^2 - 4ac}}{2a} \tag{7.2}$$

where:

$$a = 1$$
$$b = -1 \cdot ([A^-]_i + [H^+]_i + 10^{-pK_a})$$
$$c = [A^-]_i \cdot [H^+]_i$$

and then you finally find the concentrations of acid and base using:

$$[A^-] = [A^-]_i - x \tag{7.3}$$

and

$$[HA] = x \tag{7.4}$$

You also know that although there will be two roots for x, only one root is going to be physically possible. In this particular case, the correct value of x:

1. must be positive
2. must not be greater than either the starting value of the deprotonated species ($[A^-]_i$) or the amount of acid added ($[H^+]_i$)
3. must not be a complex number (i.e., no imaginary numbers)

Once you determine the correct value for x, you can use it to solve for the concentrations of the protonated and deprotonated forms, and then use Equation 7.1 to determine the $\text{p}H$. In this manner, you can predict the titration curve for any species with a *known* $\text{p}K_a$.

You figure that if you can predict the value of $\text{p}H$ as a function of acid concentration for a *known* $\text{p}K_a$, then you can use this same approach to find which value of $\text{p}K_a$ gives the best fit to experimental data for a compound with an *unknown* $\text{p}K_a$. To do so, you fit the above behavior to your data to extract the $\text{p}K_a$ using `lmfit`. This might sound complicated, but you have almost all the knowledge you need to do it!

7.1 SOLUTION

7.1.0 PLANNING

Starting a rough outline of what you want to do, you arrive at the following:

1. *Write code to predict the titration behavior using the mathematical process outlined above.* An easy way to validate this would be to simulate the titration behavior for a series of equally spaced x-values (i.e., acid additions). If you write a function to do this, you can even treat this as its own **library** and **`import`** this behavior into any code you like, such as your code to fit the experimental data.

2. *Import the data you will fit your predicted behavior to.* You are already fairly conversant with procedures for this.

3. *Fit the prediction of the titration behavior to your data* in order to extract an estimate of the $\text{p}K_a$, and the uncertainties in this estimate. You already know how to do this for models that are built-in to `lmfit`, and so you simply need to learn how to substitute in models you create yourself.

4. *Plot the results*, including the estimate of the $\text{p}K_a$ and the uncertainties in this estimate.

Looking at the list, you decide that you already know how to do the second and final steps. You know how to do most of Step 3, though there are a few new things to introduce. So the bulk of what you need to learn concerns how to code up the expected titration behavior, validate it, and import it for fitting. Learning a few new coding ideas will make this a bit easier.

7.1.1 AUTOMATICALLY RUNNING DIFFERENT CODE UNDER DIFFERENT CONDITIONS USING `if-else` STATEMENTS

Evaluating the roots of the quadratic means testing if they meet certain conditions (i.e., is the root a real number). We can do this using the `if` **control structure**. The structure is simple: we check if something is `True`; if it is, then we run the code **block** that immediately follows. However, if the value is `False`, we skip the immediate code and move on to the next block. We can illustrate this below:

```
mylist = ["True", 1, "Blue"]
if True:
    print ("This was True")
if False:
    print("This was False")
if 1 in mylist:
    print("1 is in mylist")
if 3 in mylist:
    print("3 is in mylist")
```

Running this code will output:

```
Out: This was True
Out: 1 is in mylist
```

You see "This was True" because the block after an `if` statement runs when the code to the right of `if` evaluates to `True`. Because you directly provided the **literal** `True`, Python runs the block immediately afterwards. However, when you supply the literal `False`, then it does not run, and we don't see "This was false," printed out. You do see "1 is in mylist" because `1 in mylist` evaluated to `True`, but because the list does not have three, the `3 in list` evaluates to `False`, and you don't see the results of the following block.

It is nice to be able to check for truth, but we can make the `if` control structure more powerful using `else` and `elif` (else-if)—both of which must be used in conjunction (and following) an `if` block. To see the utility of these two new additions, consider the following code that uses three `if` blocks.

```
a = 0
if a > 0:
    print("The variable is greater than 0.")

if a < 0:
    print("The variable is less than 0.")

if a == 0:
    print("The variable is 0.")
```

```
Out: 'The variable is 0.'
```

Looking at this code, you can see that you didn't really have to perform the last test: if the first and second are not true, the third must be true. Also, if the first one was `True`, then we didn't really need to check the other two. This is where `elif` and `else` help us out. The first stands for "else if" and basically is only run if the tests above it were False, while the last is run if every statement before it was false.

```
a = 0
if a > 0:
    print("The variable is greater than 0.")
elif a < 0:
    print("The variable is less than 0.")
else:
    print("The variable is 0.")

print("Done checking values.")
```

```
Out: The variable is 0.
Out: Done checking values.
```

In this example, the first if statement was `False`, so the program checked the conditional for the `elif` statement, which was `False`, so it automatically ran what was in the `else` block, then the final print statement. If the variable had a value greater than zero, the program would have run the code associated with the if statement, then skipped the `elif` and `else` blocks to go straight to the final print statement. You can play around with other examples to get a sense of how these structures work together.[1]

[1] There are other reasons to use `elif` and `else`, and you can consult Chapter F to learn more.

7.1.2 SETTING THE WORKING DIRECTORY USING `os.chdir()`

Any time you run Python code, the Python interpreter program must be active in *some* location in your **file system** (even if you are not reading or creating files). In other words, there must be some folder on your computer in which the files can be directly accessed by the Python program without having to include their **path**. That folder is known as the **working directory**. You can think of this as the specific folder on your computer where Python is currently active. You can find your working directory by typing `pwd` (print working directory) into the console in the Spyder IDE, which will give you the path associated with the working directory. If you have imported `Path` from the pathlib **library**, then you can also use `Path.cwd()` (current working directory). The latter will return a Path **object** that is the current working directory.

To change the working directory, we need to use a different library, `os`, which provides the `chdir()` function.[2] This function can accept a path object or a string object, and will set the current working directory to the directory the argument specifies, as follows:

[2] You might find code demonstrating this capability with pathlib, but at the time that this book was written, this only worked for Macs and not Windows PCs. The approach using `os` works on both.

```
In: from pathlib import Path
In: Path.cwd()
```

This will output your current working directory. We can change it to a different path

```
In: import os
In: new_path = Path('/path/to/new/folder')
In: os.chdir(new_path)
In: Path.cwd()
```

As long as the `new_path` value is different than the original working directory, you will have moved to a new working directory, as seen in the second output.

7.1.3 Importing Any Python Code from Anywhere on Your Computer

It is very convenient to use **functions** from **libraries**, because we can use the `from <library> import <component> as <alias>` construction. But, you realize that all of these libraries are just Python code! That is, they are held in a file that ends in .py! The **Numpy** library you have been using so much is just in a file called numpy.py. This must mean that we could import our own code from its own .py file. In the context of this chapter, it might be nice to store a function that simulates the titration response in a file called titration.py and then import it into our code and use it. However, Python cannot just import any .py file—largely for security reasons. There are two cases in which you can import a Python file into your script:

1. The .py file is in a folder that is part of the so-called **Pythonpath**. This is a list in which Python keeps track of where it can read .py files from. Part of installing a new library using `pip/pip3/conda install` involves insuring that library is added to the PythonPATH. When you try to import a library, Python checks the paths in it to see if the file you are trying to import are there. If so, you can import it. If not, then the next condition needs to be met.

2. The .py file you want to import is in the current **working directory**.

If neither of the above is true, then you will get an error when trying to import the Python file because Python doesn't know where to find it.

In summary, you have two options for importing your own Python code as a library: you can add the .py file to one of the **directories** that is in the PythonPATH, or you can ensure that the .py file you want to import is in the current working directory. The former is the "normal" behavior we have been using so far. The latter is the behavior we introduce in this chapter.

Why not Jupyter notebooks: revisited.
In Chapter 1 we gave a few reasons as to why we don't use Jupyter notebooks for this book. One of the reasons given was that it is hard to import code from Jupyter notebooks into other code. The reason is that a Jupyter notebook is not a .py file, and so we cannot easily import it like we will do for .py files in this book.

Another important thing to note: When importing components from an external Python file, the code in the external file will only be run the very first time you run the code. After that, you have access to the components that you specify in the `import` **statement**. For instance, if the function for simulating pH titration is called `pH_response`, we can use `from titration_sim import pH_response`. This statement will result in running the entire titration_sim.py file and then give you access to the `pH_response` function.

Finally, when you import from a Python file, the whole code will run. This may not be a desirable situation because your function could be embedded in a code that works on different data, or that takes some time to run, or that needs an input file that you don't have or don't want to give it. To avoid this, you can use a "main entry point," which is some code that defines how the script should run if it is called directly as we have normally done so far. If the same Python script is imported, then the code in the code block denoted by the main entry point will be ignored. It is denoted with a line with an `if` statement like:

```
In: if __name__ == "__main__":
```

with the code that you want in the code block executed if the expression evaluates to `True`. This will be `True` when the file is run directly, and `False` otherwise (so, `False` when run in an `import` statement). In this way, you can ensure that if you run the script directly, as we have been doing so far, it will run as expected, but when you import elements from it, the code in the main entry point will not be run.

7.1.4 Handling Errors Without Crashing Using `try-except-finally`

There are going to be times that you want to perform an operation that you know may fail sometimes. In real life, we do this often. We might attempt a new chemical synthesis knowing it could fail. However, even though the synthesis didn't work, you still want to keep doing chemistry. In Python, there are times that a step in your **script** might fail, but you don't want the failure to stop your entire script from running. Python provides a means to do this via the `try-except-finally` **control structure**.

The basic idea behind this control structure is to arrange code into two to three **blocks**. The first (required) is the `try` block. The second (required) is the `except` block. The third (optional) is the `finally` block.

When encountering this control structure, Python first attempts code in the `try` block. If this code runs without error, then the `except` block is skipped. However, if an error occurs in the `try` block, then the block is excited and the code in the `except` block is attempted. Because of this behavior, the `try` and `except` blocks cannot exist in isolation. You must have both.

The `finally` block is optional, and if it is included, then it runs after completion of either the `try` or `except` blocks.

```
var = "hi"
try:
    var  = 2 + var
except:
    print("it appears that your variable is not a number.")
finally:
    print(f"the value of var is {var}")
```

If you change this code so that var starts as a number, you will get a different output.[3]

Okay! With these concepts out of the way, you can now produce a script that will fit your titration curve and give an estimate (and uncertainty) for the pK_a.

7.1.5 PLOTTING IN A SINGLE LINE OF CODE USING `codechembook.quickPlots.quickScatter()`

The shortest code that you can produce to plot a single set of (x, y)-data in **Plotly** is three lines:

```
fig = make_subplots()
fig.add_scatter(x = xdata, y = ydata)
fig.show()
```

Notably, this code produces a figure with default formatting, which is not great for presenting scientific data, so you will at least want an additional `update_layout()` function, and if you want to include axis titles, you will also need `update_xaxes()` and `update_yaxes()`. This is starting to be a lot of work just to make a plot quickly.

The `codechembook` library contains a module, `quickPlots`, that streamlines the creation of many plots so that the same plot can be created in a single statement. As an example, consider the function `quickScatter()`: it largely mirrors the usage of `fig.add_scatter()`, but includes many keyword arguments with reasonable defaults so that

```
In: quickScatter(x = xdata, y = ydata)
```

will create and show a nice looking plot. The plot can be customized using a number of keyword arguments:

- `xlabel`, `ylabel`: accept strings containing the text for the axis labels. If these keywords are not used, the variable names passed to the `x` and `y` keyword arguments will be used as the labels.
- `name`: the name of the trace, which appears in interactive mode when the mouse hovers over a data point. This name also will appear in the legend if it is enabled, which it is by default.
- `mode`: the drawing mode, either "lines" or "markers" corresponding to the same keyword in `add_scatter()`.
- `template`: the template to use, where the default is the built-in "simple_white" template that you have been using.
- `output`: how to show the plot. By default, a .png file that will show in the Spyder Plots pane, but it will take any output option that `fig.show()` can take (e.g., "browser"), as well as `None` if you do not want the plot to show.

[3] In this usage, the `try`, `except` blocks look like another approach to error handling, but we will see that they can be used in a more flexible manner as we move through this book. You can read more details about this flow control structure in Chapter F. For now, we will use it to handle importing from a Python file that may or may not be in our current working directory.

Importantly, `quickScatter()` returns a Plotly Figure object that you can store as a variable and update using the usual Plotly Figure object methods for options that `quickScatter()` does not natively support. Thus, you can use `quickScatter()` as a quick starting point for making more complex single-subplot figures. In this case, passing `None` to the `output` keyword is convenient because the plot will not show before it is finished being formatted.

If you want to plot multiple traces simultaneously, you have three options:

1. You can use `quickScatter()` as described above, but with `output = None`, and store the Plotly Figure object as a variable. You can then use `add_scatter()` to add traces to the plot before you show it.

2. If all of your traces have the same x-data, you can pass a list of y-data arrays to the keyword `y` and each item in the list will be plotted as a separate trace. Note that all y-data arrays must be the same length as the x-data array. Also, in this case, you can pass lists of the same length containing the modes and names of the traces to the respective keywords.

3. If your traces have different lengths, you can pass lists of x-data arrays and y-data arrays to the corresponding keywords. As explained for the previous option, you can also use lists for the arguments to the `mode` and `name` parameters.

The `quickScatter()` function does not support subplots. For this, see the `quickGrid()` function in the same module. You will also find shortcuts for making bar charts, histograms, and other useful figures.

7.2 CODE

Even though you know enough now to just write a complete fitting **script**, it will save a lot of time and headache to first write and test the **function** that models the titration behavior. That way, you can make sure it works correctly before attempting to fit to it. You can do this by using it to simulate some data and check to make sure that the result is what you expect.

Thinking through the this simulation, you can see that there are four steps:

1. Set up the parameters of the experiment: the starting concentration of the deprotonated species ($[A^-]_i$) and the concentrations of the added acid for each point ($[H^+]_i$). In titrations, the latter is often expressed in terms of equivalents (let's call these `eqs`), but you can convert the equivalents to concentration using `eqs`$\cdot[A^-]_i$.

2. Solve for the change in concentration at each point. This is done using the quadratic equation, and you will find two possible values of x (call them `x1` and `x2`)

3. Determine which value of x is correct for each point.

4. Use the value of x to find the values of pH for each point and return these values.

Working from the above, you sketch out the plan (Figure 7.2):

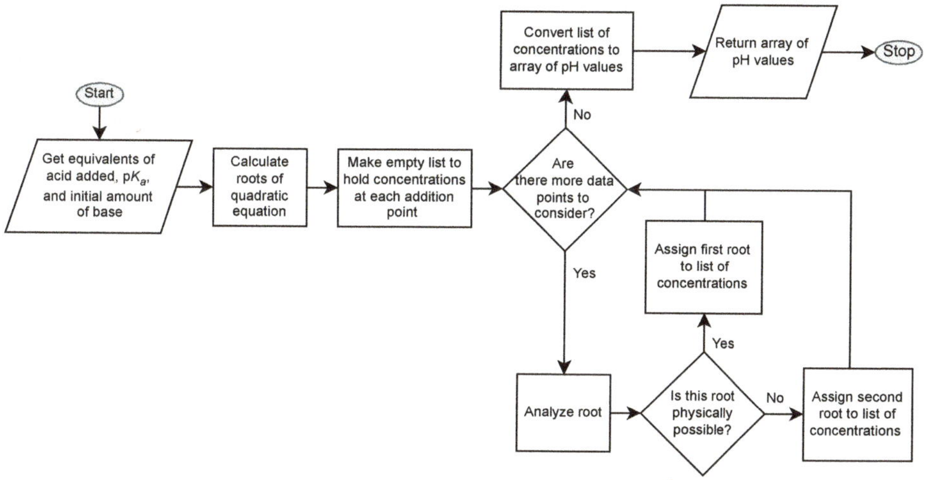

FIGURE 7.2. Flowchart for the function that will simulate titration behavior.

Your code that contains a function that performs these steps and uses the code to simulate a titration experiment and plot the results is:

```
'''
A function that returns the pH response as equivalents of acid are added to a solution
    Written by: Ben Lear and Chris Johnson (authors@codingforchemistsbook.com)
    v1.0.0 - 250210 - initial version
'''
import numpy as np
from codechembook.quickPlots import quickScatter

def pH_response (eqs, pKa = None, base_i = None):
    '''
    Calculates pH values based on equivalents of acid added, pKa, and
    starting concentration of base.

    REQUIRED PARAMETERS:
    qs (ndarray):   equivalents of acid added
    pKa (float):    the pKa of the base
    base_i (float): initialtial concentration of the base

    RETURNS:
    (ndarray of floats): the pHs at each point
    '''
    # Calculate the terms for the quadratic equation
    a = 1
```

```
    b = -1 * (base_i + eqs*base_i + 10**(-pKa)) # eqs*base_i gives concentration of acid
    c = base_i*(eqs * base_i)

    # Calculate the two roots of the equation
    x1 = (-1*b + np.sqrt(b**2 - 4*a*c))/(2*a)
    x2 = (-1*b - np.sqrt(b**2 - 4*a*c))/(2*a)

    # Make a list of the correct x-values
    x = [] # this is an empty list that will eventually hold the correct values
    for eq, e1, e2 in zip(eqs, x1, x2): # test each value
        if e1 < 0:    # can't have negative number of molecules
            x.append(e2)
        elif base_i < e1 or eq*base_i < e1:    # can't be greater than qty. added
            x.append(e2)
        elif isinstance(e1, complex):    # must be a real number
            x.append(e2)
        else:    # if none of those, then it is the correct value
            x.append(e1)

    x = np.array(x) # convert our list to a numpy array

    return pKa + np.log10((base_i - x)/(x)) # return an array, converted to pH

#  Test that the function is working, by plotting an example
if __name__ == '__main__':    # this only runs if the file is run directly
    # Make up some test numbers
    eqs = np.linspace(0.05, 0.95, 20)

    # Run the function and plot the result
    quickScatter(x = eqs, y = pH_response(eqs, pKa = 7, base_i = 1),
                 mode = 'lines+markers', xlabel = 'equivalents added', ylabel = 'pH')
```

Running this code yields the plot shown in Figure 7.3:

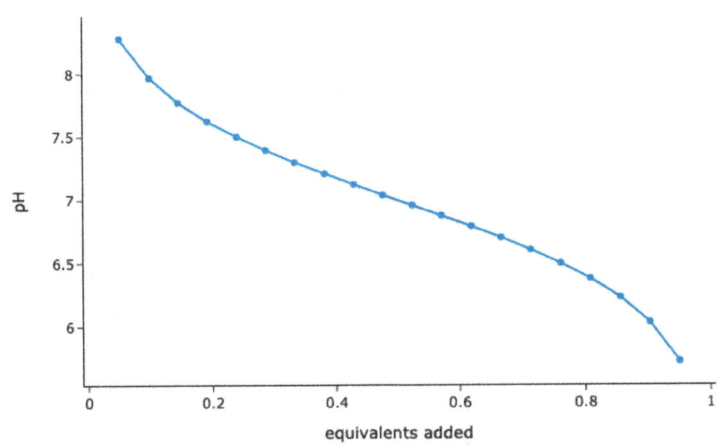

FIGURE 7.3. Results of a simulation of titration behavior. This was done using a function you wrote to simulate this behavior. This function will serve as the model for fitting experimental titration data.

You can see you have a titration curve with 20 points for a species with a pK_a of 7, but we can explain this in more detail.

7.2.0 LINE-BY-LINE EXPLANATION

6 & 7: We import **libraries** we need to use. We are introducing the `quickScatter()` function of the `codechembook` library. This provides a quicker way to generate scatter plots.

9: Define the function for simulating the titration curve. We specify the parameters that are expected. `eqs` is a positional parameter, while `pKa` and `base_i` are keyword parameters.

10–21: This is the docstring for the function. It explains the purpose of the function, the parameters that are in the function, and what is returned by the function.

23: The a-term in the quadratic formula. As long as we have a monoprotic acid, a will be equal to 1 in our quadratic equation.

24: The b-term in the quadratic formula

25: The c-term in the quadratic formula. Because `eqs` is a **Numpy array**, we will get another **Numpy array** here, with a unique value for each value of `eqs`.

28 & 29: We use the quadratic formula to calculate the two roots for x. One of these is the correct root for our problem. For both of these, because `c` is a **Numpy array**, we are also going to get a **Numpy array** back. Between these two **Numpy arrays**, we will have all the roots for all the values of `eqs`.

32: We make a list that has no items. It will eventually hold the correct roots.

33: We start a `for` loop, using `zip()` to move though the equivalents (`eqs`) and roots (`x1` and `x2`) of the quadratic equation. In this for loop, we will run a `if-elif-else` series of statements to identify the correct root, using the three conditions out outlined near the start of the chapter, and add it to the list.

34 & 35: We first check to see if the root in `x1` satisfies our first condition: that it must be positive. If it does not, then we know that the second root (`e2`) is the correct root, so we can add it to our list of correct roots.

36 & 37: If the root in `x1` passed the first condition, then we need to check the second condition: that it is smaller than both the initial concentration of base

and the added acid. The latter is done by noting the number of equivalents added, and using the `base_i` as a scaling factor to get units of concentration for the acid. Also, note that we are checking two conditions and also using `or` here. Again, if the first root fails, then the second root must the correct root and we can add it our list of correct roots.

38 & 39: If the root in `x1` passed the second condition, then we need to check the third condition: that it not be complex. If the first root fails, then the second root must the correct root and we can add it to our list of correct roots.

40 & 41: If the first root did not fail any of our criteria, then it must be the correct root, and we can add to to our list of correct roots.

43: We convert the list of correct roots to a **Numpy array**.

45: We return the **Numpy array**, converted to pH. This is the end of the function.

48: We want to include a test of this function, to ensure it is working. But we also know that we are going to want to call this function from other scripts. It will likely be the case that you do not wish the test to run each time you call the function, so we include a `if __name__ == '__main__':` block here. This block will only run if you directly run the script—which is different from reading the script during an **import** statement (which is what you would do to include this function in another script).

50: We need a series of independent variables to pass to our function. In the context in which we have developed this function, the independent variable is the equivalents of acid we have added. It makes the most sense to check equivalence just past zero (0.05) and just before one (0.95), so we use `numpy linspace` to generate 20 equally spaced points between these two limits.

53 & 54: We wish to plot the data we created. We use `quickScatter()` to do so. We supply the x- and y-data as **keyword arguments**, using the function we just wrote to create the y-values within the call to the `quickScatter()` function. We also can specify the mode we want, the x-axis and y-axis title, using the `xlabel` and `ylabel` keyword arguments. While this function does return a Plotly Figure object, the function also (by default) shows the figure. So as long as we just want to show the plot, we need only call the function, which we have done.

With a function that can simulate a titration curve, all that is left is to import that function into a fitting script that will turn this function into a `lmfit` model and use it to fit your data, as diagrammed in Figure 7.4.

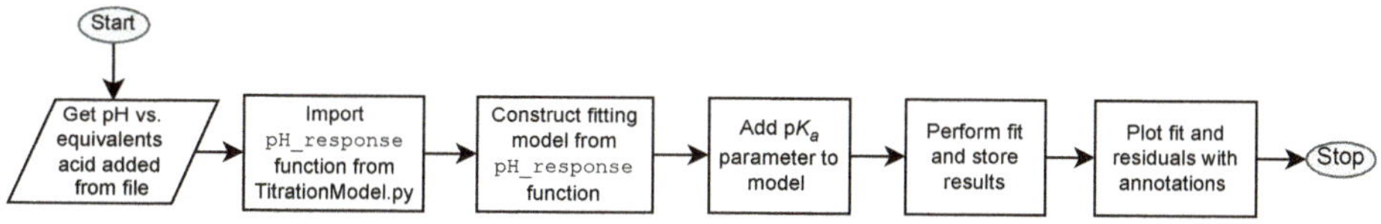

FIGURE 7.4. Flowchart for code that uses the function you wrote to simulate titration data as a model to fit the experimental titration data.

```
'''
A program to fit a titration to the Hendersson-Hassebalch equations
    Requires: pH_response from titration.py
              .csv files with col 1 as wavelength and col 2 as intensity
    Written by: Ben Lear and Chris Johnson (authors@codingforchemistsbook.com)
    v1.0.0 - 250214 - initial version
'''
import numpy as np
from pathlib import Path
from lmfit import Model
from codechembook.quickPlots import plotFit
import codechembook.symbols as cs
from codechembook.quickTools import quickOpenFilename, quickPopupMessage
import os

# Ask the user to specify a file with the data and read it
quickPopupMessage('Select the file with the titration data to fit.')
exp_eqs, exp_pHs = np.genfromtxt(quickOpenFilename(), delimiter = ',', skip_header = 1,
    unpack = True)

# Import the function we will want to use as our model
try:    # first, we attempt to import the pH_response function
    from TitrationModel import pH_response  # import the function we want direct access to
except ModuleNotFoundError:    # if import fails, ask user to find the python script
    quickPopupMessage('titration.py not found.  Please locate it using the file dialog.
    Click OK to open the file dialog.')
    script_path = quickOpenFilename(filetypes = '*.py') # locate the .py file you want to
    use
    original_path = Path('.').resolve() # record the path you are currently using
    os.chdir(script_path.parent) # change directory to the one holding the .py file
    from TitrationModel import pH_response # import the .py file
    os.chdir(original_path)  # change back to the working directory you started in

# Define a new lmfit model using the pH_response function
pH_model = Model(pH_response, independent_vars=['eqs'])  # set up the model with eqs as
    the 'x' axis

# Set up the fit parameter and non-adjustable parameter for initial amount of acid
pH_params = pH_model.make_params() # make a parameter object
pH_params.add('pKa', value = np.mean(exp_pHs)) # specifications for the parameter
base_i = 0.05 # the initial amount of acid

# Fit the model to the data and store the results
pH_fit = pH_model.fit(data = exp_pHs, eqs = exp_eqs, params = pH_params,  base_i = base_i)
```

```
41
42 # Create a figure for the fit result but don't show it yet
43 fig = plotFit(pH_fit, residual = True, xlabel = 'equivalents added', ylabel = 'pH', output
       = None)
44
45 # Add a horizontal line to highlight the pKa as determined by the fit
46 fig.add_scatter(x = [min(exp_eqs), max(exp_eqs)],
47                 y = [pH_fit.params['pKa'].value, pH_fit.params['pKa'].value],
48                 mode = 'lines', showlegend=False,
49                 line = dict(color = 'gray'))
50
51 # Add an annotation containing the best fitting pKa and its uncertainty
52 fig.add_annotation(x = max(exp_eqs), y = pH_fit.params['pKa'].value,
53                    xanchor = 'right', yanchor = 'bottom',
54                    text = f'pKa = {pH_fit.params["pKa"].value:.3f} {cs.math.plusminus} {
   pH_fit.params["pKa"].stderr:.3f}',
55                    showarrow = False)
56 fig.show('png')
```

Running this code brings up a dialog box to select the titration data to fit. You navigate to the Catalyst/Titration/pH folder and select titration.csv. The code outputs the plot shown in Figure 7.5, where the data is the orange points, the fit is the blue line, and the black line indicates the estimated value of the pK_a. Below this is shown the residuals.

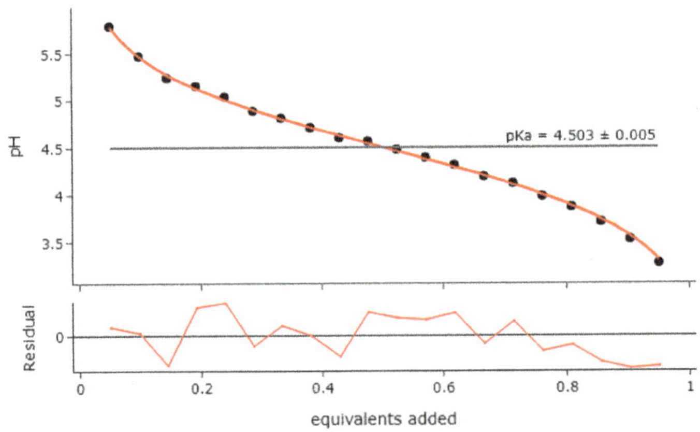

FIGURE 7.5. Results of using the function you wrote to simulate titration behavior as a model to fit the experimental data. Shown is (top) the fit to the data and (bottom) the residuals of the fit.

7.2.1 LINE-BY-LINE EXPLANATION

8–14: Import the **libraries** and portions of **libraries** we will need. Use aliases, as we like.

17: Use the `quickPopupMessage()` function to give the user instructions on what to do when the file dialog shows up.

18: Use the `quickOpenFilename()` function to point to the data we want to fit, and then pass the Path object it returns to `genfromtxt()` to extract and unpack the data into two arrays.

21–29: We wish to import the function we wrote above so that we can use it as a fit model. The process is:

21–22: We start a **`try`** block, in which we attempt to import the pH_response function we wrote from our TitrationModel.py. This will work as long as the file (TitrationModel.py) is in the current working directory. Otherwise, this will fail, which will push us to the **`except`** block.

23–29: This **`except`** block will handle the error that occurs when the current working directory does not have (TitrationModel.py).

24: We open up a popup box so that the user is aware that the import attempt failed, and that they need identify which folder contains this TitrationModel.py.

25: We use the `quickOpenFilename()` function to allow the user to identify the desired folder, using the system's native file explorer.

26: We first create a Path object pointing to the directory that is the current working directly (`Path('.')`), use the `resolve()` method of Path objects to identify where this directory is, and then store this as a variable.

27: We then change the directory to the location identified by the user on line 24.

28: We then again attempt to import the pH_response function from the TitrationModel.py, which should work now because we set the current working directory to be in the same place as this file.

29: We then change the working directory back to the place we originally were (which we stored on line 25). Thus, we imported the function we wanted, and we are still in the same working directly we started in.

32: We make an `lmfit` model object, specifying the function that provides the desired behavior, and also the fact that we will have two independent parameters. The function is the `pH_response` function we imported within our **`try-except`** control structure. The independent parameter is the equivalence values.

35: We create an `lmfit` parameter object. Initially it is empty.

36: We add the fit parameter, pK_a, to the parameter object. Here we give the parameter a name and an initial value to start the fitting process. We set the initial value to just the average of the pH values as a rough guess. Depending on the model and data you use, you might need to find other ways to estimate the initial guess for your model parameters. More information on parameters, including constraints, can be found in the online documentation for lmfit.

37: We set the value of the initial concentration, in units of molar.

40: Fit our model to the data, and pass the fit object to a variable. We need to specify the data we use, the parameters we will use, and any independent variables.

43: Using the `fitplot()` function from the `codechembook` library, we get a Figure object of the fit. Using the keyword argument `output`, we can specify that we don't want the function to show the plot. But we still return a Figure object, and assign this to a variable.

46–49: It would be nice to have a line that indicates the best-fitting pK_a value, and we construct this here and add it to the Figure object.

> 46: For the x-values, we provide just two: the minimum and maximum values of the experimental `eqs`. This ensures the line we draw spans the full range of experimental values on the x-axis.
>
> 47: For the y-values, we also provide a pair of values. In this case, both are the final value of our parameter. This value is an attribute of a dictionary entry, with the key equal to our parameter name, and this dictionary is, itself, an attribute of the fit result object (in this case `pH_fit`). We need to provide this twice because we have two x-values, and we need the same number of x- and y-values for the plot to work.
>
> 48: We want this to be a line (also, no markers), and we do not need to include it in a legend.
>
> 49: Finally, we want to change the color of our line to be black, and so we supply this information as a dictionary to the line keyword—this is how we can format the appearance of our line. The second right parentheses ")" is what ends the function call for `add_scatter()`

52–55: It would also be convenient to label the plot with the pK_a value and uncertainty that was extracted from the fit. So, we create an annotation and specify its content, format, and placement.

52: We specify the x-position of the annotation and place it at the maximum x-value. We also state that the anchor for the annotation is on the right (i.e., the x-position species the far right position of the annotation). We specify the y-position of the annotation, placing it at the best fit value for the pK_a. We access this value just as was done on line 47.

53: We also specify that the anchor is the bottom, meaning that the y-position specifies the lowest potion of the annotation.

54: We provide the string we wish to write to the `text` keyword argument. We use an f-string, adding the final fit values via the corresponding `ModelResult` object attributes, but formatted to two decimal places. We also do the same for the uncertainty.

55: Annotations default to showing a line (called: arrow) pointing to the point being annotated. We don't want this, so we set this value to False, and then end the `add_annotation( )` function call with the closing parentheses.

56: We show a static version of the figure.

CHAPTER 7 EXERCISES

TARGETED EXERCISES

Exercises for Section 7.1.1

0. Write three different functions that take as an argument a pH value and print a statement saying whether the value is "strongly acidic," "acidic," "neutral," "basic," or "strongly basic." One version can only use `if` statements, the second must use nested `if-else` statements, and the third must use `if-elif-else`. Prove that they work correctly by testing the pH values 1.0, 4.0, 7.0, 10.0, and 13.0.

1. Explain the following joke: "Your roommate is going to the store, so you ask them to pick up a gallon of milk and, if the store has eggs, to get a dozen. Your roommate returns with a dozen gallons of milk."

Exercises for Section 7.1.2

2. Change your working directory to your downloads folder. Verify using `os.pwd( )`.

3. Using the `quickSelectFolder( )` function in the codechembook library, write a line of code that will change your directory to one of your choosing.

Exercises for Section 7.1.3

4. Almost every library you use is (at least in part) just Python code, stored in a .py file. This is useful because you can learn about things that are not in the documentation by looking at the code itself. Find the folder where all of your libraries are stored, then find the folder for lmfit and open the "lineshapes.py" file. On our windows machine, the path is `C:\Users\authors\AppData\Roaming\Microsoft\Windows\Recent\lineshapes.py.lnk`, but you can also just search for this file. Explain how the `sine()` and `gaussian()` functions work.

Exercises for Section 7.1.4

5. Update the final code from in Chapter 5 to use `try-except` to ensure that the user has input a number when asked for the wavelength of the feature of interest. If the user inputs a value that cannot be converted to a float, then the code should handle the error by printing, "You must enter a number!"

6. Update the code in Chapter 2 to use `try-except` to check to make sure the data file is openable. If there is an error, inform the user and call `quickOpenFilename()` to help them find the right file. Test your code.

COMPREHENSIVE EXERCISES

7. Your friend is working up data from an amyloid formation assay using a plate reader and a chromophore that becomes absorptive when in the presence of amyloid. This gives him a set of data that contains absorption at λ_{max} versus temperature, which can be used to track the concentration of amyloid. He wants to fit this data to the following modified Sigmoid function and extract t_{50}:

$$[\text{amyloid}](t) = (m_1 t + b_1)\left(\frac{1}{\exp\frac{t_{50}-t}{\sigma_1}}\right) + (m_2 t + b_2)\left(\frac{1}{\exp\frac{t_{50}-t}{\sigma_2}}\right)$$

Here, m_1 and b_1 are the slope and y-intercept of the end of the trace, m_2 and b_2 are the slope and intercept of beginning of the trace, σ_1 and σ_2 control the shape of the curved parts of the trace, and t_{50} is the time at which half of the amyloid is formed. He provides you with a sample data file for a single well to get started. Write a script that reads the data, creates a model for this modified Sigmoid function, sets the initial guesses for the parameters, and shows the user the initial model so that the initial guesses can be updated to get close enough for the fit to work. Then, once the user determines these guesses are close, it calls the fitter, plots the results, and prints the best-fitting value of t_{50}.

CHAPTER 8

Simulating Chemical Kinetics

- ☐ **New concepts**
  - — *and their prerequisites*

- ☐ **Simulating the kinetics of one elementary reaction using** `scipy.solve_ivp()`
  - — *Section 0.1.2: Accessing capabilities beyond basic math using functions*
  - — *Section 2.1.1: Importing subsets of libraries using* `from`
- ☐ **Validating coding approaches with simple test cases**
  - — *Section 1.1.1: Operating on collections of data efficiently with Numpy arrays*
  - — *Section 0.1.7: Accessing capabilities beyond basic math using functions*
  - — *Section 3.1.1: Exploiting built-in methods to manipulate data stored in objects*
  - — *Section 2.1.10: Making appealing and interactive plots using Plotly*
- ☐ **Simulating multi-step reaction mechanisms using** `scipy.solve_ivp()`
  - — *Section 0.1.2: Accessing capabilities beyond basic math using functions*
  - — *Section 2.1.1: Importing subsets of libraries using* `from`

8.0 PROBLEM

With all the extra time you have gained automating your problem-solving with coding, you turn back to thinking about your project. It is nice to be able to do these high-throughput experiments, but how do you know what range of concentrations to start with, and how long the reaction will take to run to completion? You would like to be able to predict what the concentrations of all of the components of the reaction are, not only to understand what is happening, but to be able to design experiments to probe the mechanism in more detail.

You think, "the mechanism for this reaction is complicated, none of the solutions that I learned in p-chem class will work here." You also think about how useful it was to simulate titration behavior. "Maybe I can write a program that simulates the reaction without having to do math that may be complicated or impossible?"

The overall reaction is simple, just $A + B \xrightarrow{k_{cat}} C$. However, there are two suspected intermediates, X and Y, making the mechanism more complex:

$$\begin{aligned} \mathrm{A} + \mathrm{cat} &\xrightarrow{k_1} \mathrm{X} \\ \mathrm{X} + \mathrm{B} &\xrightarrow{k_2} \mathrm{Y} \\ \mathrm{Y} &\xrightarrow{k_3} \mathrm{C} + \mathrm{cat} \end{aligned}$$

To figure out where to even start, you check a kinetic textbook out of the library.[1] You find that chemical kinetics are an example of a initial value problem: you only know the concentrations of the species in your reaction at the beginning of the reaction. However, if you can work out the differential rate laws, then you also know how the change of the concentrations versus time depend on concentration. So, you should be able to use these rate laws to project the behavior of the reaction forward in time. The process of doing this is solving the set of differential equations, and luckily, chemical kinetics are usually represented by ordinary differential equations that can be solved straightforwardly (in principle).

[1] A good reference is Steinfeld, Francisco, and Hase's *Chemical Kinetics and Dynamics*. See a discussion about numerical solutions to kinetics problems in Chapter 2.6.

You were taught to solve kinetics problems by first writing down differential rate laws for each component. For the reaction under consideration, you come up with:

$$\begin{aligned} \frac{\mathrm{d}[\mathrm{A}]}{\mathrm{d}t} &= -k_1[\mathrm{A}][\mathrm{cat}] \\ \frac{\mathrm{d}[\mathrm{X}]}{\mathrm{d}t} &= k_1[\mathrm{A}][\mathrm{cat}] - k_2[\mathrm{X}][\mathrm{B}] \\ \frac{\mathrm{d}[\mathrm{B}]}{\mathrm{d}t} &= -k_2[\mathrm{X}][\mathrm{B}] \\ \frac{\mathrm{d}[\mathrm{Y}]}{\mathrm{d}t} &= k_2[\mathrm{X}][\mathrm{B}] - k_3[\mathrm{Y}] \\ \frac{\mathrm{d}[\mathrm{C}]}{\mathrm{d}t} &= k_3[\mathrm{Y}] \\ \frac{\mathrm{d}[\mathrm{cat}]}{\mathrm{d}t} &= -k_1[\mathrm{A}][\mathrm{cat}] + k_3[\mathrm{Y}] \end{aligned} \tag{8.1}$$

You *might* be able to make some approximations and solve this **analytical**ly in some specific cases, but you also suspect that something in the solution might be inhibiting one of the steps of the cycle and would like to be able to account for that, too. This means things will get even more complex, and it will probably be easier to just do this **numerical**ly. So, armed with your kinetics book and the internet, you set out to learn how to do it.

8.1 SOLUTION

A quick search shows that the Python **library** `scipy` can integrate systems of ordinary differential equations (ODEs) (among many other things!). You see that there are different types of ODEs and a number of **function**s to solve them built into the `scipy.integrate` **module**. Kinetics are a type of initial value problem (you know the initial conditions only), rather than a boundary value problem (in which you know the value of the resulting function at both the beginning and end of the integration). How do you map this onto a numerical solution?

The rate laws are ODEs that tell you how fast the concentrations are changing in time as a function of all concentrations. So, if you know all of the concentrations at any time, t, then you can make an estimate of the values of the concentrations at time $t + \Delta t$ by integrating the rates of change over that Δt. You can then use the concentrations at $t + \Delta t$ to integrate up to $t + 2\Delta t$. If you keep following the same process, you will have an estimate of the concentrations at a number of points in time in the future, from which you can make the familiar kinetics traces. This type of problem is known as an initial value problem (IVP). Smaller values of Δt will tend to provide better estimates of the true solution, but require more computation.

Luckily for you, you don't have to figure out how to do the integration, just how to use an integration routine provided by `scipy`!

8.1.0 SIMULATING THE KINETICS OF ONE ELEMENTARY REACTION USING `scipy.solve_ivp()`

Looking through the documentation of the `scipy.integrate` **module**, it looks like the **function** `solve_ivp()` does what you want—determine the value(s) of one or a set of quantities as a function of time given initial values and ODEs for the rates of change.[2] However, you need to translate the discussion of the usage of the function to familiar chemistry terms. It becomes clear that the y-value in the function will be your concentration, and `y0` will be the initial concentration. How to include the rate law is a little confusing, though. `solve_ivp()` has as the first required **argument** `fun`, which is a **function** that contains the time derivative of y and should be of the form `fun(t, y)`. The function `fun` then needs to return the value of $\mathrm{d}y/\mathrm{d}t$. It looks like you are required to have the `t` argument there, even though your differential equations will not explicitly depend on time. The second required argument is `t_span`, which is a two-element list or array

[2] For more information on numerical solutions of ordinary differential equations (particularly methods to evaluate accuracy when you don't already know the right answer), check out books like Hildebrand, *Introduction to Numerical Analysis*; Dover Press, 1987, or Sullivan, *Numerical Methods: An Inquiry-Based Approach With Python* (https://numericalmethodssullivan.github.io).

that contains the start and stop time for the reaction simulation. The last required argument is the initial value `y0`. You also see that there are a number of **keyword arguments** that are optional. Several appear to be notable:

- `method`: which algorithm to use to solve the IVP. They are primarily broken down into methods better used on "stiff" problems (of which many kinetics problems are examples) and "non-stiff" problems (only very simple kinetics problems qualify). The basic difference between these cases is that stiff problems have some components (reactants or intermediates) that may change very quickly with time and others that change more slowly. Stiff solvers can solve these problems without resorting to extremely small step sizes, leading to more accurate results more quickly. Typically, the default solver will be fine.

- `args`: additional arguments to pass to your differential rate law function beyond the required `t` and `y`. In your case, you don't want to hard code the rate constants into the rate laws function because you would have to change the values in the function every time. Instead, you can make your function take an argument for `k` as well and pass it whatever rates you want using this keyword.

- `options`: parameters that mostly control the speed and accuracy of the solution. Of these, the ones you are most likely to use are `rtol` and `atol`, which set the relative and absolute tolerances of the solution, respectively. Lower values of both will yield more accurate solutions but require more time to solve. Specifically, `rtol` basically controls the fractional accuracy, i.e., the percent difference, whereas `atol` controls the absolute difference. Thus, if your concentration is something like 1×10^{-3} M, then `rtol = 1e-3` will request accuracy to a part per thousand, whereas `atol = 1e-3` will request accuracy to the first decimal place.

Everything that you could want to know about the solution to the ODE, particularly the time points and the values of the function at each time, is **return**ed in a single **object**. It contains several notable **attributes**:

- `t`: time points at which the solution is known.

- `y`: values of the solution at each time point. This is a 2D **Numpy array**, where one dimension denotes the component of the reaction (each individual species present in the mechanism), and the second is corresponding concentration at each time point.

- `status`: how the solution ended up, which can be used to make sure everything went okay. This is zero if the solver finished with no issues, and 1 or -1 in the rare case that something happened along the way that caused it not to finish.

8.1.1 PLANNING

Mapping this functions usage onto your problem, you decide that you will need the following:

- Initial concentrations of all components of the reaction
- All of the rate constants
- A function containing the differential rate laws
- The total amount of time that you want to simulate the reaction

Thinking through what is required for a solution, you sketch out the four steps you need to solve the full problem:

1. *Define **functions** containing the differential rate laws that need to be integrated.*
   The function should mimic the general form of the differential rate laws:

   $$\frac{dy}{dt} = f(t, y, k), \tag{8.2}$$

   so it will need the **arguments** `t`, `y`, and `k` and **return** a number $\frac{dy}{dt}$. If you have more than one species in your reaction, this function will need to take an array of `y`s and `k`s and return an array of $\frac{dy}{dt}$s. Because of this, even if you have only one component in your reaction, you will still need to provide a one-item list or Numpy array of `y0` and will get back a one-item list with $\frac{dy}{dt}$.

2. *Set up the initial concentrations and the rates.*
   The initial concentrations will be stored as a list and passed as positional arguments to `solve_ivp()`. The rates will also be stored as a list and passed through `solve_ivp()` to the differential rate laws function using the `args` keyword.

3. *Invoke* `solve_ivp()`.
   This will just be a single line passing the needed arguments to `solve_ivp()` and assigning the results to a new variable that holds the object containing all the information about the solution.

4. *Examine and plot the results.*
   You can then unpack only the information you need from the results object. Specifically, you will want to plot all of the concentrations with respect to time!

That doesn't seem too bad, but it is probably best to start with something simple where you already know the answer, just to make sure you understand the coding part.

8.1.2 VALIDATING CODING APPROACHES WITH SIMPLE TEST CASES

You figure that the easiest thing you can do is to model a simple unimolecular reaction:

$$\mathrm{A} \xrightarrow{k} \mathrm{B}$$

with the differential rate law $\frac{\mathrm{d[A]}}{\mathrm{d}t} = -kt$. You remember that this type of reaction is described analytically by $[\mathrm{A}](t) = [\mathrm{A}_0]\exp(-kt)$. Thus, you can use `solve_ivp( )` and test the accuracy of the solutions you obtain from it against this analytical equation.

Now you can sketch out a plan for the solution (Figure 8.1):

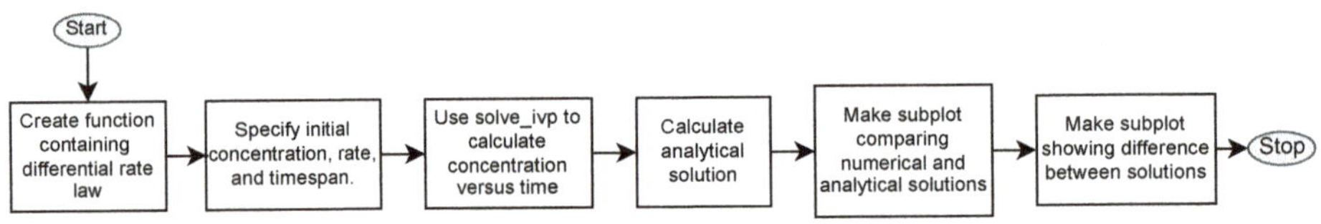

FIGURE 8.1. Flowchart for basic simulation of chemical kinetics using `solve_ivp( )` from scikit.

```python
'''
A code to numerically solve a first order kinetics problem using solve_ivp
from scipy.integrate
    Written by: Ben Lear and Chris Johnson (authors@codingforchemistsbook.com)
    v1.0.0 - 250217 - initial version
'''
import numpy as np
import scipy.integrate as spi
from plotly.subplots import make_subplots

# First we need to define a function containing our differential rate law.
def FirstOrderDE(t, y, k):
    '''
    Differential rate law for first-order kinetics

    REQUIRED PARAMETERS:
    t (float): the current time in the simulation.
               Not explicitly used but needed by solve_ivp
    y (float): the current concentration
    k (float): the rate

    Returns:
    (float): the rate of change of concentration with time
    '''
    dydt = -k * y
    return dydt

# Set up initial conditions and simulation parameters
y0 = 1.0       # concentration (mM)
k = 1.0        # rate (1/s)
time = [0, 10] # simulation start and end times (s)

```

```
33 # Invoke solve_ivp and store the result object as solution
34 solution = spi.solve_ivp(FirstOrderDE, time, [y0], args = [k])
35 
36 # Compute the known analytical solution at each simulation time point
37 anal_y = np.exp(-1 * k * solution.t)
38 
39 # Plot the numerical and analytical solutions and the difference between them
40 fig = make_subplots(2, 1)
41 fig.add_scatter(x = solution.t, y = solution.y[0],
42                 mode = 'markers',
43                 name = 'Numerical') # The data from the numerical solution
44 fig.add_scatter(x = solution.t, y = np.exp(-1 * k * solution.t),
45                 mode = 'lines',
46                 name = 'Exact') # The corresponding analytical solution
47 fig.add_scatter(x = solution.t, y = 100 * (solution.y[0] - anal_y) / anal_y,
48                 name = 'Error',
49                 row = 2, col = 1) # The percent difference between them
50 
51 # Update the plot appearance
52 fig.update_xaxes(title = 'Time (s)')
53 fig.update_yaxes(title = 'Concentration (mM)', row = 1)
54 fig.update_yaxes(title = 'Percent Difference', row = 2)
55 fig.update_layout(template = 'simple_white')
56 fig.show('png')
```

8.1.3 LINE-BY-LINE EXPLANATION

7–9: We need to import `numpy`, `scipy.integrate` (which contains `solve_ivp()`) and `make_subplots` from `plotly.subplots` to plot the results.

12–26: This is the definition of the differential rate law. `solve_ivp()` has the required parameter `t`, even though it does not appear in the function. The required parameter `y` will hold the concentration value, and `k` the rate. We simply calculate the derivative of concentration with respect to time using the equation for the differential rate law and return its value (`dydt`).

29–31: We define the values of the initial concentration, the rate, and the beginning and end of the simulation time. We don't have to do it on separate lines; we could just input their values as literals in the positional arguments and the `args` keyword when we call `solve_ivp()`. However, it is clearer to do it separately, and it makes it easier to reuse these values later in the code without having to remember to change them everywhere we use them.

34: We call `solve_ivp()` here and pass the necessary arguments and keywords to start solving the ODE. Once execution finishes, `solve_ivp()` returns an object containing arrays holding the `t` and `y` values for the solution and information to help diagnose the solution, which we store in the new variable `solution`.

37: Now we need to compute the analytical (correct) values for concentration using the known solution for first-order reactions at each time point in `t` that we have numerical values. The analytical concentrations are then stored in a new variable `analytical_y`. This will allow us to directly compute the absolute and relative errors of the numerical solution.

40–56: Finally, we make a figure with two subplots. The top subplot displays the analytical and numerical solutions for the concentration versus time, and the bottom subplot displays the relative error (percent difference) between the two solutions.

Running this code produces the plot shown in Figure 8.2.

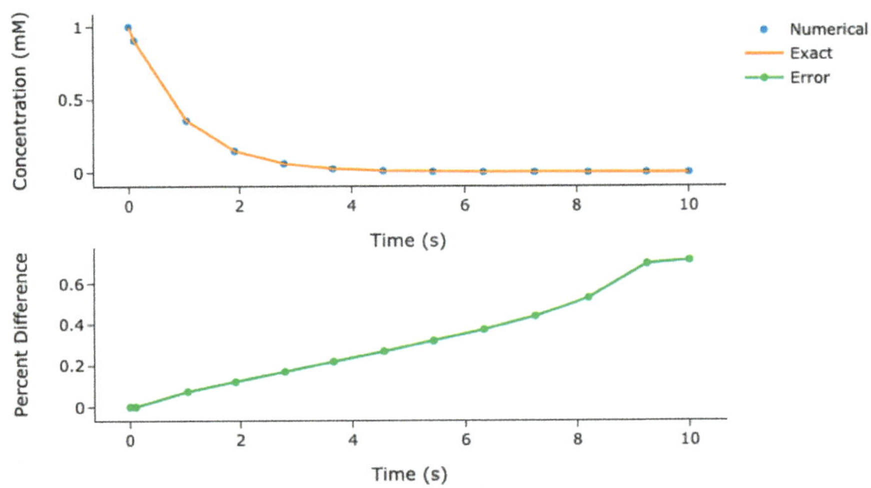

FIGURE 8.2. Result of your code to solve a simple first order reaction. (top) A plot of concentration vs. time for the numerical solution and the known analytical one. (bottom) The percent difference between the numerical and "correct" solutions.

The first thing you notice is that the time points are not evenly spaced for the numerical solution. Digging into the documentation for `solve_ivp( )`, you find that the default solver, "RK45," is an adaptive step size solver, which means that it takes as big of step sizes as it can while trying to balance accuracy. You also notice that the percent difference between the numerical solution and correct one increases with simulation time. This is likely because the concentration at long times is close to zero, so the denominator of the percent difference becomes very small, blowing up the value. Simply changing from percent difference to absolute difference (Figure 8.3) shows this is the case.

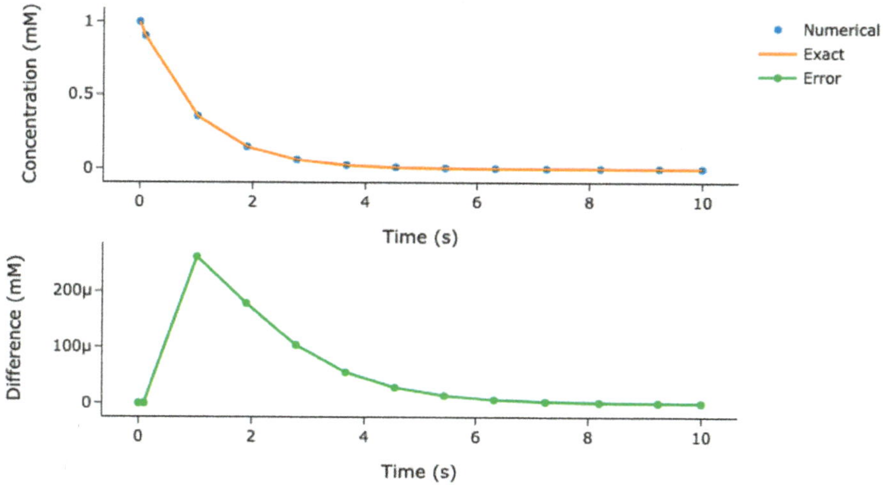

FIGURE 8.3. Like Figure 8.2, but plotting the absolute rather than percent difference.

The simulation is accurate to the order of hundreds of nM or less, not bad! You can use the keywords `rtol` and `atol` to increase the accuracy of these simulations if needed.

8.1.4 SIMULATING MULTI-STEP REACTION MECHANISMS USING `scipy.solve_ivp()`

With this example understood, you feel confident that you are ready to tackle your more complex mechanism. You realize that you can just make small modifications to your code: you need to change the function that contains the differential rate equations, you need to have six initial concentrations and three rates, and you need to change the plotting code. Other than that, you're all set.

8.2 CODE

You produce the following code:

```
'''
A code to numerically solve a reaction mechanism that is irreversible but
has two intermediates using solve_ivp from scipy.integrate
    Written by: Ben Lear and Chris Johnson (authors@codingforchemistsbook.com)
    v1.0.0 - 250217 - initial version
'''
import scipy.integrate as spi
from codechembook.quickPlots import quickScatter

# First we need to define a function containing our differential rate laws.
def TwoIntermediatesDE(t, y, k):
```

```
12      '''
13      Differential rate laws for a catalytic reaction with two intermediates:
14      A + cat ->(k1) X
15      X + B ->(k2) Y
16      Y ->(k3) C + cat
17
18      REQUIRED PARAMETERS:
19      t (float):          the cutrent time in the simulation.
20                          Not explicitly used but needed by solve_ivp
21      y (list of float): the current concentration
22      k (list of float): the rate
23
24      RETURNS:
25      (list of float): rate of change of concentration in the order of y
26      '''
27      A, B, cat, X, Y, C = y # unpack concentrations to convenient variables
28      k1, k2, k3 = k # unpack rates
29
30      dAdt   = -k1 * A * cat
31      dBdt   = -k2 * X * B
32      dcatdt = -k1 * A * cat + k3 * Y
33      dXdt   = k1 * A * cat - k2 * X * B
34      dYdt   = k2 * X * B - k3 * Y
35      dCdt   = k3 * Y
36
37      return [dAdt, dBdt, dcatdt, dXdt, dYdt, dCdt]
38
39  # Set up initial conditions and simulation parameters
40  y0 = [1.0, 1.0, 0.2, 0.0, 0.0, 0.0] # concentrations (mM) [A, B, cat, X, Y, C]
41  k = [5e1, 1e1, 5e0]                  # rate (1/s)
42  time = [0, 10]                       # simulation start and end times (s)
43
44  # Invoke solve_ivp and store the result object
45  solution = spi.solve_ivp(TwoIntermediatesDE, time, y0, args = [k])
46
47  # Plot the results
48  quickScatter(x = solution.t, # need a list of six identical time axes
49               y = solution.y,
50               name = ['[A]', '[B]', '[cat]', '[X]', '[Y]', '[C]'],
51               xlabel = 'Time (s)',
52               ylabel = 'Concentration (mM)',
53               output = 'png')
```

Running this code gives you the plot shown in Figure 8.4.

You realize you can now solve any kinetics problem for which you can write the differential rate laws, and wonder why they didn't just teach you this in p-chem class...

8.2.1 LINE-BY-LINE EXPLANATION

- Most of the code is the same as the previous one. Specific differences are:

27: Since we still only have one variable, y, we are passing the concentrations as a list. For clarity, we are first unpacking the list into variables with helpful names.

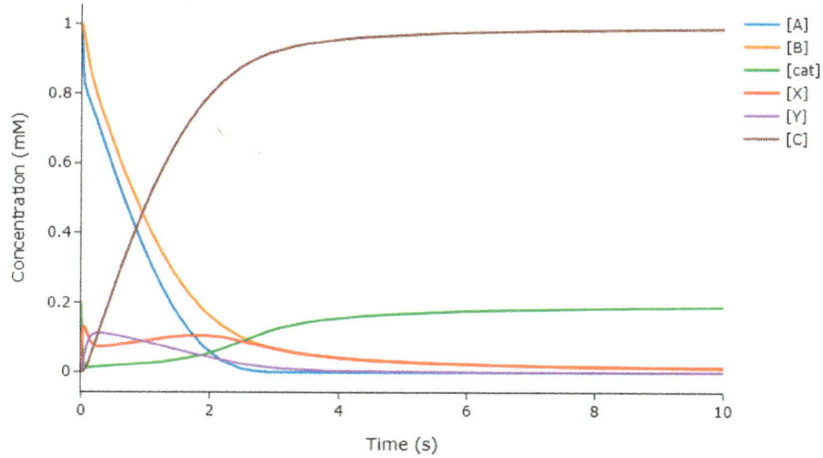

FIGURE 8.4. The output of the simulation using the code above to simulate the full kinetics of the chemical reactions outlined at the top of the chapter.

28: We are in the same situation with the rates, so we are unpacking them here, too.

30–35: Now we are defining our system of rate laws instead of just one.

37: We need to return the derivatives to `solve_ivp()` as a list (or array) in the same order as the y-values.

48–53: Because we don't have analytical solutions, there's no way to plot the error. Instead, we are just plotting all of the concentrations with respect to time using `quickScatter()`. Here, we make use of built-in functionality in `quickScatter()` whereby we can pass a list or array of y-arrays and it will automatically plot them each as a separate trace. We also pass a list as the `name` argument with names for each corresponding y-array.

CHAPTER 8 EXERCISES

TARGETED EXERCISES

Exercises for Section 8.1.0

0. Working from the code presented in Section 8.1.0, write a script that plots the maximum error in the simulation versus the relative tolerance specification within `scipy.integrate.solve_ivp()`. You can use the `rtol` keyword to set this. Use a range of `rtol` values that span 10^{-1} to 10^{-9} changing by factors of 10.

1. Working from the code presented in Section 8.1.0, use every option for the `method` keyword argument of `scipy.integrate.solve_ivp()` and compare the maximum error obtained. Try making a bar chart to compare these values, using Plotly. You can read the documentation in Chapter I for this, or use the `quickBar()` function out of the codechembook library.

Exercises for Section 8.1.4

2. Using the code presented Section 8.1.4. Adjust the reaction *rates* so that:

a) $[C]$ never levels off

b) Three of the traces appear to meet at a single point

3. Using the code presented in Section 8.1.4, adjust the initial *concentrations* so that:

a) There is only a single maximum for $[X]$

b) The bound catalyst is above a concentration of 0.5 M for most of the time

4. Change the code in Section 8.1.4 so that the results are plotted versus $\ln(t)$. This makes it easier to focus on the early timepoints.

COMPREHENSIVE EXERCISES

5. It is often difficult to work with second order processes, like:

$$A + B \rightarrow C$$

Because the reaction progress depends strongly on both $[A]$ and $[B]$ (one of which may be hard to measure). Instead, you can work in excess $[A]$ or $[B]$ and make the "pseudo-first-order approximation" to remove the explicit dependence of the differential rate law on its concentration. If, say, A is in excess, then $[A]$ will not change much during the reaction, and can be approximated as a constant. Thus, we reduce:

$$\begin{aligned} \frac{d[A]}{dt} &= -k[A][B] \\ \frac{d[B]}{dt} &= -k[A][B] \\ \frac{d[C]}{dt} &= k[A][B] \end{aligned}$$

to:

$$\begin{aligned} \frac{d[A]}{dt} &= -\kappa[B] \\ \frac{d[B]}{dt} &= -\kappa[B] \\ \frac{d[C]}{dt} &= \kappa[B] \end{aligned}$$

where $\kappa = k[A]$

It is a good idea to figure out how much excess you need for the pseudo-first-order approximation to hold. Although this could be done analytically, it is quick and still quite accurate to do numerically. To do this, we will run several trials with increasing excess of A and track how the difference between the two solutions changes. To proceed:

- Write scripts that model the second order and the pseudo-first order reactions.
- Convert these codes into functions that take as arguments the initial concentrations, rates, and reaction times and return the solution object from `solve_ivp()`.
- Write another script that creates a list of initial concentrations of A, starting with $[A]_0 = [B]_0$ and increasing logarithmically to $[A]_0 = 100 \times [B]_0$, and then solves both reactions for the same conditions.
- For each condition, compute the sum of squared differences between the second order and pseudo-first-order solutions for $[C](t)$. Note that because you are probably using an adaptive step size solver, the time points you get back by default will not be the same between the two reactions. Instead, you will need to use the `t_eval` keyword to specify common time points. A good idea would be to make sure you are specifying time points that are no farther apart than the smallest gap between times in any simulation.
- Plot the differences from above versus the ratio of concentrations for each time point. Identify the region of applicability for the pseudo-first-order approximation.

6. The iodine clock reaction is a classic chemistry demonstration. Look up a demonstration on YouTube if you're not familiar. Accurately modeling it is complex, but a simplified model can be created using the Briggs-Rauscher reaction:

$$\mathrm{A} + \mathrm{B} \xrightarrow{k_1} 2\,\mathrm{B}$$
$$\mathrm{B} + \mathrm{C} \xrightarrow{k_2} 2\,\mathrm{C}$$
$$\mathrm{C} \xrightarrow{k_3} \mathrm{D}$$

Modify our ODE solving code to work on this system. Good starting points are initial concentrations $[A]_0 = 10$ mM, $[B]_0 = 1$ mM, $[C]_0 = 1$ mM, $[D]_0 = 0$ mM, $k_1 = 0.005\,\mathrm{s}^{-1}$, $k_2 = 0.1\,\mathrm{s}^{-1}$, $k_3 = 0.01\,\mathrm{s}^{-1}$, and $t_{max} = 10000$ s. Change the initial concentrations and rates one at a time. Provide plots that show how all concentrations change with time, both for the full time range and one oscillation. Explain how this reaction works, particularly how it can oscillate even though all the reactions are forward only. Show the behavior at several limits in the rates and initial concentrations, particularly where the system no longer oscillates.

PART THREE

ANALYZING NUMERICAL DATA

CHAPTER 9

Integrating Data Numerically

- ☐ **New concepts**
  - — *and their prerequisites*

- ☐ **Integrating data using `scipy.integrate`**
  - — *Section 1.1.1: Operating on collections of data efficiently with Numpy arrays*
  - — *Section 2.1.1: Importing subsets of libraries using* `from`
- ☐ **Generating test data to evaluate the accuracy of numerical analysis**
  - — *Section 1.1.5: Operating on collections of data efficiently with Numpy arrays*
  - — *Section 5.1.4: Inserting calculation results into strings using f-strings*
- ☐ **Creating customized versions of existing functions using wrappers**
  - — *Section 0.1.7: Accessing capabilities beyond basic math using functions*
  - — *Section 0.1.6: Expressing data as readable text using strings*
  - — *Section 2.1.4: Exploiting built-in methods to manipulate data stored in objects*
  - — *Section 4.1.14: Describing how to use your functions in docstrings*
  - — *Section 7.1.1: Getting yes/no answers to questions posed by comparison statements*
  - — *Section 6.1.3: Clarifying how your code works with comments*
- ☐ **Simplifying the processing of iterable objects using comprehensions**
  - — *Section 1.1.1: Operating on collections of data efficiently with Numpy arrays*
  - — *Section 5.1.4: Inserting calculation results into strings using f-strings*
  - — *Section 3.1.1: Exploiting built-in methods to manipulate data stored in objects*
  - — *Section 3.1.11: Performing repetitive operations using* `for` *loops*
  - — *Section 4.1.4: Combining lists by concatenation*
  - — *Section 7.1.1: Getting yes/no answers to questions posed by comparison statements*

9.0 PROBLEM

The more code you write, the more you start to see the range of problems you can now solve and the benefits of automating data analysis workflows over manually working up large amounts of data in vendor-specific software. On your way into work one day, you overhear a conversation between two people in a lab down the hall from you:

"...I really need to measure the rate of this click reaction, but there's no clean peak for the reactant or product and ^{15}N-azides are too expensive for me to try nuclear magnetic resonance (NMR). I could track the disappearance of the azide by Fourier transform infrared (FTIR) spectroscopy, but the software doesn't do kinetics."

You pipe up, "You could measure the kinetics pretty easily by writing your own script to analyze the FTIR data. It is actually easier than you'd think, and probably faster than doing it by hand in some software."

You all get some coffee and talk through the problem. Tracking the azide concentration is straightforward by FTIR because its spectral signature is well-removed from the complicated fingerprint region. The peak shape is complex, so fitting it to extract the intensity is probably out of the question, but you can definitely integrate the band shape it after subtracting out the background. So, all you need is a series of spectra at different time points, and you can produce a plot of the concentration of azide versus time!

9.1 BACKGROUND

Thinking back to your calculus class, you recall that integration was introduced to you with the "rectangle" rule, where you draw a series of rectangles under the curve of the function and add up the areas of the rectangles to approximate the area under the curve. You spend a little time reading up on integration in Python and find that several **methods** for integration are provided in the SciPy **library**: the trapezoid rule, Simpson's rule, Gaussian quadrature... These are depicted graphically in Figure 9.1, but Which one should you use?

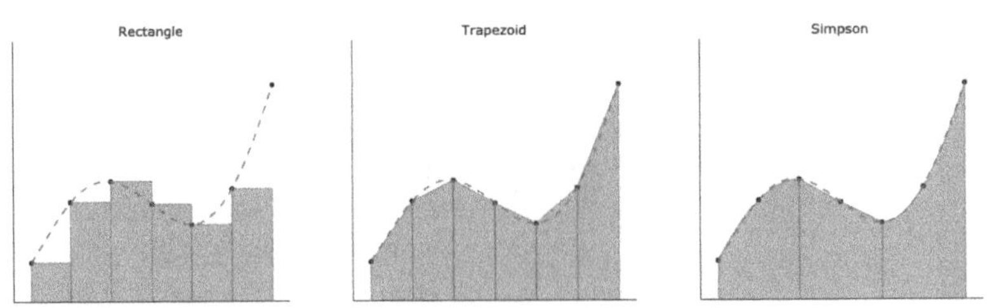

FIGURE 9.1. Different approaches to numerical integration that are commonly used.

You come across Gaussian quadrature and read that can be exactly correct for integrals of polynomials, but you are integrating experimental data with noise. It seems that it is not as frequently used for such data because noise can cause its results to be inaccurate. The rectangle rule, trapezoid rule, and Simpson's rule all basically come down to what you assume is happening to your y-values between the data points. For the rectangle rule, all the y-values are the same between any two data points. For the trapezoid, rule, you assume you can connect each x,y-pair with a straight line. For Simpson's rule, things initially seem much more complicated, but you realize that you are basically connecting every three x,y-pairs with a parabola.[1]

So, the accuracy of your result (which can rarely be perfect when working with numerical data like this, compared with integrating analytical functions) will depend on how your data changes between the point and the method that you use, and often, there is not one method that is always better. Luckily, it is easy to try different methods to see what is most accurate, in which situations.

[1] You can derive all three rules by integrating a Taylor series, a general way to estimate the value of a mathematical function at an unknown point based on knowing the value of the function and its derivatives at a known point! Many numerical methods are applications of the Taylor series up to first or second order, with higher orders being generally more accurate but slower.

9.2 SOLUTION

9.2.0 INTEGRATING DATA USING `scipy.integrate`

You already saw in the `scipy.integrate` module a function for integration of differential equations (Chapter 8), but looking at the module documentation, you noticed that you can also integrate analytical functions and numerical data with it as well. `scipy.integrate` contains built-in functions to integrate numerical data using the trapezoid rule or Simpson's rule. Apparently, the rectangle rule is too simple to bother with. So, how do you actually do it?[2]

Trapezoid rule integration is done by `scipy.integrate.trapezoid()`. This function has only one required **argument**, `y`, which is the y-values of the data to be integrated. If you provide only this argument, the **function** assumes that your data points are spaced by one on the x-axis. Usually, this is not the case, so you will be supplying the x-values using the `x` keyword. If your points are equally spaced on the x-axis and you would prefer it, you can also provide the spacing as a single number `dx` instead. If you have multidimensional data and only want to integrate one dimension, then you can specify that with the `axis` **keyword**. The function will return the integral(s) as a single value, or `ndarray` if the data is multidimensional.

[2] For more information on integration methods, check out books like Hildebrand, *Introduction to Numerical Analysis*; Dover Press, 1987, or Sullivan, *Numerical Methods: An Inquiry-Based Approach With Python* (https://numericalmethodssullivan.github.io).

Simpson's rule integration is handled very similarly by `scipy.integrate.simpson()`. In fact, you can switch from trapezoid to Simpson's just by replacing `trapezoid` with `simpson`. You decide to do a quick test, integrating $y = x$ from zero to one. You know the answer should be 0.5, so you try:

```
In: import numpy as np
In: import scipy.integrate as spi
In: x = np.linspace(0, 1, 10)
In: spi.trapezoid(x, x = x)
Out: 0.5
```

```
In: spi.simpson(x, x = x)
Out: 0.49999999999999994
```

So far so good!

The rectangle rule is not implemented in `scipy.integrate`. However, it can be performed in a single line of code. For the forward rectangle rule, where the height of the rectangle is set by the y-value of the left-side of the rectangle, the integral is simply `np.sum(y[:-1] * (x[1:] - x[:-1]))` Here, if we have N data points, then we have $N-1$ rectangles. We can compute the widths of each rectangle simply by subtracting all of the right-hand side x-values (`x[1:]`, skipping the first point) from all of the left-hand side x-values (`x[:-1]`, skipping the last point) to produce an $N-1$ array that holds the difference between each point. We can then produce an $N-1$-length slice of `y` by omitting the last data point, multiply that slice by the widths, and sum each product to get the integral. If we instead wanted to use the right-hand side y-value for the height, then we replace `y[:-1]` with `y[1:]`.

```
In: np.sum(x[:-1] * (x[1:] - x[:-1]))
```

9.2.1 GENERATING TEST DATA TO EVALUATE THE ACCURACY OF NUMERICAL ANALYSIS

Even though your azide peaks are not well reproduced by Gaussian lineshapes, the Gaussian may still serve as a good test case to evaluate the different approaches to numerical integration. If your Gaussian function is

$$f(x) = A\exp\left(-(x-x_0)_2/2\sigma^2\right), \tag{9.1}$$

where A is the amplitude, x_0 is the center, and σ is the standard deviation. Then the area can be calculated simply as follows:

$$\text{area} = A\sigma\sqrt{2\pi}. \tag{9.2}$$

So, you again have an analytical solution to for comparison. You just need some text data. You whip up a quick test script to make this data and confirm that the integration approaches will work:

```
'''
A quick test script to check the accuracy of numerical integration
    Written by: Chris Johnson and Ben Lear (authors@codingforchemistsbook.com)
    v1.0.0 - 250220 - initial version
'''
import numpy as np
import scipy.integrate as spi

# Gaussian parameters for the simulation
A = 1      # amplitude
x0 = 0     # center
sigma = 1 # width
```

```
13
14 # Create the test data and calculate the analytical value of integral
15 xdata = np.linspace(-10, 10, 11)
16 ydata = A * np.exp(-1 * (xdata - x0)**2 / (2 * sigma**2))
17 area = A * sigma * np.sqrt(2 * np.pi)
18
19 # Compute the area using the rectangle rule, the trapezoid rule, or Simpson's rule
20 area_rect = np.sum(ydata[:-1] * (xdata[1:] - xdata[:-1])) # height * width
21 area_trap = spi.trapezoid(ydata, x = xdata)
22 area_simp = spi.simpson(ydata, x = xdata)
23
24 # Print the error
25 print(f'Area = {area:8.6f}')
26 print( 'Errors:      Absolute --> % Diff')
27 print(f'Rectangle: {area - area_rect:8.6f} --> {100*(area - area_rect)/area:4.1f}%')
28 print(f'Trapezoid: {area - area_trap:8.6f} --> {100*(area - area_trap)/area:4.1f}%')
29 print(f"Simpson's: {area - area_simp:8.6f} --> {100*(area - area_simp)/area:4.1f}%")
```

9.2.2 LINE-BY-LINE EXPLANATION

6–7: We need to import `numpy` for arrays and the `scipy.integrate` module for integration.

10–12: Here we define variables holding the parameters for a Gaussian curve.

15–17: We need to compute the x,y-data points for the Gaussian curve and its area using the analytical solution.

20: We compute the rectangle rule integral. If there are N points in our curve, then there are $N - 1$ rectangles because each rectangle has to span two points. Of the two y-values for each rectangle, we will use the one with the lower x-value (left rectangle rule). We multiply this height by the width, which we calculate for each pair of x-values in case they are not equally spaced. We can do all of this in one line using a **vectorized** approach.

21: Now we compute the trapezoid rule integral by calling the `scipy.integrate` function `trapezoid()`. The first parameter is positional and takes an array holding the y-values, and the x-values are optional and are passed to a keyword parameter `x`.

21: We compute the Simpson's rule integral by calling the `scipy.integrate` function `simpson()`. Conveniently, it is used the same way as `trapezoid()`.

25-29: Print the analytically-determined area and the error of the numerical methods for comparison.

The rectangle rule calculation is so simple that it apparently wasn't worth including in Scipy! Running this script gives the following output:

```
Out: Area = 2.506628
Out: Errors:     Absolute --> % Diff
Out: Rectangle: -0.036055 --> -1.4%
Out: Trapezoid: -0.036055 --> -1.4%
Out: Simpson's: -0.522722 --> -20.9%
```

You are surprised to find that the Simpson's rule calculation was less accurate than the other two, despite being presumably a better approximation. It seems that, in this case, the parabola approximation is worse than the linear one! Because you have the script, you try several other combinations of Gaussian parameters and x-coordinates, finding similar results. Apparently, it will be fine to use the trapezoid rule for bands like this...

> **How Many Data Points Do You Need?**
> In most cases, your codes will run so fast that you can create test data with a very large number of data points and not worry about it taking long to calculate. In some cases, this is not true, and you will want to use the minimum number of points necessary to describe the phenomenon you are modeling. There are good guidelines for choosing the number, or frequency, of data points to use that come from signal analysis. A good place to start is the Nyquist-Shannon Sampling Theorem, which links the spacing between points to the maximum rate of change of the data. In essence, you do not want your data points to be so far apart that there could be an important shape in the data that falls completely between two data points and is, therefore, not represented at all by the data points.

9.2.3 Creating Customized Versions of Existing Functions Using Wrappers

Now that you know how to integrate, you will need to define the range over which the integral should be evaluated. You note that the Scipy integration routines just integrate the whole array, and provide no direct means to set the limits of integration. You figure that you can just use **masking** to select a range of data, much like you did when fitting the UV-vis spectra.[3] This seems like a pain to code every time, and you could see yourself using such a function to integrate high-performance liquid chromatography (HPLC) and gas chromatography (GC) chromatograms, mass spectra, and maybe even NMR data, so you decide to make your own function that calls the Scipy integration functions that you call `IntegrateRange`. These kinds of functions turn out to be common, and are called **wrappers** because they wrap an existing function with some helper code to make the function more convenient to use. You could, of course, include the extra code

[3] See Chapter 6.

in your **script**, but if you use the **function** often enough, the wrapper will likely save you considerable time.

```
1 import numpy as np
2 import scipy.integrate as spi
3
4 def integrateRange(y, x, limits, method='trapezoid'):
5     '''
6     Integrate a numeric function over a range less than the full extent of
7     the function.
8
9     REQUIRED PARAMETERS:
10    y (ndarray): the y points for integration
11    x (ndarray): the x points for integration (need not be evenly spaced)
12    limits (list of numeric): the lower and upper limit of integration
13
14    OPTIONAL PARAMETERS:
15    method (string): which approach to use (trapezoid (default), rectangle, simpson)
16
17    RETURNS:
18    (float): the value of the integral
19    '''
20    # Sort the limits so they are in the order [lower, upper]
21    limits.sort()
22
23    # Find the indicies of the range that we want to integrate
24    int_range = (x >= limits[0]) & (x < limits[1])
25
26    # Decide which integration method the user wants
27    if method == 'simpson':
28        return spi.simpson(y[int_range], x[int_range])
29    elif method == 'rectangle':
30        return np.sum(y[int_range[:-1]] * (x[int_range[1:]] - x[int_range[:-1]]))
31    else:    # the user typed 'trapezoid' or an invalid string - default
32        if method != 'trapezoid':
33            print('Invalid method specified, defaulting to trapezoid')
34            print('Please choose rectangle, trapezoid, or simpson')
35        return spi.trapezoid(y[int_range], x[int_range])
```

9.2.4 LINE-BY-LINE EXPLANATION

1–2: We need to import `numpy` and `scipy.integrate` to get `trapezoid()` and `simpson()`. Because `y` and `x` must be Numpy arrays, we will not have to import `numpy` separately.

4: Define the wrapper function. It requires the y-values, the x-values, and a list of the limits of integration. We include a keyword parameter for a string specifying which method we want to use (default: trapezoid).

21: We sort the list so that it doesn't matter in which order the user provides it.

24: We determine the elements of the x- and y-arrays that are between the limits of integration using **mask**ing as we did in Chapter 6.

27–35: Depending on which method the user specifies, calculate and return the appropriate integral.

9.2.5 Simplifying the Processing of Iterable Objects Using Comprehensions

Sometimes all you want to do with a `for` **loop** is perform the same simple operation on each element of an **iterable object**. For instance, suppose you have a list of pH values, and you want to make a list of formatted strings that express those pH values. You could use a `for` loop:

```
pH = [5.5, 5.7, 5.9, 6.2, 6.5, 7.1, 8.6]
pH_text = []

for trial in pH:
  pH_text.append(f'The pH is {str(trial)}.')
```

This is clear and simple, but requires writing a fair amount of code. You'd rather do something that only takes a single line, and some searching leads you to quickly discover that Python has a way to do that called a **list comprehension**. The basic format of the list comprehension is [<statement producing new element> for <element> in <iterable variable>], which evaluates to a new list that can be assigned to a new variable. For the example above, you instead write:

```
pH = [5.5, 5.7, 5.9, 6.2, 6.5, 7.1, 8.6]
pH_text = [f'The pH is {str(trial)}.' for trial in pH]
```

You further discover that list comprehensions also allow you to construct new lists that use only certain elements from an original list using a form like [<statement producing new element> for <element> in <iterable variable> if <conditional test based on element>]. The simplest example you can think of is:

```
In: pH_basic = [trial for trial in pH if trial > 7.0]
```

which yields the list

```
Out: [7.1, 8.6]
```

This looks to you to be the list equivalent to the Numpy concept of masking.

You realize that you can combine these concepts. Before, if you wanted to make a new list for only the trials in the above example where the solution was basic, you would write:

```
pH = [5.5, 5.7, 5.9, 6.2, 6.5, 7.1, 8.6]
pH_text = []

for trial in pH:
  if trial > 7.0:
    pH_text.append(f'The pH is {str(trial)}.')
```

but instead you write a very terse list comprehension:

```
pH = [5.5, 5.7, 5.9, 6.2, 6.5, 7.1, 8.6]
ph_text = [f'The pH is {str(trial)}.' for trial in pH if trial > 7.0]
```

The advantage of list comprehensions is that they do simple, boring things without taking up too much space, making it easier to see more important parts of your code. The disadvantage is that their **syntax** is not terribly intuitive, particularly to new coders or coders not familiar with Python, making your code in some ways potentially harder to read. However, they are considered "**Pythonic**" so you are likely to encounter them even if you do not personally use them in your codes. You can also see that they could be a great way to work with dictionaries stored in lists.

9.3 CODE

Now you feel like you know have you need to solve the full problem. You come up with a plan:

1. Read the data files and the associated times that they were recorded.
2. Background subtract each spectrum.
3. Integrate the azide peak for each spectrum and store the result.
4. Plot the relative concentration versus time.
5. Fit the data to obtain a rate.

For each spectrum, you have several pieces of data: the wavenumber array, the absorbance, the background-subtracted absorbance, the time the spectrum was recorded, and the value of the integral. You figure that it would be best to store all of these values as a dictionary for each spectrum to keep them organized, and then make a list of these dictionaries. When thinking about the plot, you also think it might be nice to display three subplots: one that shows the full IR spectrum (so you can remember what it looks like), one that is a zoom-in on the azide (so you can see where the integrated values are coming from), and one that is the results of the numerical integration (so you can

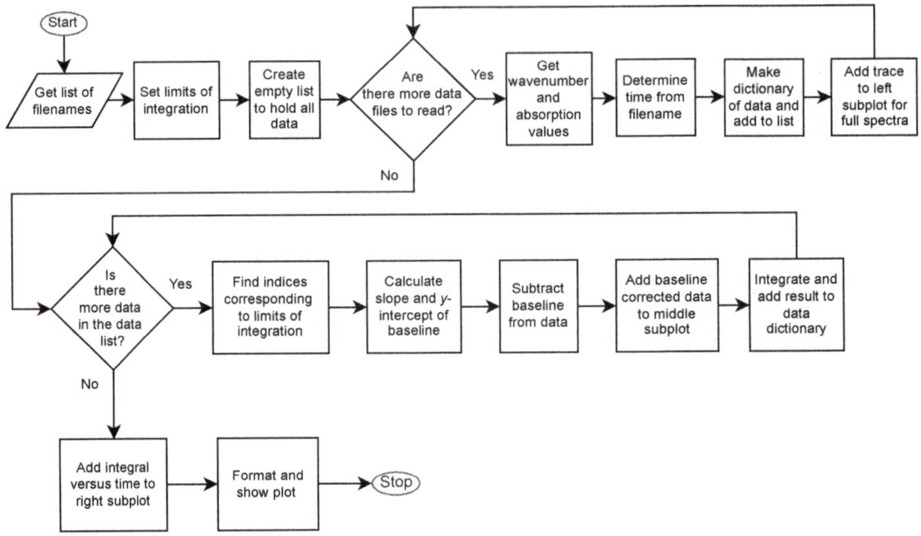

FIGURE 9.2. Flowchart for calculating the integrals of the azide band in several FTIR spectra and constructing a kinetic trace showing how their concentrations vary with time.

verify that the results you obtain meet your expectations). Sketching out the work flow, you produce Figure 9.2.

Now you are finally ready to solve the problem. You whip up the following code:

```
'''
Integrate the intensities of azide peaks in an FTIR and
produce a kinetics plot
    Requires: FTIR spectra in .csv format for each time point
              filename must include the time the spectrum was taken
    Written by: Chris Johnson and Ben Lear (authors@codingforchemistsbook.com)
    v1.0.0 - 250220 - initial version
'''
import numpy as np
from plotly.subplots import make_subplots
from codechembook.quickTools import quickOpenFilenames, quickPopupMessage
from codechembook.quickPlots import sampleColorScale
from codechembook.symbols.chem import wavenumber as wn
from codechembook.numericalTools import integrateRange

# Get and sort the spectrum file names
quickPopupMessage(message = 'Select CSV files containing the FTIR spectra.')
data_files = sorted(quickOpenFilenames(filetypes = 'CSV files, *.csv'))

# Set the upper and lower wavenumbers for the integration
integration_limits = np.array([1950, 2150])

# Make a color scale for the different traces in the figure
colors = sampleColorScale(len(data_files)) # Get samples of a continuously changing color
    scale

# Loop through files, read data and times, put them in a dict, and add to the plot
```

```
data = []
fig = make_subplots(rows = 1, cols = 3)
for c, file in zip(colors, data_files):
    wavenumber, absorption = np.genfromtxt(file, delimiter = ',', unpack = True)
    time = float(file.stem.split('_')[1]) # get the time from the file name

    # Create a dictionary with everything we need for this file, add it to the list
    data.append(dict(wavenumber = wavenumber, absorption = absorption, time = time))

    # Add the trace to the plot
    fig.add_scatter(x = wavenumber, y = absorption,
                    line = dict(color = c),
                    name = time, row = 1, col = 1)

# Loop through the spectra to background subtract and integrate
for c, spec in zip(colors, data):
    # Find the indicies of the points at the limits of integration
    xmin_index = np.argmin(abs(spec['wavenumber'] - integration_limits[0]))
    xmax_index = np.argmin(abs(spec['wavenumber'] - integration_limits[1]))

    # Calculate line that connects the x and y values at the limits of integration
    m = (spec['absorption'][xmax_index] - spec['absorption'][xmin_index]) / (spec['wavenumber'][xmax_index] - spec['wavenumber'][xmin_index])
    b = spec['absorption'][xmin_index] - m * spec['wavenumber'][xmin_index]

    # Subtract the baseline from the data and add it to the figure
    spec['back corr'] = spec['absorption'] - (m * spec['wavenumber'] + b)
    fig.add_scatter(x = spec['wavenumber'], y = spec['back corr'],
                    line = dict(color = c),
                    showlegend = False, row = 1, col = 2)

    # Integrate the baseline-corrected data and store it in the dict
    spec['integral'] = integrateRange(spec['back corr'], spec['wavenumber'], integration_limits)

    # Calculate the maximum y value for this spectrum section
    spec['ymax'] = np.max(spec['back corr'][xmin_index:xmax_index])

# Find the max y height to determine the zoom range for the y-axis
ymax = max([spec['ymax'] for spec in data])

# Plot the kinetic trace
tdata = np.array([spec['time'] for spec in data]) # get array of times
intdata = np.array([spec['integral'] for spec in data]) # get array of integrals
fig.add_scatter(x = tdata, y = intdata, name = 'kinetic trace',
                line = dict(color = 'black'), row = 1, col = 3)

# Format the plot
fig.update_xaxes(title = f'wavenumber / {wn}', row = 1, col = 1)
fig.update_yaxes(title = 'absorption', row = 1, col = 1)
fig.update_xaxes(title = f'wavenumber / {wn}', range = integration_limits, row = 1, col = 2)
fig.update_yaxes(title = 'absorption', range = [-0.1 * ymax, 1.1 * ymax], row = 1, col = 2)
```

```
77 fig.update_xaxes(title = 'time /s', row = 1, col = 3)
78 fig.update_yaxes(title = f'intensity / AU{wn}', row = 1, col = 3)
79 fig.update_layout(template = 'simple_white')
80 fig.show('browser')
```

Your new collaborator provides you with a series of csv files named "azide_XXXX.csv" where XXXX is the time in seconds after the start of the reaction at which the spectrum was recorded. You place the files in a new folder in your research directory called "Azide" and run the code. It asks you to specify these files, and then churns away for a second or two and produces the plots in Figure 9.3.

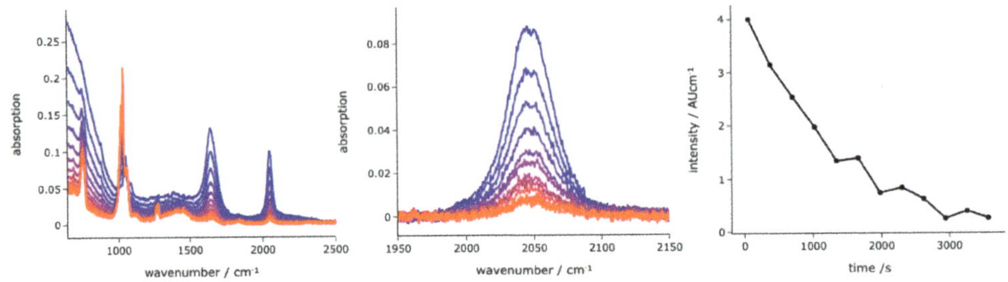

FIGURE 9.3. (left) An overview of the spectra. (middle) The feature of interest after baseline subtraction. (right) The kinetics of the azide formation at short times.

You take your computer down the hall to show your colleague, who says, "I can't believe that this wasn't produced by some professional software—I could just directly use this as a figure in my paper! Maybe you can show me how to do this..." You say, "Sure, let's meet for lunch tomorrow and get started. Bring your laptop!"

9.3.0 LINE-BY-LINE EXPLANATION

9–14: Import `numpy` and `make_subplots`. We will also import `quickOpenFilenames` and `quickPopupMessage` from the `codechembook` library for convenient file selection and user notifications.

17 & 18: We give the user a message instructing them to pick the csv files that they want to analyze, and then open the file dialog box so that they can find and select them. This process returns a list of `Path` variables, which we immediately sort and store as a new variable. There is one new import here, `customColorList`, which will be used on line 24.

21: Set the limits over which we want to integrate.

24: We want to make a color series that smoothly varies from one color to another. We will use a simple function from the `codechembook.quickPlots` module that helps us do that. The function `sampleColorScale( )` will start with a Plotly color scale that is designed to continuously vary through colors and will pick out N evenly spaced values from the scale, where N is the number of colors you want. This value is a required parameter, and there is one optional parameter that specifies the name of the color scale (default: 'bluered'). It returns a list of rgb values.

27: We define an empty array to hold the data we are about to read.

28: We want to make a plot that shows the original spectra, a zoom in of the peak we care about after baseline subtraction, and the kinetic trace, so we will make a 1 $\times$ 3 subplot array.

29–39: We loop through each file one at a time to to read and plot them:

- 29: Zip the color scale and the filename list together so that each file is associated with a different color.
- 30: Read the wavenumber and absorption data into arrays.
- 31: Get the time point for each spectrum from the file name, similar to our approach in Chapter 6. The files are named `azide_*.csv` where `*` is the reaction time point at which the spectrum was recorded. We split the stem of the filename at _ and take the second part, which just has the text for the time.
- 34: Create a dictionary that holds the wavenumber, absorption, and time point of each spectrum so that we don't have to manage several arrays later.
- 37–39: Add this spectrum to the plot of the full spectra.

42–61: Loop through all of the files again, this time to perform the background subtraction and integration:

- 44 & 45: We need to figure out the array indices closest to our limits of integration. Subtract the limit of integration from each x-value and look for the one closest to zero by taking the absolute value of the difference and finding the smallest one. This is done with `numpy.argmin( )`, which returns the index of the smallest value of an array.

48 & 49: Now we need to get the slope and intercept of the background. This can be done algebraically by solving first for the slope using $\frac{rise}{run}$, then plugging that slope into $b = y - mx$ for one of the points to find the y-intercept.

52–55: Here, we subtract the baseline to create a new array and enter it into the dictionary. Next, we and add this trace to our second panel of our subplot with the corresponding color.

58: We can integrate now. We use the `integrateRange()` function from this chapter to simplify the process. We store the value of the integral as a new entry in our dictionary.

61: We will also store the maximum y-value for this spectrum to help us determine the y-range of the subplot that shows a zoom in to the peak we are integrating.

64: Now we want to find the maximum y-value for all the spectrum to set the y-range for the subplot. Here we will use a list comprehension to build a list of the maximum y-values for each spectrum. Then, we use `max()` to get the maximum of these values.

67–70: We need to build the plot of the integral versus time. Create arrays for time and integral value using list comprehensions to pulhe data out of the list of data dictionaries, and plot these arrays in the third subplot.

73–80: Format the plot areas and show the plot.

CHAPTER 9 EXERCISES

TARGETED EXERCISES

Exercises for Section 9.2.0

0. The cumulative integral is the integral from the beginning of the data up to a given data point, plotted against the value of the last data point integrated. Write a code that plots the cumulative integral of a Gaussian function with an amplitude of 1, a center of 5, and a width of 1, from 0 to 10 with 100 points on the x-axis.

Exercises for Section 9.2.1

1. Write two functions, one that returns a Lorentzian distribution and one that returns a log-normal distribution. You can either write them yourself or import these functions from LMFIT now that you know how to find them in the library files (see

Exercise 4 of Chapter 7). These functions should take as arguments the x-data points, the center, the sigma (width), and the amplitude. Generate four different x-data arrays spanning 0 to 10 with 10, 50, 100, and 1000 points, respectively. For each, of these x-data arrays:

- Generate the corresponding y-values for a mean of 5, a width of 0.5, and an amplitude of 1. Because you have two functions, you will have eight y-value arrays.
- Numerically integrate these eight y-value arrays.
- Make a log-log plot of the difference between the numerical solution and the known analytical solution (1, since they are normalized) as a function of the spacing between the points on x-value arrays.
- Increase the width to 1, 2, and 4 and integrate each the same way over the same range. You will notice that the accuracy decreases substantially. Why?

Exercises for Section 9.2.3

2. Write a wrapper that combines `quickReadCSV( )` and `quickOpenFilename( )` from `codechembook.quickTools` so that you can quickly read a csv file using a dialog box to identify the file.

Exercises for Section 9.2.5

3. Write a list comprehension that accomplishes each of the following tasks, in a single line of code:

   a) Add the suffix "`_old`" to each name in a list of names (as strings). You could also use this approach to rename, for example, files in a list of files.

   b) Cerate a sine wave with amplitude 1 and 10 points spanning 0 to 2π. Create a list of points that are positive or zero and a second list of the points that are negative.

   c) Redo what you just did, but now change the value of the numbers failing the test to zero instead of omitting them from the list.

   d) Create a new list that contains the three-point moving average (the current point, the point prior, and the point after) of your list for positive values from part c. Your new list will have to have two fewer items in it than the original list. You might use this to attempt to smooth data—although we will address much ways to smooth data in the next chapter.

   e) Calculate the slope between adjacent points for the same list. Your new list will have to have one less points in it than the original list. This might be used to take the numerical derivative of your data. This idea will be explored much more thoroughly in Chapter 13.

For each of these, invent some test data and use it to verify that your solution works.

COMPREHENSIVE EXERCISES

4. Write a code where you can supply a lists of arbitrary numbers of initial and final integration limits, and then the code will integrate data over these limits and output the results for each range.

CHAPTER 10

Smoothing Data Responsibly

- ☐ **New concepts**
  - — *and their prerequisites*

- ☐ **Understanding why implementing smoothing requires caution**
- ☐ **Comparing smoothing approaches using test data**
- ☐ **Deciding when smoothing is appropriate**
- ☐ **Generating noisy test data using random numbers from `numpy.random`**
  - — *Section 1.1.5: Operating on collections of data efficiently with Numpy arrays*
  - — *Section 2.1.1: Importing subsets of libraries using* `from`
- ☐ **Smoothing by moving averages using `scipy.signal.convolve()`**
  - — *Section 1.1.1: Operating on collections of data efficiently with Numpy arrays*
  - — *Section 2.1.1: Importing subsets of libraries using* `from`
- ☐ **Smoothing using local regression with `scipy.signal.savgol_filter()`**
  - — *Section 2.1.1: Importing subsets of libraries using* `from`

10.0 PROBLEM

In the process of working up the FTIR data in the last chapter, you and your classmate notice some small peaks that appear to come and go in the spectra. Your classmate remarks, "I wonder if these are signatures of the intermediates in this reaction? I've always wanted to get some more detail about the mechanism. We could integrate these, too, and see if they track the kinetics of an intermediate species." Looking closer, you think, "They're so small that it's hard to tell if they are there at all—much less to know where the center of the peak is, what the limits of integration should be, or how to subtract the solvent background."

You say, "We can integrate these peaks with the same basic code as the azide peak before, but we will probably have to smooth the data first to make it easier to set up the parameters." Your classmate looks nervous. "My advisor always complains about people smoothing data. He says there's rarely a good reason to do it, and it can make you see things in the data that aren't really there."

This is not a bad point, but you need to do something because you'd just be making educated guesses with the noise you have in the data as recorded. Not only that, but because the solvent background is so much more intense than the actual peak, any error in background subtraction will be a large error in the estimate of the integral. This also seems bad.

"Let's give it a shot. Your advisor is correct to be cautious about smoothing, but maybe it will make sense in this case, and maybe we can also prove to your advisor that it's okay in this particular case. We can still integrate the raw data, we just need some well-founded numbers to put into this script."

10.0.0 UNDERSTANDING WHY IMPLEMENTING SMOOTHING REQUIRES CAUTION

How bad can it possibly be to smooth? And how do you even do it? Looking through the documentation of the Scipy **library**, there are a bewildering number of smoothing options to choose from. You also don't see the most common option you can think of, the moving average. What is going on?

You track down your labmate and ask, "What do you know about smoothing? And how do you do it in Python."

"Oh, smoothing is super interesting!" They stand tall and gives you their best noble look. " 'With great power comes great responsibility.' Why do you want to smooth something? Hopefully not to 'make it look better!' "

You explain what you want to do, but then get curious: "What is wrong with making it look better?"

"Let me show you something." They take your laptop, turn it around so that you can't see the screen, and in a minute, turn it back around to show you Figure 10.1.

"How many carbonyl stretches do you see in this spectrum?"

You squint at the screen. "Four?"

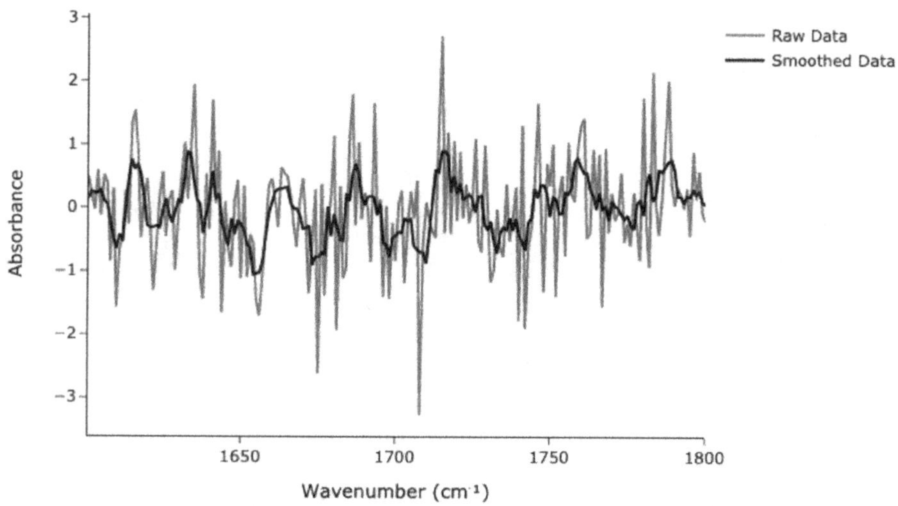

FIGURE 10.1. Data that your labmate shows you, wondering how many peaks there are.

"Nope! This is just an array of normally-distributed random numbers that has been smoothed. There's nothing there. That's the issue."

You are slightly irritated, but the point is taken. "I can see how it can go wrong, and we will have to keep this in mind, but it must still be valuable in some cases. So how do you do it?"

"Your labmate explains that the easiest and most common ways to smooth are moving weighted averages and local regressions. Moving weighted averages basically take each y-point and average them together with a set number of adjacent y-points. The moving average that you are familiar with is actually a special case of a more general averaging approach. In the moving average, every point in the average has the same weight, but that need not be the case. You can weight the points however you like: using a binomial distribution, a normal distribution, an exponential, or really anything you can imagine."

"Local regression-type methods are a little more complicated, but basically boil down to fitting a subset of your data to a function that ignores the noise, and then using that function to produce the new smoothed y-point for each x-point. In fact, one of the most common of these methods, Savitzky–Golay smoothing, was invented by chemists who wanted to smooth noisy HPLC chromatograms!"

"There are many other approaches, but moving averages and local regression should cover most of your bases without having to learn a lot of new math."

"So how do we know which one we should use?"

Your labmate says that more smoothing will reduce noise, but it can shorten and broaden your peaks. Different methods will cause different amounts of peak distortion. Luckily, you can just make up test data to see how these effects would impact your real data.

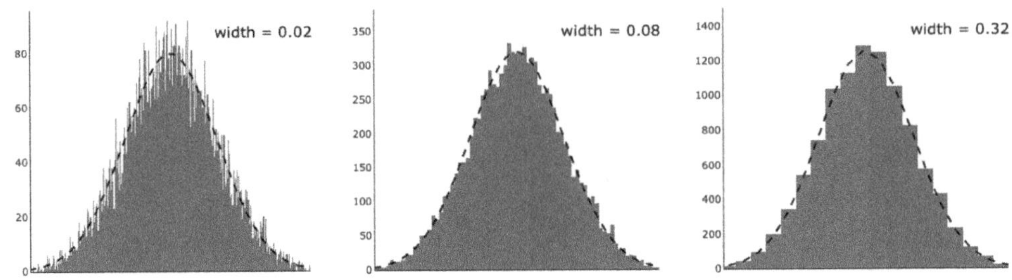

FIGURE 10.2. As we increase the binning on the histogrammed data, the apparent noise decreases. Note that as the bin width increases, the number of points that fall within each bin increases, so the y-axis scale increases.

10.0.1 COMPARING SMOOTHING APPROACHES USING TEST DATA

10.0.1.1 Binning Data to Preserve Meaningful Statistics

The simplest approach to smoothing is to simply combine adjacent points into one point—so-called "binning." This has the advantage of not fundamentally altering your data: you are essentially zooming out from it, letting fluctuations cancel one another out and seeing the broader trends. It also has the advantage of being mathematically simple, which allows for relatively straightforward rigorous error propagation. The problem with binning is that the "smoothing" effect is not as significant as with other methods that actually *change* your data. Because binning is essentially summing multiple measurements, signal to noise improvements with binning scale as $\sqrt{N}$, where N is the number of points being binned together.[1] Figure 10.2 shows the effect of increased binning on data.

[1] See, for instance, Bevington, P. and Robinson, D.R. *Data Reduction and Error Analysis for the Physical Sciences*, 2002.

Binning is most naturally applied to distributions of data that could be plotted as a **histogram**; that is, data that could be visualized in a bar chart with a continuous independent variable. In these types of charts, we often use as the dependent variable the "frequency" that a measurement appears between the edges of an individual bar or bin, or the total number of measurements between the edges. Thus, we can simply make the bin limits wider to include more measurements, thereby reducing the apparent noise. Spectra, which are just histograms (the probability of the sample absorbing photons between 405 and 406 nm, for instance), x-ray powder diffraction patterns, and images can also be straightforwardly binned even though we do not typically think of them as histograms.

10.0.1.2 Understanding the Common Moving Average

The moving average is likely the smoothing technique with which you are most familiar. In a simple moving average, N adjacent points are averaged together to determine the value of a single point. This is similar to the binning process outlined above, except for two important points: the moving average has the same number of points as the original signal, and the moving average *changes* the data in a way that may make it complicated to derive the uncertainty in the smoothed data from the uncertainty in the measured data. For the simple N-point moving average, our smoothing function is just an array of N points that is normalized such that its sum is one, i.e., each point is simply weighted by $1/N$.

Most moving average smoothing methods can be cast as convolution problems. Convolution as a mathematical and numerical concept is ably covered in many resources online, and thus will only be briefly outlined here for the specific context of smoothing. Figure 10.3 graphically depicts the process of convolution. Essentially, convolution for smoothing is the process of taking a signal that you would like to smooth, multiplying it by a function that you would like to smooth by (for a simple moving average, that function is just a flat top function that is constant for the data points to average and zero everywhere else), and integrating the resulting product. Translation of the smoothing function against the original signal specifies the range of data over which you are averaging.

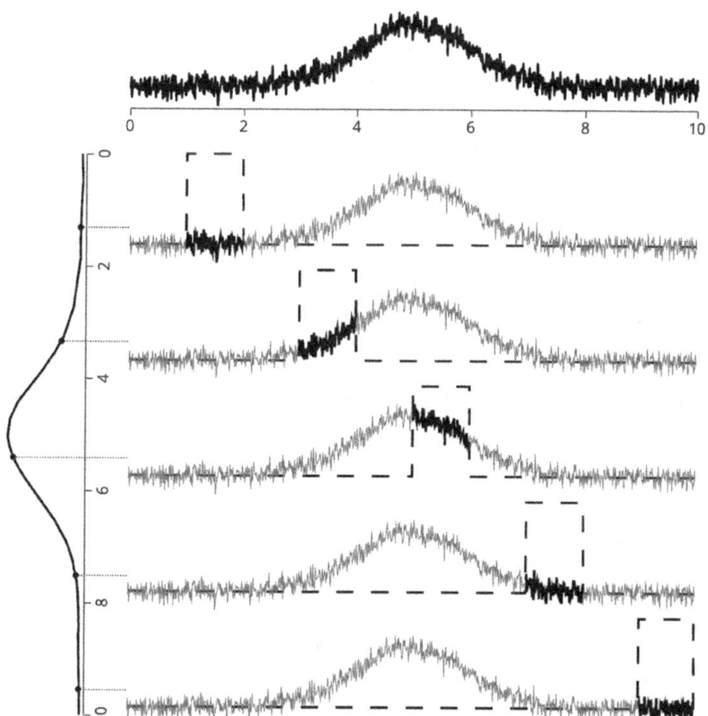

FIGURE 10.3. Graphical representation of simple moving averages. The dashed lines represent the smoothing function. The smoothing function is moved one point at a time across the noisy signal (top), and after each movement, multiplied by the signal. This product is then averaged, and the average is used as the new data point (plotted vertically to the left). Each new point corresponds to a shift of the smoothing function across the noisy data.

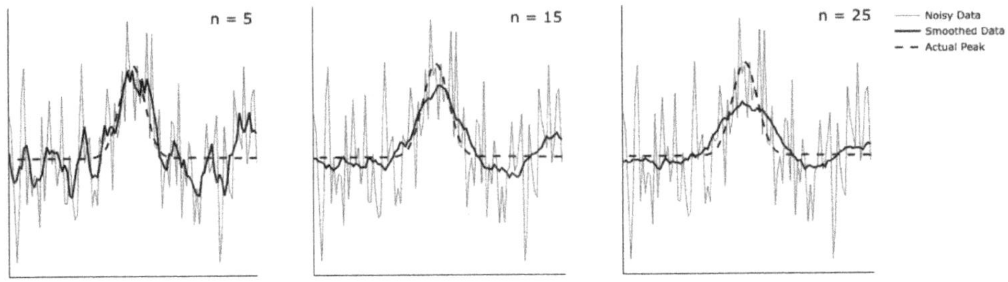

FIGURE 10.4. As we increase the number of adjacent points to average, peak widths increase and amplitudes decrease.

You may say, "So what if moving average smoothing changes the data?" One particularly problematic example is the smoothing of relatively narrow peaks. Figure 10.4 shows an example of this. As we increase the number of points to average, the data does indeed get smoother. However, we can also plainly see that the peak is reduced in intensity and is significantly broader. If you require high resolution, like in FTIR or NMR spectra, or if it is important that peak intensities are quantitative, then this smoothing approach is probably not the best one.

Of course, you are not limited to simple moving averages; you can choose any weighting of points for smoothing when you use the convolution approach. It is common to weight points such that the current point (middle point of the collection of points being averaged) carries the most weight, and then each nearest neighbor carries less and less weight. Binomial and Gaussian distributions of weights can often be used to minimize the broadening effects described here.

10.0.1.3 Smoothing by Local Regression

The basic problem with the class of moving average smoothing approaches that we just covered is that they do not explicitly attempt to describe the underlying trend in the data; the trend is just revealed by averaging away noise. We could instead attempt to fit the data to some function,[2] which explicitly treats the underlying trend in data with a model described by a well-chosen mathematical function. However, this function often does not exist, and we often do not even have a well-formed have a numerical function to use as the fit model.[3]

We can instead choose a relatively simple function, only try to fit a small region of the data to it, and use the best-fitting function to calculate the smoothed y-value for that specific x-value. We can then shift the small region of data to fit to the next point, and do it all again. This is the idea behind **local regression**-type smoothing algorithms. Figure 10.5 shows a simplified, intuitive view of this process for a linear local regression. Clearly, this smoothing approach also changes the data.

One of the most well-known (and chemistry-relevant) methods of local regression smoothing is the Savitzky–Golay filter.[4] This method was developed by Savitzky and Golay to smooth noisy HPLC data and remains to this day a very commonly used

[2] As we did in Chapters 5 and 6.

[3] Like we did in Chapter 7.

[4] See Savitzky, A.; Golay, M. J. E. Smoothing and Differentiation of Data by Simplified Least Squares Procedures, *Anal. Chem.* **1964**, *36*, 1627–1639.

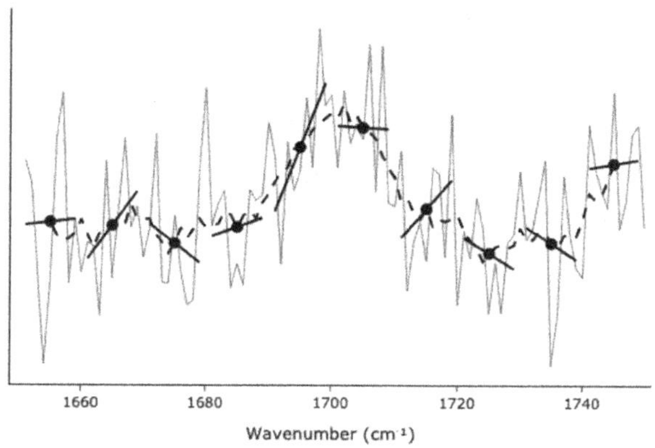

FIGURE 10.5. Simplified view of local regression smoothing. At each point, a function is fit to a range of surrounding points. The best fitting function is then used to calculate the "smoothed" value for that point. This example shows a linear function, but low-order polynomials are also often used.

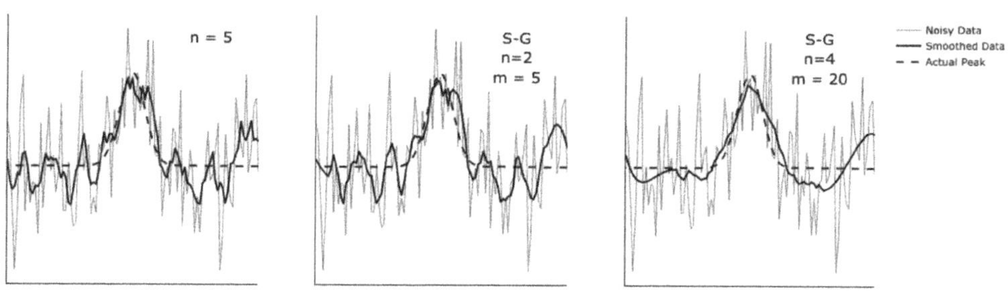

FIGURE 10.6. Comparison of a five-point moving average (left) with Savitzky–Golay smoothing using two different sets of n and m (center and right). With Savitzky–Golay smoothing, you can typically achieve more noise reduction while maintaining peak widths and amplitudes better.

method. The approach is to fit a polynomial of order n (where the order is the largest power in the polynomial, i.e., two for a parabola) to an m-point-wide span of the data. There are a wide range of possible values of n and m. As m gets larger, more points are used, and there is a larger smoothing effect. As n gets larger, more complex curvature can be reproduced, which is a double-edged sword: on one hand, more significant changes in the data can be handled, like smoothing narrow peaks in a spectrum, but this can also make specific noise patterns look like real peaks by accident. In practice, the choice of m and n can be made by modeling artificial data in which the underlying curve is known, as shown in Figure 10.6.

More generally, these kinds of approaches are classified as locally estimated scatterplot smoothing (LOESS) or (if the data is weighted during fitting) locally weighted scatterplot smoothing (LOWESS) methods. These methods are computationally more intensive than moving averages, and also make uncertainty estimation even more difficult (except for in some special cases with analytical solutions), but they are often quite effective once validated.

10.0.1.4 Smoothing by More Advanced Methods

You have only scratched the surface of smoothing here, but beyond these methods, you mostly find approaches that are mathematically much more complex or are only used in niche cases. A quick survey reveals that two more approaches are quite common.

Spectral methods make use of the fact that a signal often lies in a specific frequency range (very often, below a specific cut off frequency). Because noise spans all frequencies that can be represented in the data, it is possible to filter out the high-frequency part of the noise to achieve a smoother curve. These approaches are usually classified as digital filters. Their implementation typically leverages fast Fourier transforms, which are commonly used in chemical instrumentation like FTIR and NMR spectrometers.[5]

Smoothing splines are similar in concept to (and in some cases, a special case of) LOESS or Savitzky–Golay smoothing, but are often separately implemented (for instance, in the Scipy **library**). These methods usually use cubic functions for fitting because cubic spline–based numerical techniques are particularly well developed.[6]

[5] This class of approaches is mathematically more complex and beyond the scope of this book.

[6] You will see another use of splines in Chapter 11.

10.0.2 Deciding When Smoothing is Appropriate

Now that you have more background on smoothing, you still don't quite know when you should use it. You go back to your labmate to pick their brain, and they explain, "Usually, you should not smooth your data. Smoothing data typically fundamentally changes the data, sometimes in a way that will lead you to make qualitatively wrong conclusions. You probably should **not smooth** in the following situations:

1. *To make your data look better.* If your data is too noisy to find the trend/peak/feature in it that you expect, then smoothing may create a trend/peak/feature out of pure noise as we saw in Figure 10.1. If you are worried that other researchers will not believe your data, it is better to prove them wrong using rigorous statistical methods!

2. *Before fitting.* You should essentially never fit smoothed data. Just fit the noisy data; the math still works fine! Because smoothing changes your data, your fit result will likely not be the same. Furthermore, typical goodness-of-fit metrics will be misleading, giving you more confidence than you should have in the statistical relevance of your results.

3. *Before integrating.* Again, smoothing changes your data. Just integrate the noisy data. If the noise is actually random, it will mostly cancel out during integration because you should have a roughly equal number of points that are too high as you do peaks that are too low.

So why do we spend all of this time learning how to smooth? There are times where smoothing is a very good (or maybe the only) idea:

1. *As an intermediate step in data processing.* Sometimes you need to determine parameters, limits, and models for exploratory data analysis. Smoothing can give you an idea of starting points for these analyses, then you can do the full analysis on the unsmoothed data. Furthermore, when you are automating data analysis, your automated algorithm may be fooled by noise. Smoothing data first, then applying automated methods to set up initial guess, to find peaks, and so on, can get your algorithm moving in the right direction. Then, it will be more likely to be robust when applied to the unsmoothed data.

2. *When you want derivatives of noisy data.* If your data is noisy, then derivatives of the data will show significantly more noise due to the fact that the slopes between adjacent noisy peaks are much larger than those of the underlying signal. Particularly when you do not have a model to fit to your data, smoothing can allow you to more reliably estimate the value of the derivatives of your data.[7]

[7] To learn more about derivatives, see Chapter 13.

3. *When you have a pattern of variations in the data that you already understand and want to remove.* As long as you know where the noise is coming from, and its frequency (or period), then you can smooth it away to more easily see the underlying trend. However, it is usually still better to model the noise, if you can, than to smooth it out.

These are both not exhaustive lists, but will provide you food for thought as you run into situations in which you think smoothing might be an effective tool."

10.1 SOLUTION

What you need to do is pretty clear:

1. You will modify your **script** for integrating FTIR peaks to smooth the data first.[8]

[8] See Chapter 9

2. From the smoothed data, you will determine the background and the limits of integration.

3. You will take the background you got from the smoothed data and subtract it from the raw data.

4. You will then perform the integration on the background-corrected unsmoothed data in order to avoid any numerical errors from the smoothing.

First, however, you need to figure out what is the right smoothing approach. You decide to test different smoothing methods on simulated data. You can accomplish this by:

1. Creating the x- and y-values of a simulated absorption peak.
2. Adding noise to each y-value.
3. Smoothing the noisy data.
4. Comparing the smoothed peak height and width to the "clean" one.

Creating the "clean" peak will be no problem; you have already calculated a Gaussian curve.[9] How do you make it noisy?

[9] See Section 9.2.1.

10.1.0 GENERATING NOISY TEST DATA USING RANDOM NUMBERS FROM `numpy.random`

In its purest form, noise comes from random errors in measurement and should be characterized by a random normal variance (a.k.a. Gaussian, bell curve, and so on) from the "true" value. You can simulate noise by having the computer generate random numbers. For a measurement like FTIR, most of your noise is in the y-values, so you can simulate noise by adding positive or negative random numbers to your y-data. Numpy provides a suite of random number generators that will do the trick.

You notice that the *standard* random number generator in Numpy produces numbers *evenly* distributed across the range $0 \leq x < 1$. This won't do because you need numbers that (1) are greater than and less than zero, and (2) have a probability distribution described by the Gaussian probability distribution. A little more digging shows that Numpy has functions that allow for this. In particular, the `normal()` function of `numpy.random` will produce the random numbers you want. This function accepts arguments that allow you to specify the center and width of the distribution as well as the size of the array that it returns. You give it a shot:

```
In: noise = np.random.normal(loc = 0, scale = 0.2, size = 100)
```

and check to see if it gives you the expected results:

```
In: np.mean(noise)
Out: 0.004253909771452216 5
```

```
In: np.std(noise)
Out: 0.20823387191838444
```

Not exactly the numbers you gave it, but hey, these are random! You try a couple more times and see that the numbers are different each time, but close to the values you expect.[10]

So, all you have to do is add a random number to each of your y-values and you have a rough estimate of noise.[11] Now that you can make a noisy peak, it is time to figure out how to smooth it!

[10] Sometimes, you are testing some code and want the *same* "random" numbers every time it runs. For this, you can set a "seed" the random number generator (which normally is seeded by a clock on your computer's processor). See the documentation for the Random Generator in Numpy for more details.

[11] In reality, noise often deviates somewhat from a perfect Gaussian distribution due to the presence of noise sources with specific frequencies and filtering effects of data acquisition electronics, but a Gaussian distribution will get the job done most of the time.

> **Random Numbers and Monte Carlo Techniques**
> A number of numerical problems cannot be efficiently solved using standard numerical techniques. This is especially true for problems with large numbers of dimensions, like the structure of liquids or proteins, where creating a grid of positions of all atoms leads to too much data and too many calculations to do in a reasonable amount of time on a computer. In some of these cases, so-called Monte Carlo techniques, in which a problem is solved indirectly by statistical analysis of a large number of solutions for individual "trial" conditions (for structure, a given set of atomic coordinates), can be used.

10.1.1 SMOOTHING BY MOVING AVERAGES USING `scipy.signal.convolve()`

You could compute the moving average explicitly by working out the algebra, but you will instead use a more general approach that employs convolution. The convolution approach also avoids a tedious feature of algebraically computing the moving average: dealing with edge cases. For instance, if we are doing a five-point moving average, then obviously the first point does not have two points ahead of it. Thus, we are forced to deal separately with these points. This is not a problem, but it is somewhat of a hassle.

Luckily, the convolution-based smoothing process is quite simple to code! Scipy's `signal` **module** contains the `covolve()` **function**, which does everything that you need. It takes two required **arguments**, the data that you would like to convolve, and the function with which you are convolving (which should be normalized, i.e., the array should sum to 1). There is one important thing to note, however: mathematically, when we convolve an n-length **Numpy array** with an m-length convolving function, we will get an $(n+m)$-length array as a result. This is usually inconvenient, so `scipy.signal.convolve()` features a helpful **keyword argument** `mode`, which when set to `'same'`, produces a resulting array that is equal in length to the original data. So, a simple five-point moving average can be done in a single line (assuming we have already run `import scipy.signal as sps` and `import numpy as np`:

```
In: smoothed_y = sps.convolve(data, np.array([.2, .2, .2, .2, .2]), mode = ↩
      ↪'same')
```

or more generically[12]

[12] See Section 1.1.4 for details on np.ones( ).

```
In: smoothed_y = sps.convolve(data, np.ones(5)/5, mode = 'same')
```

10.1.2 SMOOTHING USING LOCAL REGRESSION WITH `scipy.signal.savgol_filter()`

Implementing the Savitzky–Golay filter is straightfoward. The **function** `savgol_filter()` has three **positional arguments**:

1. `x`: an array holding (confusingly) the y-values that you want to smooth.
2. `window_length`: the number of points that you want each polynomial to span.
3. `polyorder`: the order of the polynomial to use for smoothing (i.e., three for cubic).

There are a number of **keyword arguments** that you are unlikely to use. `savgol_filter` returns a **Numpy array** of the same length as `x` containing the smoothed y-values.

Now that you have a survey of some smoothing methods, you are ready to write the code to smooth the weak peak and determine the baseline to subtract from the integral.

10.2 CODE

You decide that you want to use Savitzky–Golay smoothing, but you can choose among any combination of n (polynomial order) and m (number of points). What to do? You write a quick test script that will allow you to try different combinations of n and m to see how the data responds. You sketch out a quick flowchart (Figure 10.7) to organize your thoughts:

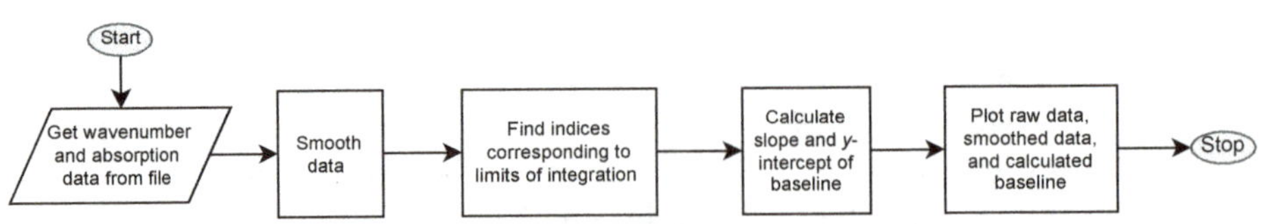

FIGURE 10.7. Flowchart for code that will read in spectral data and then apply smoothing, so that a baseline can be established. Much of this solution is the same as in Chapter 9 and is omitted here for clarity.

From this, you produce the following code:

```
'''
Find the baseline of a small peak by smoothing the data first
    Requires: An FTIR spectrum in .csv format
    Written by: Chris Johnson and Ben Lear (authors@codingforchemistsbook.com)
    v1.0.0 - 250226 - initial version
'''
import numpy as np
import scipy.signal as sps
from codechembook.quickPlots import quickScatter
from codechembook.quickTools import quickOpenFilename, quickPopupMessage
from codechembook.symbols.chem import wavenumber as wn

# Get the spectrum file names
quickPopupMessage(message = 'Select a CSV file containing the FTIR spectrum.')
data_file = quickOpenFilename(filetypes = 'CSV files, *.csv')

# Read the file
wavenumber, absorption = np.genfromtxt(data_file, delimiter = ',', unpack = True)

# Apply Savitzky-Golay smoothing with a window of 100 and an order of 3
smooth_absorption = sps.savgol_filter(absorption, 100, 3)

# Get the baseline connecting the limits of integration
base_min, base_max = 1790, 1870 # the limits
abs_min = smooth_absorption[np.argmin(np.abs(wavenumber - base_min))]
abs_max = smooth_absorption[np.argmin(np.abs(wavenumber - base_max))]
m = (abs_max - abs_min) / (base_max - base_min)
b = abs_min - m * base_min
baseline = m * wavenumber + b

# Plot all of the spectra to allow the limits of integration to be determined
fig = quickScatter(x = wavenumber, y = [absorption, smooth_absorption, baseline],
                   name = ['raw data', 'smoothed data', 'baseline'], output = None)

# Modify each trace to have the appearance we want
fig.update_traces(selector = 0)
fig.update_traces(selector = 1, line = dict(width = 4))
fig.update_traces(selector = 2, line = dict(width = 4, dash = 'dash'))

# Find the y axis range to show just the data we want to see in the plot
int_range = (wavenumber > 1750) & (wavenumber < 1900)
ymin, ymax = np.min(absorption[int_range]), np.max(absorption[int_range])

# Format the plot area
fig.update_yaxes(title = 'absorption', range = [ymin - 0.1*(ymax-ymin), 1.1 * ymax])
fig.update_xaxes(title = f'wavenumber /{wn}', range = [1750, 1900])
fig.update_layout(template = 'simple_white')
fig.show('png')
fig.write_image(format = 'png', file = 'SmoothIR.png')
```

You run the code, it asks you for a csv file with an FTIR spectrum, and you select the first FTIR time point "azide_0062.csv" in your "Azide" folder. It produces the plot shown in Figure 10.8.

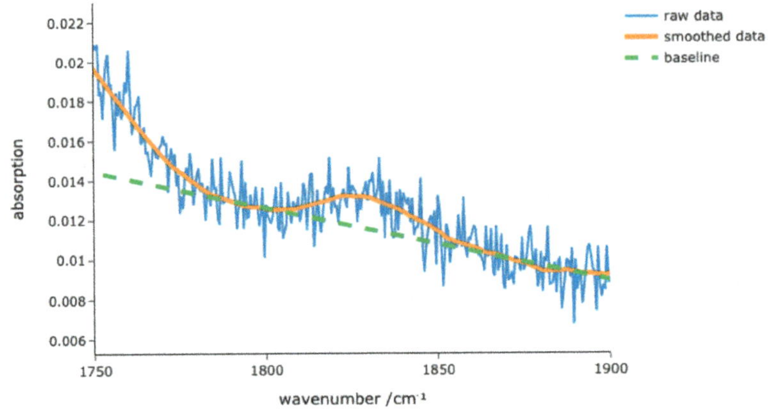

FIGURE 10.8. The results of your test code for smoothing. The small peak that you wanted is centered at 2520 cm^{-1}. You can see that your baseline estimate appears to be reasonable, so you feel confident that the procedure works.

You try several combinations of n and m, finding that including more points in the smoothing (increasing m) leads to the small peak and the large peak staring to merge, which means your baseline would be incorrect. Using fewer points adds more noise. Changing n to 2 does not make the smoothing much more effective, but changing it to 4 degrades the smoothness, again raising the question of whether the baseline will be correct. You decide that $n = 3$ and $m = 100$ seems like a happy medium.

10.2.0 LINE-BY-LINE EXPLANATION

7–11: Import `numpy` to get `genfromtxt()`, `scipy.signal` to get `savgol_filter()`, and `quickScatter()`, `quickOpenFilename()`, `quickPopupMessage()`, and the `symbols` module from `codechembook`.

14 & 15: Get the name of the csv file that we want to smooth.

18: Read the data file.

21: Smooth by Savitzky-Golay using 100 points and a third-order polynomial. You can use this script to try out different combinations of these values to see what seems reasonable.

24–29: Compute the baseline:

- 24: First, set the integration limits for the peak, i.e. where you expect the peak to return to the baseline. We do this in one line using unpacking.

25 & 26: Now, determine the y-values at these points, much like in the last chapter.

27–29: Finally, compute the slope and intercept using the x- and y-values determined above and calculate the baseline points.

32 & 33: We will make the figure using `quickScatter()` to save time, but don't want to show the figure yet, so we'll use `output = None`. We make use of the fact that we can pass `quickScatter()` a single array as the argument for `x` and a list of arrays as the argument for `y` and it will figure it out.

36–38: There's no easy way to set trace attributes using `quickScatter()` (hey, it's supposed to be *quick*!) So, we'll update them here.

41 & 42: The only minor detail here is that we want to zoom in only on this peak by setting the x-axis range, but the y-axis range will still be the full range of the spectrum. So, we have to calculate the maximum and minimum y-values for just the x-values of interest.

45–49: Format the figure and show it.

Looking at the integration code that you already wrote for the azide peak,[13] you realize that it will be very easy to adapt to this smaller intermediate peak. All you have to do is `import scipy.signal as sps`, create a smoothed version of the spectrum before the baseline determination on lines 47–49, use this smoothed spectrum to determine the baseline, and change the limits of integration, and you're finished. So, you replace lines 46–48 with the following lines of code:

[13] In Chapter 9.

```
# Smooth the spectrum
smooth_absorption = sps.savgol_filter(spec['absorption'], 100, 3)

# Determine the baseline that connects the data at the limits of integration
m = (smooth_absorption[xmax_index] - smooth_absorption[xmin_index]) / (spec['
wavenumber'][xmax_index] - spec['wavenumber'][xmin_index])
```

then add `import scipy.signal as sps` and change the limits of integration to 1790 to 1870 cm^{-1} and add the import statement. Satisfyingly, the code just works!

You take the results (Figure 10.9) to your colleague, who says, "Well, it looks like it is some small intermediate that forms relatively quickly. This will really change the way that we think about the mechanism! I'll have to go look for it in the NMR and see what it is telling me. In any case, thanks! I have some clear data tell my advisor and include in the paper."

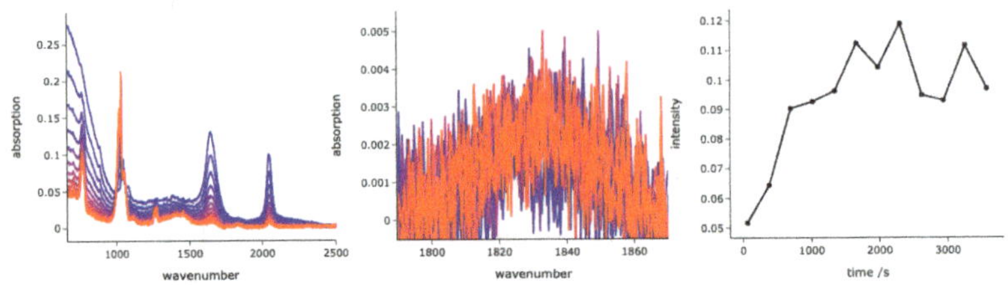

FIGURE 10.9. The spectra that you are analyzing (left), a zoom of the small spectral component in question (center), and its kinetics (right). It appears to be a byproduct of the reaction, rather than an intermediate.

CHAPTER 10 EXERCISES

TARGETED EXERCISES

Exercises for Section 10.0.1

0. Imagine you have data with uncertainties for the dependent variables. You are considering smoothing using either binning or moving averages. The bin width and the moving average window would be the same size. After smoothing, which approach results in the largest absolute uncertainty? Which approach results in the largest relative uncertainties? You can, of course, use code to answer this question, if you want.

Exercises for Section 10.1.0

1. We are going to explore what 'random' means in Python. Use the following code, which generates two arrays using a Numpy random number generator and then subtracts one from another:

```
import numpy as np
rng1 = np.random.default_rng(42) # sets a 'seed' for random number generation
rand1 = rng1.random(10) # get 10 random numbers from the generator`
rng2 = np.random.default_rng(42) # set up random number generation with the same seed as above`
rand2 = rng2.random(10) # get another 10 random numbers, using the second generator`
rand1 - rand2
```

   a) Run the above code. What does it mean that `rand1 - rand2` gave an array of all zeros?

   b) Now, repeat the above, but change the value given to `default_rng()`, the *second* time you use it (you can also supply no argument to this function).

   c) What does it mean that `rand1 - rand2` did not give an array of all zeros?

   d) Repeat the calculation again, just as above. Do you get the same difference as before?

   e) What does all of this tell you about random number generation in Numpy?

2. Perform a linear fit to a series of 10 random normally-distributed y-values (a mean of zero and a standard deviation of one) over a range of x-values from 0 to 10. Do 10 trials and compute the standard error of the slopes, and compare that with the uncertainties of the slopes in each individual fit. The expected slope is 0.

3. Repeat Exercise 2 using 100, 1000, and 10,000 points. What do you notice about the uncertainty of the slope and intercept?

4. Repeat Exercise 2 using standard deviations of 2, 3, and 4. What do you notice about the uncertainty in the slope and intercept?

5. Reproduce Figure 10.2 using your own code. Use 10,000 random numbers. You can make a histogram using the `add_hist()` method of a plotly figure, or the `quick_hist()` function of quickPlots from the codechembook library.

6. There are some general guidelines for understanding how large to make bins. One common one is the Friedman–Diaconis rule, but there are others, such as Sturges's formula or the Rice rule. Modify the final code from Exercise 3 to plot the results from each of these formulas.

Exercises for Section 10.1.1

7. A common bandshape to encounter in vibrational spectroscopy of condensed phases is a Voigt profile. This profile is a convolution of a Gaussian and Lorentzian profiles, and can be generated using the `scipy.signal.convolve()` function. However, many times people use pseudo-Voigt profiles instead, which are the sum of Gaussian and Lorentzian profiles. The LMFIT module has some documentation of this, but it is also well described on Wikipedia. In this problem, generate a Voigt and 'equivalent' pseudo-Voigt profile. Plot them against one another. Describe the differences you notice.

Exercises for Section 10.1.2

8. We explored several different smoothing approaches in this chapter, but did not directly compare them. Take the bin centers and counts from the data in Exercise 3, apply the following methods, and then plot them all on one plot:

   a) binning

   b) moving averages

   c) Savitzky–Golay

COMPREHENSIVE EXERCISES

9. The appearance of binned data can change drastically as the width of the bins change. Using the particle size measurement data from our website (sizes.csv),

explore the effect of different bin sizes. Vary the number of bins between the fourth-root of the number of measurements and one-half-the number of measurements. For each choice, fit the histogram to a log-normal distribution. How much smoothing can you do before the best-fitting values change by more than their uncertainty compared using the Freedman-Diaconis rule, as introduced in Exercise 6?

10. Using the convolution function `scipy.signal.convolve( )`, write a wrapper function that performs weighted moving average smoothing of the X-ray powder diffraction data provided on the website for this book. This function should take as arguments: the data to be smoothed and an array that specifies the shape of the smoothing function. For the latter, use:

    a) square function
    b) parabolic function
    c) triangular function
    d) sine function
    e) Gaussian function
    f) Lorentzian function

    Compare the results using Plotly and comment on which methods appear to give the best noise reduction with the minimum peak broadening.

11. Generate a Gaussian with $\sigma = 1$ and $x_0 = 0$ on an x-axis from -5 to 5 with 20 points. Then, add noise by adding normally-distributed random numbers with a standard deviation of 0.1. Fit the a Gaussian profile to this data using `lmfit`, then smooth it using a simple seven-point moving average and fit it again. Compare the fit of the original data versus the fit of the smoothed data: do the best fitting parameters agree within uncertainty? Has the uncertainty changed? What do you think about fitting smoothed data, rather than original noisy data?

CHAPTER 11

Interpolating Between Data Points

☐ **New concepts**
— *and their prerequisites*

☐ **Implementing linear interpolation using** `numpy.interp`
— *Section 1.1.1: Operating on collections of data efficiently with Numpy arrays*

☐ **Implementing cubic spline interpolation using** `scipy.interpolate.CubicSpline`
— *Section 2.1.1: Importing subsets of libraries using* `from`

11.0 PROBLEM

One evening, you are chatting with a friend in the lab next to yours while you're waiting for a reaction to finish. Your friend complains that they have a problem with data analysis. A collaborator sent them a sample of nanoparticles for your friend to add a new ligand to. Your friend explains that they characterized the resulting particles using Raman spectroscopy and really needs to plot the difference spectrum for a paper so that they can highlight the changes.

"Normally, subtracting two spectra would be simple," your friend says, "but the spectrum our collaborator took used a 532-nm laser and our instrument uses a 785-nm laser and different gratings. So all of the wavenumber values are different! Also, I used up all the original nanoparticles in my synthesis and because nanoparticle syntheses are so variable, I'll never get the exact same particles as the starting ones! Like, maybe we can get ones that are close, but it would be super nice to just subtract the spectra I have already—it would be more correct and take less time and money!"

You think that this seems like a problem that could easily be solved with Python. "I bet we could write a quick program that uses the old data to calculate what the data points should be for each wavenumber value in the new data, and then subtract that from the new data."

"That would definitely make my month!"

11.1 BACKGROUND

But where to start? The simplest thing to do would be just shifting the x-value of each data point in the old spectrum to the nearest x-value of the new data, but this doesn't seem right to you. For instance, it could easily be the case that the locations *and* spacing between points is different, so that there won't be an even shift in the points, and some points may have two options of where to shift! Sketching out the problem, you get something like Figure 11.1.

You realize the core problem: *All the data you have is at the points you collected...* What happens in between them? If you could figure this out (or take reasonable guesses) maybe you could subtract that behavior.

Thinking more about it, you consider that a spectrum is a continuously varying thing with values that change continuously over a continuous set of wavelengths or energies. How can you deal with such a thing in a computer? We could keep adding more and more points, but we would never reach the continuous limit, our computer would run out of storage space, and our programs would take longer and longer to run. So, the data that you are used to working with is a discrete set of points that will *represent* these continuously changing phenomena. You go down a rabbit hole on signals and find that this process is known as sampling; each (x, y) pair (or (t, y), or whatever) is like a measurement of the value of the continuous phenomenon, much like a time series that you would record in an experiment. When you sample it, you create a discretized function.

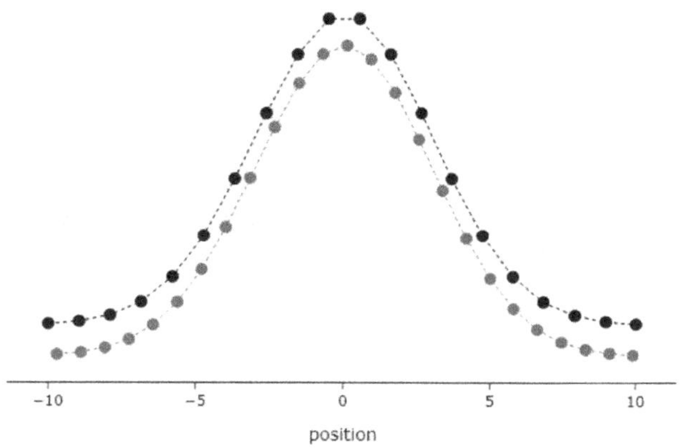

FIGURE 11.1. A plot of two sets of data that have different locations of the data (circles) and different spacing between data points. Even though they both appear to be Gaussian lineshapes, there is no way to subtract one from the other, since the data points never line up. Note: the dotted lines are not data, they simply connect the data points.

Thinking about it a little more, you realize that this situation is the opposite: you are starting with a discretized function and trying to figure out the continuous phenomenon that it samples. If you could propose a functional form, then you could fit it, but what could be the functional form be for a Raman spectrum? Many of the peaks look asymmetric, what function could you use? This seems like a problem, so maybe trying to reproduce the data is the wrong approach. Maybe you can do what you would just do visually: look at how the data is changing point by point and try to fill in the gaps. For instance, in Figure 11.1, the points are drawn connected by straight lines. Could you use that? Or perhaps another function?

A quick search for "estimate value of data between data points" quickly makes it clear that what you want to do is called interpolation. Digging into interpolation, you unsurprisingly find that there are a number of different approaches from which to choose: cubic splines, linear interpolation, constant interpolation, and others. In fact, this is reminding you a lot of the situation you found with numerical integration. There, you could link each method with the corresponding order of complexity in a Taylor series, with each order (in principle) giving you a more accurate estimate. You remember that the whole point of the Taylor series, when you first about learned it, was that you could use the values and the rates of change of a function at a known point to estimate the value of that function at an unknown point, which is exactly what you want to do here. So, linear interpolation is just using the first order of the Taylor series, a line with constant slope, to estimate the value of the function between the data points. A cubic spline uses a cubic equation, the third order, to do the same, which allows it to take into account some of the curvature of the function between the points.

You think, "this seems like the kind of thing that Numpy or Scipy would be able to do." Sure enough, you find that Numpy has a 1D interpolation **function**, `interp( )`, and Scipy has a whole **module**, `scipy.interpolate`.

You decide to write a quick **script** to see how they work.

11.2 SOLUTION

You sketch out a plan to test the interpolation approaches that you want to try:

1. Pick an analytical function that will give you exact values at any point, so you can test the accuracy of different interpolation methods. Peaks in spectra are often well described by Gaussian functions, so that seems like a good place to start.

2. Pick a set of x-values to operate as the "data." These should be relatively sparsely spaced.

3. Pick a different set of x-values to operate as the points at which you would like interpolated data.

4. Do the interpolation.

5. Plot the analytical function, the data, the interpolated data, and the difference. Compute the Root-Mean-Squared Error (RMSE) between the interpolated data and the analytical curve.

11.2.0 Implementing Linear Interpolation Using `numpy.interp`

The only part of this that is new to you is actually doing the interpolation. Inspecting the documentation for `numpy.interp( )`, you see that this is simply a **function** that takes as **positional arguments** the x-values at which you want interpolated points and the x- and y-values of the data from which you want to interpolate.[1] This function simply returns an array that holds the y-values for any x-value where you want an interpolation to occur

[1] For more information on interpolation methods, check out books like Hildebrand, *Introduction to Numerical Analysis*; Dover Press, 1987, or Eric Sullivan's *Numerical Methods: An Inquiry-Based Approach With Python* (https://numericalmethods sullivan.github.io).

In:
```
import numpy as np
```

In:
```
y_interp = np.interp(x_interp, x_data, y_data)
```

assuming that you already have the arrays `x_data` and `y_data`.

11.2.1 Implementing Cubic Spline Interpolation Using `scipy.interpolate.CubicSpline`

Looking at the documentation for `scipy.interpolate.CubicSpline`, you find that the situation is more complex. `CubicSpline` is a **class** and the constructor function only takes the x- and y-values of the data from which you will interpolate—it does not need to know the x-values at which you would like interpolated y-values. All that it is doing in the background is computing the coefficients of all of the cubic splines. Once an instance of the

`CubicSpline` class is created, the resulting **object** has a default **method** that takes a **positional argument** that is the x-value(s) at which you would like interpolated y-values. So, you figure the usage is two steps instead of the single step for `numpy.interp`:

```
In:     import scipy.interpolate as spi
In:     cs = spi.CubicSpline(x_data, y_data)
In:     y_interp = cs(x_interp)
```

Now you are ready to make a test **script** to see how well these approaches do. It turns out to be pretty simple. You want to:

1. Define an analytical function so that you know the exact y-value of the function at any x-value. Because the Gaussian is so prevalent in science, you will just use it.

2. Create some sparsely-separated data points to serve as the "fake" data that you will use interpolate from.

3. Create the exact x- and y-values for the points that you would like to interpolate to. This will be a good way to compare the accuracy of the interpolated points.

4. Use linear and cubic spline interpolation to calculate interoplated data points at each exact x-value. Calculate the root-mean-squared error between the interpolated and exact data.

With this plan in hand, you quickly bang out a script to play with:

```
'''
A quick test of the accuracy of linear and cubic spline interpolation

    Written by: myself (email@domain.com)
     Changelog: 240606 - v1.0.0 - initial version
'''
import numpy as np
import scipy.interpolate as spi

# Define a function that returns the y-values for a Gaussian function
def Gaussian(x, A, x0, sigma):
    return A * np.exp(-1 * (x - x0)**2 / (2 * sigma**2))

# Define the parameters of the Gaussian
A = 1
x0 = 0
sigma = 2

# Create the x- and y-values for the fake data
x_data = np.linspace(-5, 5, 8)
y_data = Gaussian(x_data, A, x0, sigma)

# Create the x- and (exact) y-values for the points at which we would like interpolation
x_interp = np.linspace(-5, 5, 1000)
```

```
y_interp_exact = Gaussian(x_interp, A, x0, sigma)

# Do the interpolation.  First we will do a linear interpolation
y_interp_linear = np.interp(x_interp, x_data, y_data)

# Do the cubic spline interpolation - first create the cubic spline object
cs = spi.CubicSpline(x_data, y_data)

# Now generate the array of interpolated points using the cubic spline object
y_interp_cubic = cs(x_interp)

# Compute and print the RMSE
print(f"Linear Interpolation RMSE: {np.sqrt(np.sum((y_interp_linear - y_interp_exact)**2)):4.2f}")
print(f" Cubic Interpolation RMSE: {np.sqrt(np.sum((y_interp_cubic - y_interp_exact)**2)):4.2f}")
```

which gives you the output:

```
Out:     Linear Interpolation RMSE: 0.73
Out:      Cubic Interpolation RMSE: 0.13
```

Clearly the cubic spline interpolation is more accurate, but you want to really see what is going on, so you create some plots:

1. A plot of the exact analytical curve, the data points, and the interpolated curve.

2. The error in the interpolated y-values, defined as $y_{error} = y_{interp} - y_{exact}$.

3. The relative error in the interpolated y-values, defined as $y_{relerror} = y_{interp} - y_{exact}/y_{exact}$.

The plots for linear and cubic spline interpolation are given in Figure 11.2.

Again, the cubic spline is more accurate than the linear interpolation. So you wonder why anyone would use linear interpolation, but then think, maybe it doesn't handle noise as well. So, you add noise to your simulation and get the following results:

```
Out:     Linear Interpolation RMSE: 5.48
Out:      Cubic Interpolation RMSE: 6.46
```

In this case, the linear interpolation approach is slightly less affected by noise and would probably be preferable. Plotting the linear and cubic spline interpolation results similarly for the same Gaussian peak with noise, shown on the right in Figure 11.2, shows the same story, with the cubic spline interpolation tending to be off by more in the areas between data points despite being off by the same amount at the data points. Based on this, you figure that cubic splines will generally do better if you don't have noise and if your data is reasonably closely spaced—basically, if you have a few data points on each side of your peak.

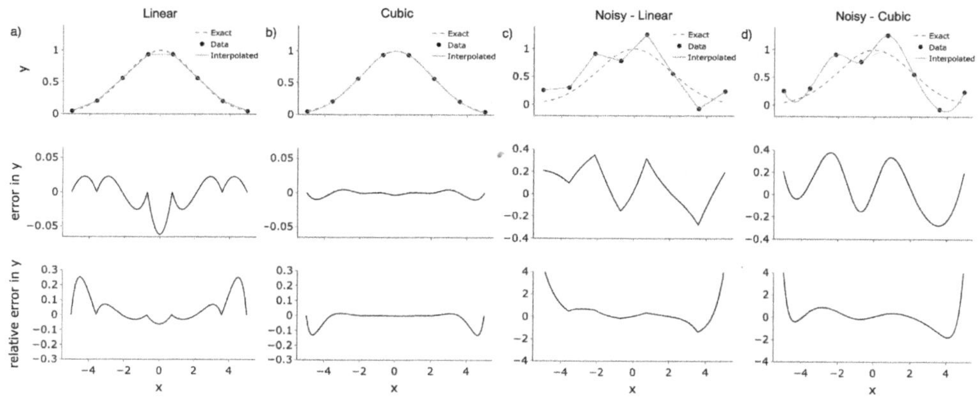

FIGURE 11.2. Example of the performance of (a) linear and (b) cubic spline interpolation for a generated Gaussian peak without noise, compared to the effects of noise on the same methods (c and d, respectively).

With the different approaches validated, writing the script for your friend will be quick. All you need to do is read both data files, interpolate one spectrum using the x-values of the other, and save the interpolated spectrum. You reason that it would be better to use the data file with more closely spaced x-values (the data collected with the 532 nm laser) to define the common x-values.

11.3 CODE

The flowchart for the desired process is given in Figure 11.3, from which you work to produce the code.

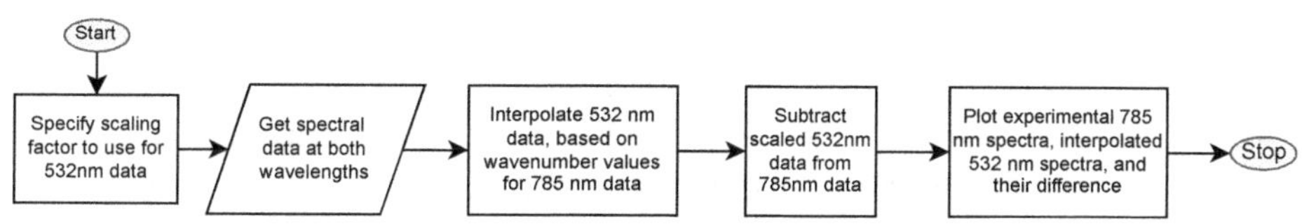

FIGURE 11.3. Flowchart for code that will take two spectral data files with different x-values, interpolate one, and then subtract one spectrum from the other.

```
'''
Produce an interpolated 785 nm Raman spectrum with data points that
match an old 532 nm spectrum
    Requires: one 532 nm and one 785 nm Raman spectrum
    Written by: Chris Johnson and Ben Lear (authors@codingforchemistsbook.com)
    v1.0.0 - 250304 - initial version
'''
```

```
8  import numpy as np
9  from plotly.subplots import make_subplots
10 from codechembook.quickTools import quickSelectFolder, quickPopupMessage
11 from codechembook.symbols.chem import wavenumber as wn
12
13 # Scaling factor for 532 data
14 scale_532 = 2.95
15
16 # Get the folder containing the files to process
17 quickPopupMessage(message = 'Select the folder with the Raman spectra.')
18 folder_name = quickSelectFolder()
19
20 # Read the data: 785 is the new data that has a contaminant, 532 is the old data
21 x532, y532 = np.genfromtxt(folder_name/'oldNPs.csv', delimiter = ',',
22                            skip_header = 2, unpack = True)
23 x785, y785 = np.genfromtxt(folder_name/'newNPs.csv', delimiter = ',',
24                            skip_header = 2, unpack = True)
25
26 # Interpolate to the 532 spectrum because 785 has the larger span of x points
27 y785_interp = np.interp(x532, x785, y785)
28
29 # Normalize the 532 data and subtract it from the 785 data
30 y_delta = y785_interp - scale_532 * y532
31
32 # Plot the 785 spectrum, the 532 spectrum, and the difference spectrum
33 fig = make_subplots(2, 1)
34 fig.add_scatter(x = x532, y = y785_interp, name = '785 nm', row = 1, col = 1)
35 fig.add_scatter(x = x532, y = scale_532 * y532, name = '532 nm', row=1, col=1)
36 fig.add_scatter(x = x532, y = y_delta, name = 'Subtracted', showlegend = False, row = 2,
       col = 1)
37 fig.update_xaxes(title = f'wavenumber /{wn}')
38 fig.update_yaxes(title = 'intensity')
39 fig.update_layout(template = 'simple_white', font_size = 18, width = 3 * 300,
40                   height = 3 * 300, margin = dict(b = 10, t = 30, l = 10, r = 10))
41 fig.show('png')
42 fig.write_image('raman.png')
```

You run this script, select the folder "Nanoparticles" in your directory, and it produces the plots in Figure 11.4.

You show this result to your friend, who says, "That was so easy for you! I couldn't figure out any way to do this with any of the programs that our lab uses. Now we can be really quantitative about the effects of the ligand exchange reaction we did. Can I use this in my paper? We will totally plug you in the acknowledgments!"

11.3.0 LINE-BY-LINE EXPLANATION

8–11: We need `numpy`, `make_subplots()`, `quickSelectFolder()` and `quickPopupMessage()` from `codechembook.quickTools`, and the wavenumber symbol from the `codechembook.symbols.chem` module.

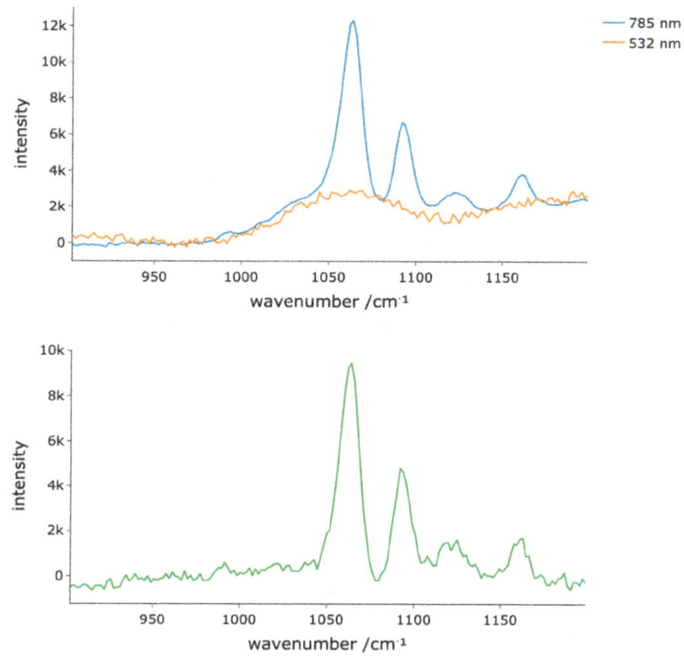

FIGURE 11.4. Results of interpolation and subtraction of the spectra.

14: The two spectra have different y-scaling, so we need a normalization factor before we subtract. Put it here so that it is easy to find and change.

17 & 18: Get the path to the folder that holds the two Raman spectrum csv files.

21–24: Read both files and store them in separate variables.

27: Do the actual interpolation. Here x532 is the x-values at which we want interpolated y-values, and x785 and y785 are the x- and y-values of the data we want to interpolate. We need to interpolate the 785-nm spectrum onto the x-values of the 532-nm spectrum because the 785-nm spectrum has x-values that fall outside of the range of the 532-nm spectrum and thus cannot be interpolated!

30: Here we subtract the 532 nm spectrum from the interpolated 785-nm spectrum. We need a normalization factor for the 532-nm spectrum because it is lower in intensity than the 785-nm spectrum. You can try to make this fancy or you can just guess and check a few times to get very close.

33–42: Nothing fancy here, we're just plotting the two spectra in one subplot and the difference between them in another.

CHAPTER 11 EXERCISES

TARGETED EXERCISES

Exercises for Section 11.2.0

0. Consider a Gaussian distribution with $x_0 = 0$, $\sigma = 2$, and amplitude of 1.

a) Create 1000 x-values between -10 and 10, and use this to create y-values with the above Gaussian parameters. This will represent the 'true' Gaussian shape.

b) Repeat the above, but with 5, 6, 7, 8, 9, 10, 11, 12 points, each from -10 and 10. These will represent sparsely sampled points.

c) For each set of points generated in the second step, perform a linear interpolation using `numpy.interp( )`. This accepts three positional arguments: an array of x-values you wish to generate interpolated points for, the original x-values, and the original y-values used for interpolation.

d) Make a Plotly plot holding a separate subplot for each of the eight sparse arrays. On this plot, have the 'true' Gaussian shape, the sparsely sampled points, and the interpolated values.

e) Comment on how the interpolation improves with sample points.

f) Comment on what regions of the Gaussian seem to be reproduced well, and which are not.

Exercises for Section 11.2.1

1. Repeat Exercise 0, but use a cubic spline for interpolation.
In addition to answering all parts from Exercise 0, comment on the differences between the results from the linear interpolation, versus the cubic interpolation.

COMPREHENSIVE EXERCISES

2. Modify the final code from Chapter 8 to interpolate the kinetic traces so that they have evenly spaced points. Use both linear interpolation and cubic splines. Do you notice any obvious inaccuracies for either method?

3. At the end of this chapter, you guessed you should use the data with the more close spaced x-values to define the common x-values after interpolation. Check if there are differences in the result if you do the opposite. If there are differences, which do you think is better, and why?

CHAPTER 12

Finding Peaks

☐ **New concepts**
   — *and their prerequisites*

☐ **Finding local maxima using** `scipy.signal.find_peaks()`
   — *Section 2.1.1: Importing subsets of libraries using* `from`

☐ **Looping until some conditions are met with** `while` **loops**
   — *Section 2.1.3: Representing answers to yes/no questions using booleans*
   — *Section 3.1.11: Performing repetitive operations using* `for` *loops*

12.0 PROBLEM

You need to do a series of cyclic voltammetry experiments on your catalysts under a number of different conditions. You will need to evaluate different scan rates, different catalyst derivatives, and different substrates, adding up to more than a *hundred* individual experiments. You want to figure out the $E_{1/2}$ for each of them, but with so many experiments, that will take a lot of slow, hands-on data analysis. You are not looking forward to quantifying all these results by hand, so you start thinking about how to automate it using your growing coding skills.

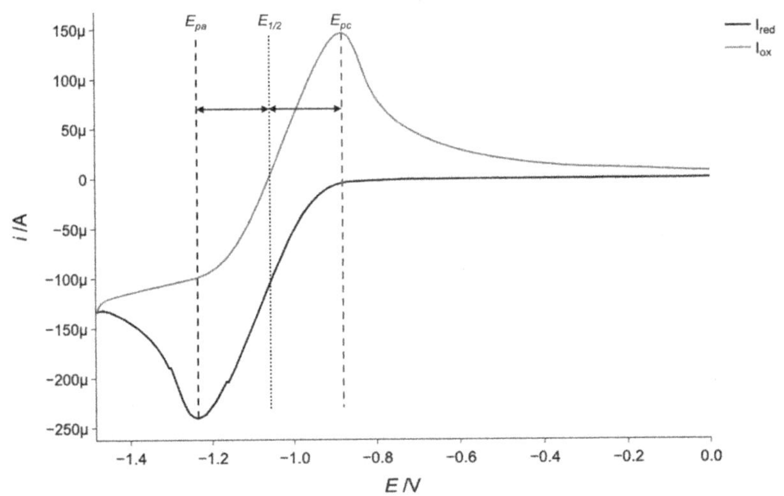

FIGURE 12.1. A diagram of your task—you need to figure out the value of $E_{1/2}$, the potential halfway between the anodic and cathodic peak potentials.

12.1 BACKGROUND

You think about how you would find a peak just by looking at a cyclic voltammogram (CV) like the one shown in Figure 12.1—you do it without even really thinking about it. It's just the highest data point in a range of data, or in other words, the data point that has a higher value than any other point (at least in its vicinity). So, it seems like the easiest thing to do is just finding the index of the *global* maximum potential value in the reduction wave, and the minimum potential value in the oxidation wave, and using it to index the potential and current arrays. Thinking about this in very general terms, you draw something with peaks and valleys (Figure 12.2) and immediately realize this would only find the leftmost peak in your drawing.

How to handle multiple peaks? Then finding the maximum or minimum is will only find one of them. Time for a more sophisticated approach that finds *local* maxima

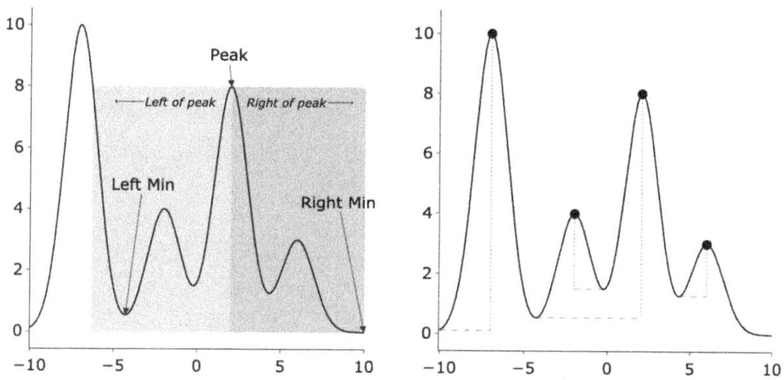

FIGURE 12.2. (Left) Example of how prominence is determined for a given peak. (Right) The prominence of a series of peaks.

and minima. Again, as a human, you can usually easily spot them, but how do you get a computer to do it?

You realize that there are a few more possible problems. Your catalyst is not very soluble in all the solvents you will use in your experiment. This means that the resulting data could be fairly noisy in those solvents. Thus, the local maximum is likely not going to be the actual maximum of the underlying signal, but rather a point close to the maximum of the underlying signal that just happens to have a particularly large contribution from noise. This is like what is shown in Figure 12.3. What can you do to mitigate these problems?

You could fit the peak to a function that has the peak center as a parameter, if your peak follows a well-behaved analytical function or if you just fit part of the peak (say, to a parabola). However, this would probably require a lot of user intervention and it also seems like it might take a bit of tuning of parameters, so maybe there is a more generic method.

Another approach could be to simply find every point that is higher than the points adjacent to it. This would pick way *more* peaks than you want, but you could filter out

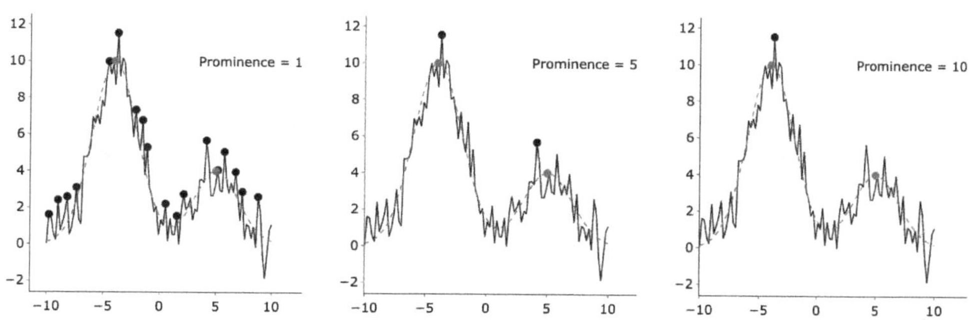

FIGURE 12.3. Examples of peak filtering with difference prominence requirements.

the ones you don't want with a number of different filters until you have narrowed your list down to only the ones you want. Talking with your labmate , you learn there are a number of different filters that you could imagine applying that would cover a range of peaks and a range of different signal-to-noise ratios. They point you to the `scipy.signal` signal module.

12.2 SOLUTION

12.2.0 Finding Local Maxima Using `scipy.signal.find_peaks()`

In `scipy.signal`, you find a function, `find_peaks()`, which appears to do what you want. It seems to just go through your data and look for points that meet several criteria, the most basic of which is that its value is higher than those of its nearest neighbors. It will return to you an array of **indices** of the peaks. However, because this could find way too many peaks if your data is noisy, there are other more complex criteria by which you can filter out most or all of the spurious peaks that aren't real or that you don't actually want. You can filter by the y-value a number of different ways using a few options: `height` is the minimum y-value (or a range of y-values if a two-element array is specified) of allowable peak heights, `threshold` is similar but specifically refers how much higher than its nearest neighbors it must be, and `prominence` is the difference in y-values between the peak height and the highest valley between it and the next nearest peak.

So, if you know that the peak you want is much higher than any neighboring peaks, you can use `threshold` to filter out baseline noise or `prominence` to filter out noise that might be causing multiple peaks to appear near the top of the peak that you want. In particular, `prominence` seems a useful parameter in many chemistry contexts.

You can also filter on properties related to the x-axis, with `distance` setting the minimum number of points between neighboring peaks and `width` setting the minimum (if a single value is given) or a range (if two values are given) of peak widths that are acceptable; `width` is particularly useful if you have an expectation of the widths of your peaks—often noise is much narrower than signal features.

12.2.1 Looping Until Some Conditions are Met with `while` Loops

The above seems like a solid plan, but you wonder how to handle the fact that the cathodic trace is going to have a negative peak? You realize that you can just take the cathodic scan, and multiply the y-values by -1, so that you can treat them as positive, but this raises another question... how do you know when you are dealing with points on the cathodic scan? You would need a way to loop through all the values of potential in the scan, and then determine when you moved from the anodic to the cathodic scan.

You already know how to use **for loops**, but the problem with them is that for loops will run for exactly as many iterations as the length of the object that is being iterated over. This means you need to have an object that is exactly as long as the number of time you want the for loop to run. Sometimes, there is no way to know how many times you need the loop to run ahead of time: either you are depending on input from a user, data from a source outside of the computer, or simply executing some analysis in which it is not clear how many iterations you will need. In the case of the CV experiment we are considering, you don't know how many iterations there will be until you move from the anodic to the cathodic scan.

A while loop addresses this case. A while loop will keep iterating until some specific logical expression evaluates to False. In Python, while loops are written like this:

```
while <conditional expression>:
    <code to execute repetitively>
<rest of script>
```

As long as <conditional expression> evaluates to False, <code to execute repetitively> will continue to loop. The while loop will terminate once <conditional expression> evaluates to False. This evaluation is performed only at the end of the loop. If something changes in the middle of an iteration, the iteration will complete, and

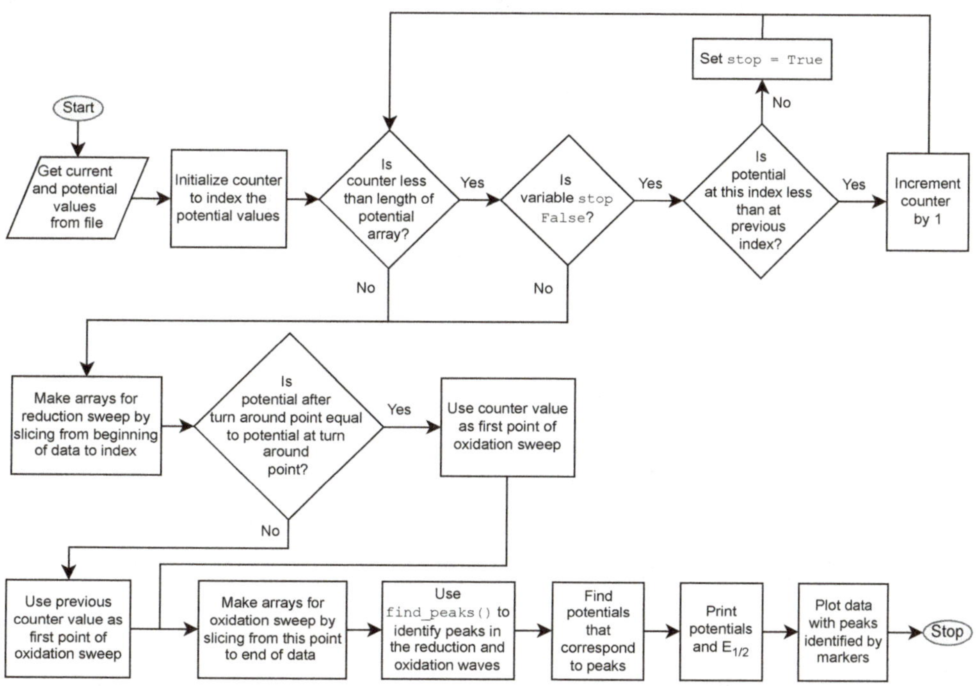

FIGURE 12.4. Flowchart for code that reads in cyclic voltammetry data, splits the data into anodic and cathodic scans, and then identify the locations of peaks in the anodic and cathodic scans.

then the loop will exit to move on with the rest of the main code. If the <conditional expression> evaluates to False even before the first iteration, the code in the loop will never run at all.[1]

[1] See Chapter F.5 for additional information.

12.3 CODE

The flowchart for this process (Figure 12.4) looks a bit complicated, but each box requires only a simple bit of code as demonstrated in the script below.

```
'''
Find the peaks of the reduction and oxidation waves in a CV experiment
and calculate E_1/2
    Requires: CV data in a .csv file with reduction first then oxidation
    Written by: Chris Johnson and Ben Lear (authors@codingforchemistsbook.com)
    v1.0.0 - 250322 - initial version
'''
import numpy as np
import scipy.signal as sps
from plotly.subplots import make_subplots
from codechembook.quickTools import quickOpenFilename, quickPopupMessage
from codechembook.symbols.greek import Delta
from codechembook.symbols.typesettingHTML import textsub, textit

# Find the data file
quickPopupMessage(message = 'Select a file with CV data.')
filename = quickOpenFilename(filetypes = 'CSV files, *.csv')

# Read the data - reduction and oxidation waves are included here
E, I = np.genfromtxt(filename, delimiter = ',', skip_header = 1, unpack = True)

# Find the index of the scan turnaround to separate the oxidation and reduction waves
i = 1 # a counter variable
stop = False # will change to true to stop the loop
while i < len(E) and stop == False: # loop until we reach the end or find the turnaround
    if E[i] < E[i-1]: # as long as E is reducing, then we are on reduction
        i += 1
    else: # E started to increase so this must be the turnaround
        stop = True

# Get separate arrays for reduction and oxidation
E_red, I_red = E[:i], I[:i] # reduction is simple
if E[i] == E[i+1]: # check special case where two points are the same at the turnaround
    E_ox, I_ox = E[i:], I[i:]
else: # only one point at the turnaround
    E_ox, I_ox = E[i-1:], I[i-1:] # we want to include the turnaround point too, thus i-1

# Find indicies of the peak currents.  Use prominence of 10% of max to avoid noise
E_pa_index, E_pa_dict = sps.find_peaks(-1*I_red, prominence = .1*np.max(-1*I_red))
E_pc_index, E_pc_dict = sps.find_peaks(I_ox, prominence = .1*np.max(I_ox))

```

```
42 # Find the potentials at the peak current
43 E_pa = E_red[E_pa_index[0]]
44 E_pc = E_ox[E_pc_index[-1]]
45 
46 # Print the results
47 print(f'E_pa = {E_red[E_pa_index[0]]:.3f} V')
48 print(f'E_pc = {E_ox[E_pc_index[0]]:.3f} V')
49 print(f'{Delta}E = {np.abs(E_pa - E_pc):.3f} V, E1/2 = {(E_pa + E_pc)*.5:.3f} V')
50 
51 # Plot the results
52 fig = make_subplots()
53 fig.add_scatter(x = E_red, y = I_red, name = f'I{textsub("red")}',
54                 line = dict(color = 'red'))
55 fig.add_scatter(x = E_ox, y = I_ox, name = f'I{textsub("ox")}',
56                 line = dict(color = 'blue'))
57 fig.add_scatter(x = E_red[E_pa_index], y = I_red[E_pa_index],
58                 name = f'E{textsub("pa")}', mode = 'markers',
59                 marker = dict(color = 'red', size = 10))
60 fig.add_scatter(x = E_ox[E_pc_index], y = I_ox[E_pc_index],
61                 name = f'E{textsub("pc")}', mode = 'markers',
62                 marker = dict(color = 'blue', size = 10))
63 
64 # Format the plot and display
65 fig.update_xaxes(title = f'{textit("E")} /V', tickformat = '0.1f')
66 fig.update_yaxes(title = f'{textit("i")} /A')
67 fig.update_layout(template = 'simple_white', font_family = 'arial',
68                   font_size = 18, width = 3 * 300, height = 2 * 300,
69                   margin = dict(b = 10, t = 30, l = 10, r = 10))
70 fig.show('png')
71 fig.write_image('CVpeaks.png')
```

You run this code, navigate to the eChem subfolder of your Catalyst folder, and select the file 'CVdata.csv' to use as test data for this script. In a flash, the following command line output appears:

```
Out:    Epa = -1.236 V
Out:    Epc = -0.885 V
Out:    delta E = 0.351 V, E1/2 = -1.061 V
```

along with the plot shown in Figure 12.5.

This looks great! You just saved yourself a lot of time opening and analyzing CV data by hand.

12.3.0 LINE-BY-LINE EXPLANATION

8–13: We need `numpy` and `make_subplots()` like usual, as well as `scipy.signal` to get `find_peaks()`. We'll also import the capital delta symbol from the `codechembook.symbols.greek` module and separately import the HTML formatting functions `textsub()` and `textit`. We could just import the whole symbols module, but the code looks cleaner this way compared with typing out `cs.typesettingHTML.textsub()` each time we want to use it.

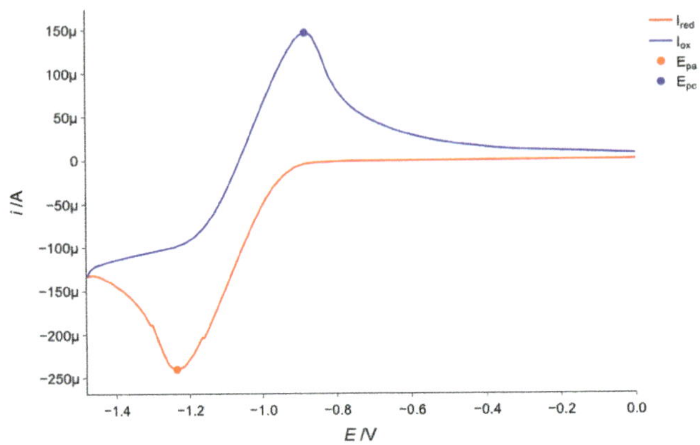

FIGURE 12.5. The CV data you analyzed along with the peaks marked.

16 & 17: Get the filename for the CV data csv file by notifying the user what to do and then opening the file dialog.

20: Read the raw CV data. This data has both the oxidation and reduction waves in it, so we will have to separate them next.

23–29: We want separate arrays for the reduction and oxidation waves. First, we need to find the index at which the potential starts going positive (after going negative for the first part of the scan). We do this by looking for where the difference between the last and the current potential switches signs:

23: We want to count data points so that we know the index. First we need to initialize the count. We set it at one instead of zero because we are computing the direction of the change of the voltage by subtracting the current point from the point behind it, and there is no point behind the point at index 0, so we have to ignore it.

24: We also initialize the stop condition to `False` because the loop will terminate when it becomes `True`.

25–29: Now we loop until we reach the end of the voltage array or the stop condition is no longer `False`. If we don't include the first condition, the counter could become greater than the length of the voltage array, causing an error. If the voltage at this index is less than the voltage at the previous index, then we are still scanning to the negative direction, and we increment the index and continue the loop. If this is not true, then we have stopped scanning in the negative direction and have found the turnaround point, so we set `stop` to be `True`, which terminates the loop.

32: Now we can split the CV data into reduction and oxidation waves. We will slice from the beginning to the turnaround point to get the reduction wave.

33–36: The oxidation wave is slightly more difficult. It is possible that there are two points that are exactly the same at the turnaround: the last voltage for the negative scan and the first voltage for the positive scan. We use an `if-else` statement to test for this case and slice accordingly.

39 & 40: Now, we will use `find_peaks()` on the reduction and oxidation waves, looking for any peak that has at least 10% prominence (i.e., 10% greater than any surrounding base). Because `find_peaks()` only takes the prominence in terms of absolute difference between the base and the peak, we can calculate this in terms of a fraction of the maximum value of the wave. `find_peaks()` returns the indices of the peaks as an `ndarray` and a dictionary of the peak properties, so we need to accept both of them on the left-hand side (LHS) of the assignment operator.

43 & 44: We don't want to know the indices of the peaks, just the E values, so we use the peak indices to index the data and return the E value of each peak.

47–49: Print out the results to the command line.

52–71: Plot the data with the found peaks marked on top.

CHAPTER 12 EXERCISES

TARGETED EXERCISES

Exercises for Section 12.2.0

0. Go to the book's companion website (https://www.codingforchemistsbook.com/) and get the following files: Raman UIO66.csv, HDI IR.csv, and AuNP Raman.csv. Using `scipy.signal.find_peaks()` find the peaks in these files. Experiment with the `prominence`, `height`, and `width` keyword arguments to successfully only find one point per peak in each set of data.

Exercises for Section 12.2.1

1. Modify the final code from Chapter 5 so that it will keep asking the user for the location of the feature of interest until the user provides a valid response (something that can be converted to an integer and that is present in the x-data).

2. Create a game that selects a random number between 0 and 10, ask the user to guess the number, and tells them if they are correct. Make sure that the user can guess the number until the user is correct. Include functionality so that it runs until you type "quit." Also, ensure that it keeps track of the number of valid guesses.

3. Create a Wordle-like game that selects a random five-letter word from a list, and then lets the user guess the word, reporting back how many letters were correct, and the position of these letters.

COMPREHENSIVE EXERCISES

4. Often, just finding the highest point in data is not an accurate determination of the center position of a peak in a spectrum or diffraction pattern. This is especially true if your data has noise that is noticeable. In this case, it is best to use the peak finding step not as the final step, but as a way to automatically generate a reasonable initial guess for a fitting process. In the case of a CV, we don't have a great analytical function to fit the full data to, but we can fit just the area around the peak to a function like a Gaussian or parabola to get a better estimate of the true peak. Try this for the positive going and negative going peaks in the CV data presented in this chapter (file 'CVdata.csv' on the book's companion website https://www.codingforchemistsbook.com/) and see how different it is from the answer you get using `find_peaks()`. Another potential benefit from fitting data is that you can get estimates of errors in the parameters, and so consider these as well, when comparing your results.

5. Modify your code above to add more noise to the data by adding normally distributed random numbers as we did in Section 10.1.0 and see how the difference between the `find_peaks()` result and the fitting result increases. Keep making the noise larger and comment on how this continues to change.

6. In the final code for Chapter 5, we had to ask the user which wavelength to use to extract the absorbances that we wanted to fit. Modify that code to automatically finds the peak for the initial parameter guess rather than ask the user. Note that the peak you use for the absorbances is not the highest peak in the spectrum, so you will have to get the correct peak from the list returned by `find_peaks()`.

7. Return to Exercise 6 from Chapter 8. This exercise dealt with simulating kinetics for an iodide clock. It turns out that the iodide clock only works for a narrow range of reactant concentrations, and the ones supplied there may not be the optimal ones. In this exercise, simulate the iodide clock reaction for different concentrations of B. Make the code automatically determine the amplitude of the first oscillation as a function of the concentration of B. Find the concentration for which the amplitude of this oscillation is largest. Plot the concentration of B versus amplitude and then find the peak in this data.

8. Using the HCl FTIR spectrum from the book's companion website (https://www.codingforchemistsbook.com/) for this book (HCl-spectrum.csv), find all peaks greater than 15 standard deviations above the baseline. Automatically determine the average and standard deviation for the baseline by selecting part of the data that clearly has no peaks in it.

9. In Exercise 5, you found the peaks in the spectrum, but these points are not necessarily the values we would report for the center of each peak. We would like to get a more accurate estimate of the peak centers by fitting. We can use the peaks we found as initial guesses for our fit:

   - Create a model that is the sum of 47-ish Gaussians. Do not hard code the number of peaks; if the number of peaks changes, you will have to change your code. We want this to work for any number of peaks. The easiest way to do this is with a for loop.
   - Start with an empty model, and for each iteration, add a `GaussianModel` with its own prefix. At the same time, use `make_params` to get initial parameters for the current `GaussianModel`, with setting amplitude to be the y value from your peak finder, center to be the x value, and sigma to be something like 0.25. Hint: use the `update()` method for the Parameters class in LMFIT.
   - Plot your data initial guess to show that your initial guess is not crazy.
   - Now that you have the model and parameters, you should be ready to fit. Do the fitting, plot the results of the fit using `plotFit()`, and print the fit report.

CHAPTER 13

Computing Numerical Derivatives

- ☐ **New concepts**
  - — *and their prerequisites*

- ☐ **Refactoring code to turn it into functions you can reuse**
  - — *Section 0.1.7: Accessing capabilities beyond basic math using functions*
  - — *Section 4.1.13: Specifying the result of function evaluation using the* `return` *keyword*
  - — *Section 4.1.14: Describing how to use your functions in docstrings*
- ☐ **Importing objects from Python files in arbitrary locations using** `codechembook.quickTools.importFromPy()`
  - — *Section 2.1.7: Adding* `codechembook` *to your Python installation*
  - — *Section 7.1.3: Giving your computer instructions using Python commands*
- ☐ **Computing derivatives using** `numpy.gradient()`
  - — *Section 1.1.1: Operating on collections of data efficiently with Numpy arrays*

13.0 PROBLEM

You just wrote a **script** that found the peaks of the reduction and oxidation waves of a CV experiment and used them to calculate the redox potential, $E_{1/2}$. You expected your advisor to be impressed, but instead they only had a series of questions, the main one being how you can know if the redox event are reversible without *looking* at the waves.

After some discussion, you both come to the conclusion that you need to automatically calculate the peak oxidative and reductive currents. So now you need to figure out how to automatically extract these values and compare them to see if they're "close enough" to establish reversibility. You know how to find the peak positions (Chapter 12), but you cannot simply read out the y-values at these points. This is because CV scans have a capacitive charging component that makes a significant contribution to the observed currents. You will need to figure out what this capacitive component is, and then subtract it from the peak currents to find the current due to the chemical redox event.[1]

[1] If you care to know more, we recommend the open-access tutorial article *A Practical Beginner's Guide to Cyclic Voltammetry* by Elgishi et al. in *J. Chem. Ed.*

The capacitive charging portion of the CV wave is typically treated as a linear extrapolation of the straight part of the scan at higher potential than the peak for oxidation and at lower potential than the peak for reduction (Figure 13.1). Normally, people will fit a line to the linear regions and then use the fit to compute the capacitive contribution to the total current at E_{pa} or E_{pc}. Subtracting the capacitive contribution from the total current gives the electrochemical current. For a reversible process, I_{pa} and I_{pc} should be equal.

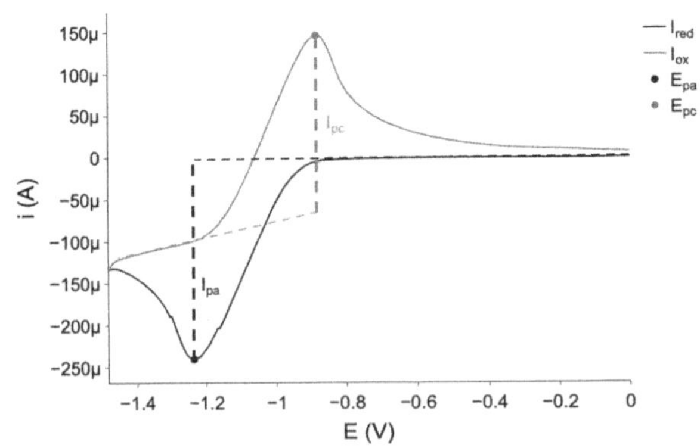

FIGURE 13.1. A diagram of your task—you need to figure out the slope of the capacitive parts of the CV and extrapolate them to the peak to determine I_{pa} and I_{pc}.

It is not immediately apparent how to use Python to do this. You can certainly use the lmfit **library** to do the fitting, but how do you automatically figure out the right *part* of the wave to fit, so that you can extrapolate for the capacitive components?

Because the redox potential of each sample will likely be somewhat different, and your sweep rates are likely to also be different, then you can't just always fit the same range of potentials for each file. "I need to figure out where in the wave the slope is basically constant," you mutter to no one in particular. So, you need to calculate the slope and look for regions where it is essentially the same. "Better yet, I can look for when slope of the slope—the second derivative—is zero, then I know the slope is not changing!" Now you need to figure out how to calculate derivatives of data.

Remembering back to your calculus class, you recall a number of rules for derivatives, like

$$\frac{\mathrm{d}}{\mathrm{d}x}x^n = nx^{n-1}$$

or

$$\frac{\mathrm{d}}{\mathrm{d}x}\cos x = -\sin x.$$

How do you apply these rules to numerical data? Thinking back a little farther, you remember the definition of a derivative at any point x:

$$\frac{\mathrm{d}}{\mathrm{d}x}f(x) = \lim_{h\to 0}\frac{f(x+h)-f(x)}{h} \tag{13.1}$$

In analytical math, you can take this limit, but with discrete data, there is a minimum spacing between points, and as such, h can only get as small as $x_{i+1} - x_i$, where i is the index of an array of data points. So, if you define x as x_i and replace h with $x_{i+1} - x_i$, then you get

$$\frac{\mathrm{d}}{\mathrm{d}x}f(x)|_{x_i} \approx \frac{f(x_{i+1})-f(x_i)}{x_{i+1}-x_i}. \tag{13.2}$$

This is just the old formula for the slope that you learned in algebra!

$$m = \frac{y_2 - y_1}{x_2 - x_1} \tag{13.3}$$

Is it really true that numerical differentiation is as easy as finding the slope?

By this definition, you can sketch out what your derivative process should look like, as shown in Figure 13.2. Although this makes sense initially, when you start to think about it, you notice some issues. First, and most importantly, your sketch has seven data points but only six derivatives! This makes sense because the derivative connects every pair of points, but it seems pretty inconvenient. Secondly, the derivative you calculate is the slope between two adjacent points, but this doesn't actually tell you the value of the derivative at any given point. Perhaps this tells you the value of the derivative at some unknown point between the two data points, like in interpolation,[2] but that isn't necessarily what you want.

[2] See Chapter 11.

"Maybe I can average the derivative before the point with the derivative after the point to get a better estimate of the derivative at the point?" Working out the math, you reason:

$$\frac{\mathrm{d}}{\mathrm{d}x}f(x)|_{x_i} \approx \frac{\frac{f(x_{i+1})-f(x_i)}{x_{i+1}-x_i} + \frac{f(x_i)-f(x_{i-1})}{x_i-x_{i-1}}}{2}. \tag{13.4}$$

At least when your x-values are equally spaced, this simplifies to:

$$\frac{\mathrm{d}}{\mathrm{d}x}f(x)|_{x_i} \approx \frac{f(x_{i+1})-f(x_{i-1})}{x_{i+1}-x_{i-1}} \tag{13.5}$$

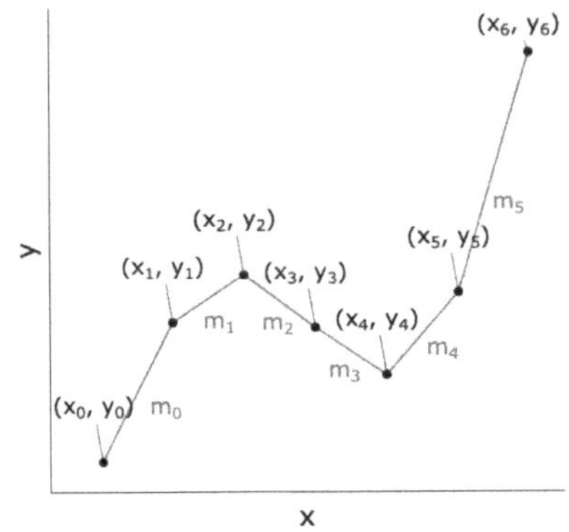

FIGURE 13.2. Basic sketch of numerical derivatives, in this case, the "forward derivative."

which is just the slope of the line connecting the point just before each point to the point just after it.

You realize that this can provide you with a solution for the slope for all points except the first and last point, so now you are two points short for the number of derivative points compared with the number of data points. However, you figure that you can probably get the first point's derivative by computing the slope to the next point, and the last point's derivative by computing the slope to the previous point, and that will give you a value for the derivative at each data x-value.

Finally, because the second derivative is just the derivative of the derivative, you can perform this same procedure over and over to achieve whatever order you need. Of course, while you know you could code this, you also think that taking numerical derivatives is common enough that there must be a library for it. A quick Google search for "numerical derivatives Python" shows that Numpy—that old familiar standby—has functions for this.

13.1 SOLUTION

Your task is fairly simple when looking at it broadly. Starting with your peak finding code,[3] you can break the task up into six major activities: reading the data from the file, splitting it into reduction and oxidation waves, finding the peaks, subtracting the capacitive charging contribution, printing the results, and plotting a figure. This is summarized by the flowchart in Figure 13.3.

[3] See Chapter 12.

You could just add the capacitive charging part to the code that you already have, but you think that would not be a great option because you do these experiments a lot and would likely end up with many slightly different copies of the same code on your computer. This could lead to headaches and errors down the road.

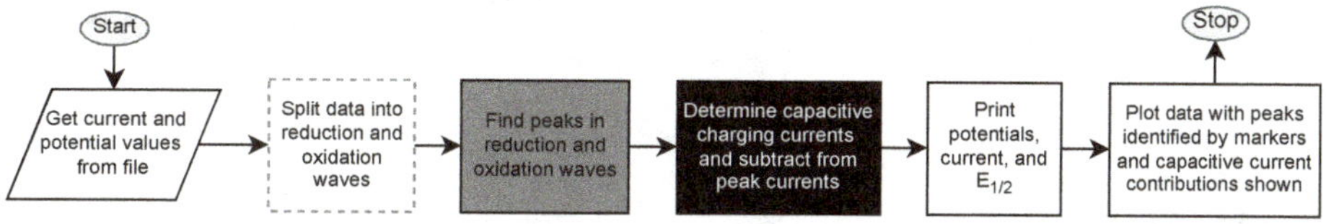

FIGURE 13.3. Flowchart showing the big picture of what you are trying to accomplish when expanding your CV analysis code to include correcting the peak currents for capacitive charging effects The boxes with the dashed outline and grey coloring correspond to regions in Figure 13.4, while the box with and black coloring corresponds to the flowchart in Figure 13.8.

Because you plan to keep doing these experiments and analyses, you would like to break down the steps into a set of tools that not only are easy to maintain, but allow you to do the analysis interactively if needed. You realize that your flowchart provides a clear way to break up the task into logically related subtasks! You could write a function that splits the data, another function that finds the peaks, and a third function that determines the capacitive charging contribution. You already have the code that would need to be in the first two functions, they're just in a script instead of their own functions. You could rewrite them to be a function, which will make them more convenient to use here. Reading a little about programming, you realize that what you want to do is called **refactoring**, so you dig in to it more.

13.1.0 REFACTORING CODE TO TURN IT INTO FUNCTIONS YOU CAN REUSE

Looking at your flowchart for your CV peak finding code, you see that it is easy to find which steps go with splitting the data and which steps go with finding the peaks (Figure 13.4).

The process for refactoring this code will be to identify the logically related code segments, move them to a new function that needs to be defined, and determine the parameters that need to be passed and the variables that need to be returned. Looking at the code in Chapter 12, you see that lines 22–36 deal with finding the turnaround point of the potential and creating the oxidation and reduction waves, so you create a new function called `SplitWaves()` and move those lines into its body. This code only needs the potential and current arrays to function, so you include them in the function definition as **positional arguments**. It produces four new arrays, the current and potential for reduction and oxidation, so you include those in a `return` statement at the end. Finally, you write a helpful **docstring**. Now you have a function that does the work, and you can replace all of the code that you just moved with a simple function **call** to it in the main code! Following the same logic, you create a function called `AnalyzeCV()` to do the peak finding, calculate $E_{1/2}$, and determine the indices of the peaks. Replacing this code with a call to the new function gets your code working again, but now in a refactored form. You can import these functions at the command line if you want to work with data

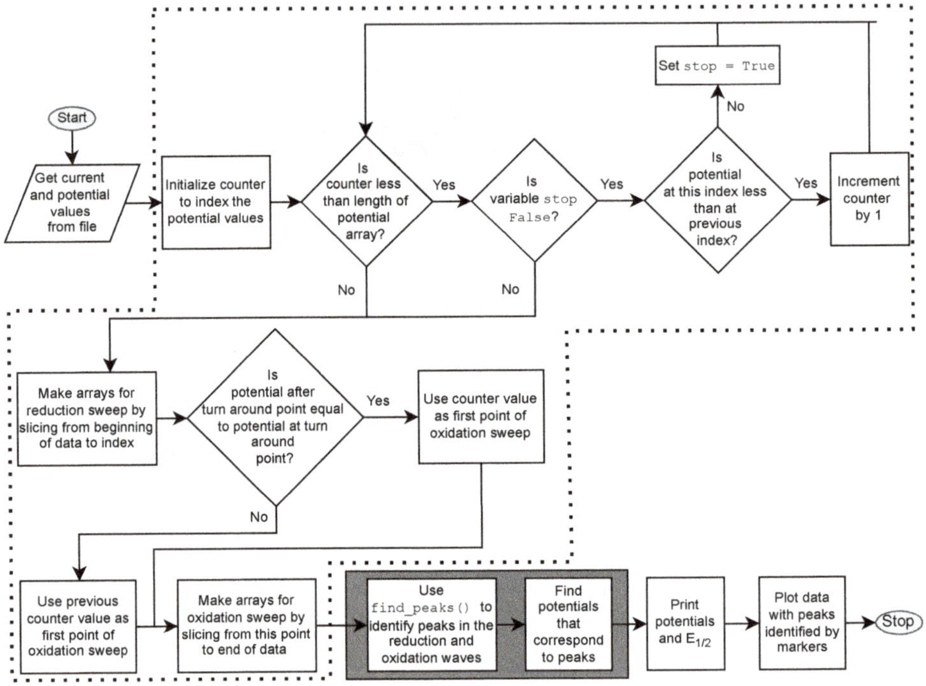

FIGURE 13.4. Your flowchart from Chapter 12, annotated to identify the logical blocks to split the oxidation and reduction waves (dotted box) and the peak finding (gray shaded box).

interactively and save yourself a ton of time. You can also import the functions into a new code that performs the capacitive charging correction. The rest of the code, which handles reading files and output of results, you leave in a block at the end of the script. This block is denoted by an `if __name__ == '__main__':` statement that makes sure it only runs if you run the script directly instead of importing the functions.[4] The code you come up with is:

[4] See Section 7.1.3.

```
'''
Find the peaks of the reduction and oxidation waves in a CV experiment
and calculate E_1/2
    Requires: CV data in a .csv file with reduction first then oxidation
    Written by: Chris Johnson and Ben Lear (authors@codingforchemistsbook.com)
    v1.0.0 - 250322 - initial version
    v1.1.0 - 250323 - refactored to make AnalyzeCV function importable
'''
import numpy as np
import scipy.signal as sps
from plotly.subplots import make_subplots
from codechembook.quickTools import quickOpenFilename, quickPopupMessage
from codechembook.symbols.greek import Delta
from codechembook.symbols.typesettingHTML import textsub, textit

def SplitWaves(E, I):
```

```
    '''
    Split E and I arrays into separate reduction and oxidation E and I arrays

    REQUIRED PARAMETERS:
    E (ndarray): potential data points
    I (ndarray): current data points

    RETURNS:
    (ndarray): E_red, reduction wave potential data points
    (ndarray): I_red, reduction wave current data points
    (ndarray): E_ox, oxidation wave potential data points
    (ndarray): I_ox, oxidation wave current data points
    '''
    # Find the index of the scan turnaround to separate the oxidation and reduction waves
    i = 1 # a counter variable
    stop = False # will change to true to stop the loop
    while i < len(E) and stop == False: # loop until we reach the end or find the turnaround
        if E[i] < E[i-1]: # as long as E is reducing, then we are on reduction
            i += 1
        else: # E started to increase so this must be the turnaround
            stop = True

    # Get separate arrays for reduction and oxidation
    E_red, I_red = E[:i], I[:i] # reduction is simple
    if E[i] == E[i+1]: # check special case where two points are the same at the turnaround
        E_ox, I_ox = E[i:], I[i:]
    else: # only one point at the turnaround
        E_ox, I_ox = E[i-1:], I[i-1:] # we want to include the turnaround point too, thus i-1

    return E_red, I_red, E_ox, I_ox

def AnalyzeCV(E_red, I_red, E_ox, I_ox):
    '''
    Find the peaks of reduction and oxidation waves and get E_1/2

    REQUIRED PARAMS:
    E_red (ndarray): reduction wave E points
    I_red (ndarray): reduction wave I points
    E_ox (ndarray):  oxidation wave E points
    I_ox (ndarray):  oxidation wave I points

    RETURNS:
    (float): E_pa, the peak of the anodic wave
    (float): E_pc, the peak of the cathodic wave
    (float): delta_E, the difference between the peaks
    (float): E_1/2, the reduction potential
    (int):   E_pa_index, the index of the peak of the anodic wave
    (int):   E_pc_index, the index of the peak of the cathodic wave
    '''

    # Find peaks in reduction and oxidation regions
    E_pa_index, E_pa_dict = sps.find_peaks(-1*I_red, prominence=0.1*np.max(E_red))
```

```
    E_pc_index, E_pc_dict = sps.find_peaks(I_ox, prominence=0.1*np.max(I_ox))

    # Calculate E_pa and E_pc
    E_pa = E_red[E_pa_index[0]]
    E_pc = E_ox[E_pc_index[-1]]

    # Calculate delta E and E1/2
    delta_E = np.abs(E_pa - E_pc)
    E1_2 = (E_pa + E_pc) * 0.5

    return E_pa, E_pc, delta_E, E1_2, E_pa_index, E_pc_index

if __name__ == '__main__':
    # Find the data file
    quickPopupMessage(message = 'Select a file with CV data.')
    filename = quickOpenFilename(filetypes = 'CSV files, *.csv')

    # Read the data - reduction and oxidation waves are included here
    E, I = np.genfromtxt(filename, delimiter = ',', skip_header = 1, unpack = True)

    # Get separate E and I arrays for oxidation and reduction
    E_red, I_red, E_ox, I_ox = SplitWaves(E, I)

    # Call the analysis function
    E_pa, E_pc, delta_E, E1_2, E_pa_index, E_pc_index = AnalyzeCV(E_red, I_red, E_ox, I_ox)

    # Print the results
    print(f'E_pa = {E_red[E_pa_index[0]]:.3f} V')
    print(f'E_pc = {E_ox[E_pc_index[0]]:.3f} V')
    print(f'{Delta}E = {np.abs(E_pa - E_pc):.3f} V, E1/2 = {(E_pa + E_pc)*.5:.3f} V')

    # Plot the results
    fig = make_subplots()
    fig.add_scatter(x = E_red, y = I_red, name = f'I{textsub("red")}',
                    line = dict(color = 'red'))
    fig.add_scatter(x = E_ox, y = I_ox, name = f'I{textsub("ox")}',
                    line = dict(color = 'blue'))
    fig.add_scatter(x = E_red[E_pa_index], y = I_red[E_pa_index],
                    name = f'E{textsub("pa")}', mode = 'markers',
                    marker = dict(color = 'red', size = 10))
    fig.add_scatter(x = E_ox[E_pc_index], y = I_ox[E_pc_index],
                    name = f'E{textsub("pc")}', mode = 'markers',
                    marker = dict(color = 'blue', size = 10))

    # Format the plot and display
    fig.update_xaxes(title = f'{textit("E")} /V', tickformat = '0.1f')
    fig.update_yaxes(title = f'{textit("i")} /A')
    fig.update_layout(template = 'simple_white', font_family = 'arial',
                      font_size = 18, width = 3 * 300, height = 2 * 300,
                      margin = dict(b = 10, t = 30, l = 10, r = 10))
    fig.show('png')
    fig.write_image('CVpeaks.png')
```

13.1.1 IMPORTING OBJECTS FROM PYTHON FILES IN ARBITRARY LOCATIONS USING `codechembook.quickTools.importFromPy()`

It is often the case that you want to use functions or variables defined in Python scripts located in some folder of your file system other than the current working directory (CWD). We discussed this already in Section 7.1.3, but briefly, our options are typically:

1. Copy the file to the CWD. This is not ideal because you now have multiple copies to keep track of, making it more likely that you have multiple different versions of the code to deal with.

2. Add all the locations to the PythonPATH. The code to do this is clunky to use in every script that needs this capability.

The `codechembook.quickTools` module features a function to simplify this process,[5] `importFromPy()`. This function has a required parameter to specify the name of the file that has the code we want to import, and then a series of arbitrary positional parameters that are the names of the objects (functions, variables) in that code that you want to import. When you run the function, the specified Python code will run, and the specified objects will then be available to the rest of the code that calls `importFromPy()`.

[5] Technically, what this function is doing is adding the requested object's variable names to the "namespace" of the Python kernel. This allows you to use the variable names, just as you normally would. Exploring this more is well beyond the scope of this book, but we mention it here, so that you have terms to search, if you want to learn more.

13.1.2 COMPUTING DERIVATIVES USING `numpy.gradient()`

Turning your attention back to derivatives, you realize that computing these slopes is very simple, particularly if your points are evenly spaced. You could easily code this up yourself, but why do that when you could use `numpy.gradient()`. Reading the documentation, you see that this **function** performs derivatives in multiple dimensions (including one) and has some useful features built in.[6] You find that this function is very simple—it has one required **argument**, the y-values, although the second **positional argument** is the x-values, which will generally be useful if your points are not spaced by the default of one or are unequally spaced. The second argument could also be a single value per dimension representing the spacing between points, but only if that spacing is constant in each dimension. It also has one **keyword argument**, `edge_order`, that specifies how the first and last points are handled. The default, 1, treats it as described above, whereas 2 represents a higher-order approximation that is often more accurate. You decide to do some quick tests by creating some data representing a function whose derivative is known analytically, $\cos(x)$ whose derivative is $-\sin(x)$. You can then test the approach to see how it works. You whip up a quick script, which produces Figure 13.5:

[6] For more information on numerical derivatives, check out books like Hildebrand, *Introduction to Numerical Analysis*, Dover Press, 1987, or Eric Sullivan's *Numerical Methods: An Inquiry-Based Approach With Python* (https://numericalmethodssullivan.github.io).

```
'''
Testing numerical derivatives by taking the derivative of a cos function
    Written by: Chris Johnson and Ben Lear (authors@codingforchemistsbook.com)
    v1.0.0 - 250427 - initial version
'''
```

```
import numpy as np
from plotly.subplots import make_subplots

# Sample a cos function and its analytical derivative, sin, over one period
x = np.linspace(0, 2*np.pi, 20)
y = np.cos(x)
dy_analytical = -1 * np.sin(x)

# Compute the numerical derivative using 1st and 2nd order edges
dy_gradient_1 = np.gradient(y, x, edge_order = 1)
dy_gradient_2 = np.gradient(y, x, edge_order = 2)

# Plot all the curves to compare
fig = make_subplots()
fig.add_scatter(x = x, y = y, name = 'cos(x)', line = dict(color = 'black'), )
fig.add_scatter(x = x, y = dy_gradient_1, name = 'Gradient, order 1',
                line = dict(color = 'gray', width = 12),)
fig.add_scatter(x = x, y = dy_gradient_2, name = 'Gradient, order 2',
                line = dict(color = 'black', dash = 'solid', width = 8),)
fig.add_scatter(x = x, y = dy_analytical, name = 'd(cos(x))/dx',
                line = dict(color = 'lightgrey', dash = 'dot', width = 4), )

fig.update_yaxes(title = 'y')
fig.update_xaxes(title = 'x')

fig.update_layout(template = 'simple_white', font_size = 18,
                  legend = dict(x = 1, y = 0, xanchor = 'right'),
                  width = 3 * 300, height = 2 * 300,
                  margin = dict(b = 10, t = 30, l = 10, r = 10))
fig.show('png')
fig.write_image(format = 'png', file = 'CosDerivativeExample.png')
```

You note that the derivatives are fairly accurate and that, in this case, the second-order edge calculation works much better. As would be expected, increasing the number of points that you use to sample substantially decreases the error.

With this understanding, you are now ready to write your code.

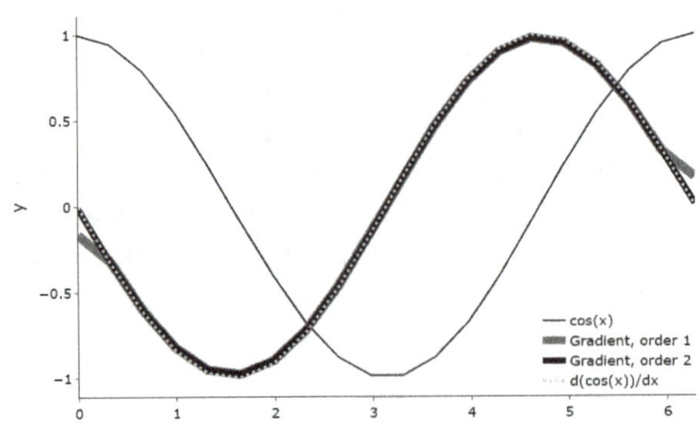

FIGURE 13.5. Results of a test of the accuracy of numerical derivatives, comparing the known analytical solution to numerical solutions using first- and second-order edge corrections.

13.2 CODE

You start with the refactored peak finding code you wrote earlier as a template. You can write another function that separately calculates the currents so that your code is equally modular for current as it is for potentials. Your plan of attack is:

1. Calculate the first and second derivatives (easy now).
2. Find the part of each wave where the second derivative is nearly zero.
3. Fit the wave where the second derivative is nearly zero to a linear model.
4. Use the linear fit parameters to calculate the capacitive current at the peak potentials.
5. Subtract those currents from the peak currents.

Part 2 is really the only part that you don't know how to do. Naively, you would think that you can just find the parts of the second derivative that are zero, but even clean CV data has some noise that will make it so the second derivative is usually not *exactly* zero. Your first idea is to look for where the second derivative is less than some threshold value. It would be good to actually visualize what the second derivative is, so you quickly whip up a code to plot it and obtain the plot shown in Figure 13.6.

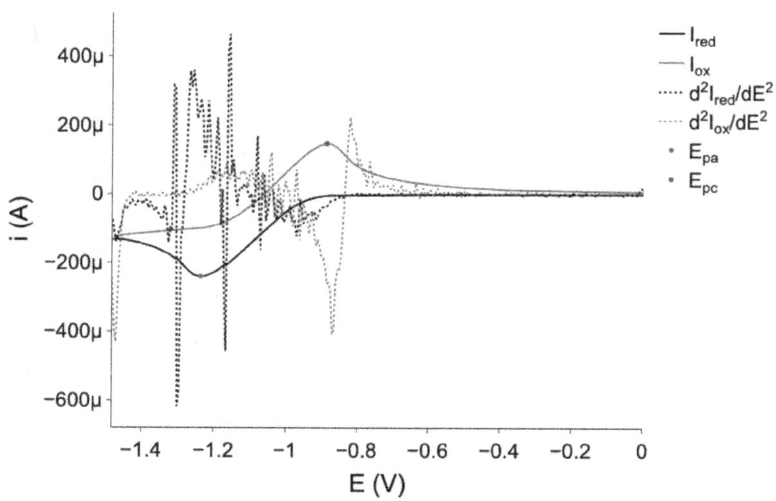

FIGURE 13.6. Plots of the CV, the first derivative, and the second derivative, as well as the peak locations.

You see the problem: the second derivative is also nearly zero by accident in many different parts of the wave because the noise is amplified by taking the derivative. So, if you just look for where the second derivative is near zero, you will get points all over the place, and trying to fit them will give you wrong slopes. So, you need to narrow down the points somehow. One thing you know is that for the reduction wave, the linear part should be somewhere between the beginning of the scan and the peak, and for the oxidation wave, it should be somewhere between the turnaround point and the peak. This is helpful, but still doesn't solve the problem. One thing you could do is find the largest contiguous range of points that are all nearly zero. This should filter out all of the points where random noise made the value accidentally zero. You could also smooth the data, or smooth the derivatives, to try to reduce this noise some. All of this would require a lot of user intervention, however, for different files with different noise levels. You would like something that is relatively immune to noise.

Derivatives from Local Regression Smoothing
For particularly noisy data, it can be useful to obtain derivatives directly from the smoothing process rather than from the raw or smoothed data. Local regression smoothing techniques use analytical functions that have analytical formulae for their derivatives, and `scipy.signal.savgol_filter()` can return derivatives via the `deriv` keyword.

After thinking about it for a while, you come to a realization. You can account for the noise automatically by using the standard deviation. When the second derivative is really approximately zero, essentially all of the variance from zero will be noise. When the second derivative is changing, then you will get an additional contribution to the standard deviation from the fact that the underlying signal is also changing. This will necessarily increase the standard deviation. So, you could do the following (Figure 13.8).

1. Approximate the standard deviation at each point by computing the standard deviation in a moving window of points. The standard deviation of each point will then be the standard deviation in the range of points around it, say ± 10 points.

2. Find the minimum standard deviation of each wave.

3. Start at the beginning of each wave and find the first point where the standard deviation drops below some multiple of the minimum standard deviation.

4. Keep going until you find the last contiguous point at which the standard deviation remains below that multiple of the minimum standard deviation.

5. This range will be the range that you fit.

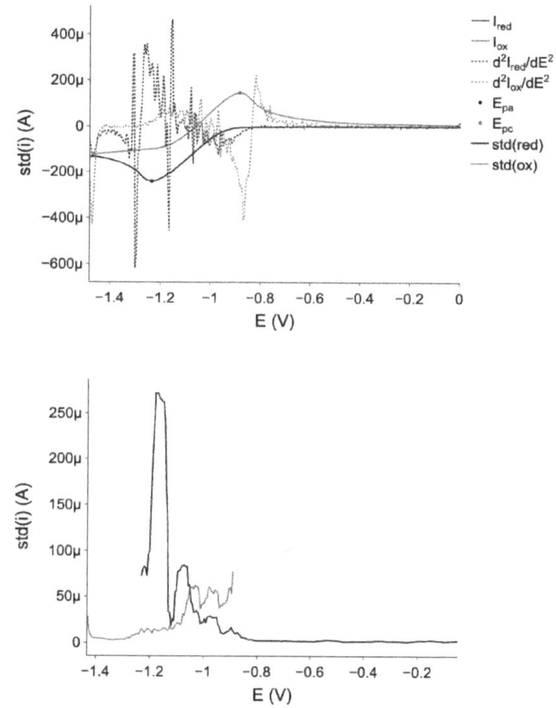

FIGURE 13.7. Plots of the standard deviations of the reduction and oxidation waves with respect to potential.

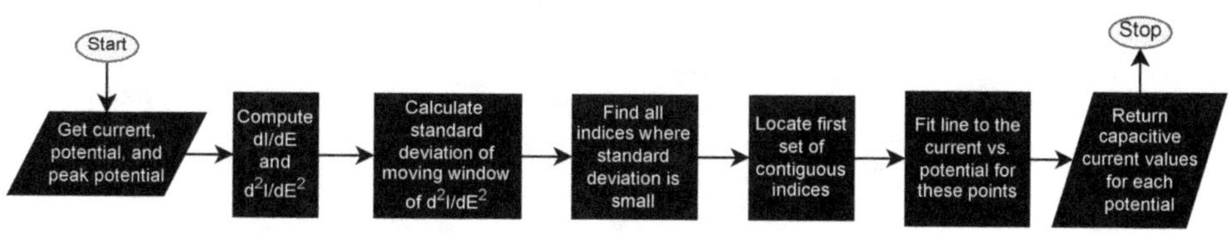

FIGURE 13.8. Flowchart for code that will determine the capacitive charging component of the current at any potential point.

You modify your code to also plot the standard deviation and obtain the plots shown in Figure 13.7.

You can now clearly see the parts of the curves that you want to fit!

Finally, you put all of these workflows together and get the code working on a single file:

```
'''
Get E_1/2, i_pa, and i_pc for a CV experiment
    Requires: CVPeakFind_refactored.py, CV data in a .csv file with
              reduction first then oxidation
    Written by: Chris Johnson and Ben Lear (authors@codingforchemistsbook.com)
    v1.0.0 - 250427 - initial version
'''
```

```
import numpy as np
from plotly.subplots import make_subplots
from lmfit.models import LinearModel
from codechembook.quickTools import quickOpenFilename, importFromPy, quickPopupMessage
from codechembook.symbols.greek import Delta
from codechembook.symbols.typesettingHTML import textsub, textit
importFromPy('CVPeakFind_refactored.py', 'SplitWaves', 'AnalyzeCV')

# Get the capacitive contribution to the current
def getCapCharge(I, E, E_peak):
    '''
    Compute the capacitive charging component of the i vs. E wave

    REQUIREDPARAMS:
    I (ndarray):    the current data points
    E (ndarray):    the potential data points
    E_peak (float): the potential of the peak in the wave

    RETURNS:
    (ndarray): cap, the linear extrapolated capacitive current at each E
    (ndarray): dI, the derivative of the current
    (ndarray): d2I, the second derivative of the current
    (ndarray): std_d2I, the standard deviation over a subset of E
    (ndarray): std_valid, the points at which the std is evaluated
    '''

    # Take the first and second derivatives
    dI = np.gradient(I, E)
    d2I = np.gradient(dI, E)

    # Estimate the standard deviation of the second derivative of the current
    # at each point by calculating it in a moving window
    window = 5
    std_d2I = np.array([np.std(d2I[i-window:i+window]) for i in np.arange(10, E_peak)])

    # Find the indicies of points at which the standard deviation is less than twice the minimum
    std_valid = np.array([i for i in range(len(std_d2I)) if std_d2I[i] < 2*np.min(std_d2I)])

    # Find the first contiguous range of points with a low standard deviation
    # First we calculate the difference between the indicies calculated above
    # and the indicies we would expect if every point was contiguous
    std_valid_shift = std_valid - np.arange(np.min(std_valid),
                                            np.min(std_valid) + len(std_valid))

    # Select only the indicies where the difference is zero i.e. the range was contiguous
    deriv_indicies = [val for val, test in zip(std_valid, std_valid_shift) if test == 0]

    # Fit that range to a linear model
    fit = LinearModel()
    fit_results = fit.fit(I[deriv_indicies], x = E[deriv_indicies])

    # Compute the capacitive charging component
    cap = fit_results.best_values['slope'] * E + fit_results.best_values['intercept']
    return cap
```

```

# Find the data file
quickPopupMessage(message = 'Select a file with CV data.')
filename = quickOpenFilename(filetypes = 'CSV files, *.csv')

# Read the data - reduction and oxidation waves are included here
E, I = np.genfromtxt(filename, delimiter = ',', skip_header = 1, unpack = True)

# Get separate E and I arrays for oxidation and reduction
E_red, I_red, E_ox, I_ox = SplitWaves(E, I)

# call the analysis function
E_pa, E_pc, delta_E, E1_2, E_pa_index, E_pc_index = AnalyzeCV(E_red, I_red, E_ox, I_ox)

# Get the capacitive currents for both waves
cap_red = getCapCharge(I_red, E_red, E_pa_index[0])
cap_ox = getCapCharge(I_ox, E_ox, E_pc_index[0])

# Print the results
print(f'Epa = {E_red[E_pa_index[0]]:.3f} V, ipa {I_red[E_pa_index[0]] - cap_red[E_pa_index
    [0]]:9.3e} A')
print(f'Epc = {E_ox[E_pc_index[0]]:.3f} V, ipc {I_ox[E_pc_index[0]] - cap_ox[E_pc_index
    [0]]:9.3e} A')
print(f'{Delta}E = {np.abs(E_pa - E_pc):.3f} V, E1/2 = {(E_pa + E_pc)*.5:.3f} V')

# Plot the results
fig = make_subplots()
fig.add_scatter(x = E_red, y = I_red, name = f'I{textsub("red")}',
                line = dict(color = 'red'))
fig.add_scatter(x = E_ox, y = I_ox, name = f'I{textsub("ox")}',
                line = dict(color = 'blue'))
fig.add_scatter(x = [E_red[E_pa_index]], y = [I_red[E_pa_index]],
                name = f'E{textsub("pa")}', mode = 'markers',
                marker = dict(color = 'red', size = 10))
fig.add_scatter(x = [E_ox[E_pc_index]], y = [I_ox[E_pc_index]],
                name = f'E{textsub("pc")}', mode = 'markers',
                marker = dict(color = 'blue', size = 10))
fig.add_scatter(x = E_red, y = cap_red,
                showlegend = False, line = dict(color = 'red', dash = 'dash'))
fig.add_scatter(x = E_ox, y = cap_ox,
                showlegend = False, line = dict(color = 'blue', dash = 'dash'))

# Format the plot and display
fig.update_xaxes(title = f'{textit("E")} /V', tickformat = '0.1f')
fig.update_yaxes(title = f'{textit("i")} /A')
fig.update_layout(template = 'simple_white', font_family = 'arial',
                font_size = 18, width = 3 * 300, height = 2 * 300,
                margin = dict(b = 10, t = 30, l = 10, r = 10))
fig.show('png')
fig.write_image('CVpeaks.png')
```

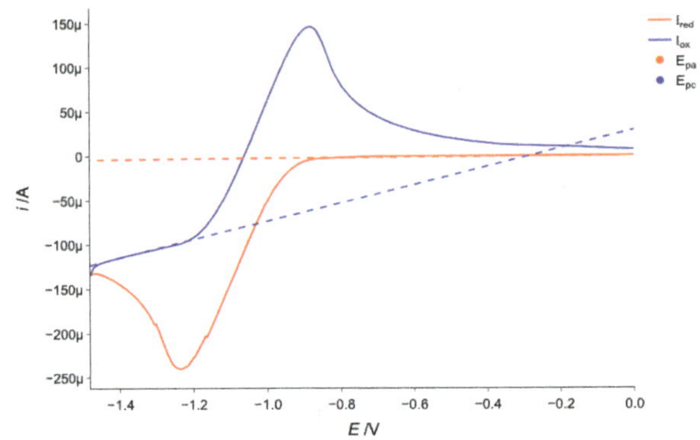

FIGURE 13.9. Results of applying the above code. (top) The data visualization that is produced and (bottom) the text output in the **console**.

You can run this code the same way as the previous version,[7] and it outputs the plot shown in Figure 13.9.

[7] See Chapter 12.

The following text is also output to the console:

```
Out: Epa = -1.236 V, ipa -2.371e-04 A
Out: Epc = -0.885 V, ipc 2.081e-04 A
Out: delta E = 0.351 V, E1/2 = -1.061 V
```

The currents are close enough to call this reversible. You'll have to test this code a bit more on some additional data files to make sure that the process is robust, but you are even more sure than before that you have saved yourself a ton of time!

13.2.0 LINE-BY-LINE EXPLANATION

8–13: We will need `numpy` and `make_subplots()` as usual. We also need `LinearModel` from `lmfit` for the fitting and extrapolation. We'll use `quickOpenFilename()`, `quickPopupMessage()`, and `importFromPy()` from the `codechembook.quickTools` module. Finally, we'll import the capital delta symbol from the `codechembook.symbols.greek` module and, in addition, separately import the HTML formatting functions `textsub()` and `textit`. We could just import the whole symbols module, but the code looks cleaner this way compared with typing out `cs.typesettingHTML.textsub()` each time we want to use it.

14: We use `importFromPy` to get the `SplitWaves()` and `AnalyzeCV` functions from the **refactor**ed version of the code in Chapter 12.

17–61: Compute the capacitive charging component of the current. Because we need to do this for the reduction and oxidation wave, we just create a function that we can call twice rather than reproduce the same code again.

35 & 36: Take the first and second derivatives using `numpy.gradient` with default parameters.

40 & 41: Find the standard deviation of the second derivative at each relevant point between the start of the wave and the peak. We will ignore anything after the peak because it definitely should not be fit to determine the charging. At each point, we define a window of points before and after it to use to compute the standard deviation.

44: Now we want to get the indices where the second derivative is nearly zero. We will compute this by using list comprehensions to find the points at which the standard deviation of the second derivative is less than twice its minimum value.

49: We want to find the first contiguous range over which the standard deviation is nearly zero. There are many ways to do this, but a simple way here is to compare the array of indices where the standard deviation is nearly zero to an array of indices that starts at the same first index and just increases by one. At some data point, the standard deviation will be more than twice the minimum and that index will not be included in the first array. Subtracting the second array from the first one thus gives you the difference in indices. This should be zero for the contiguous part and non-zero at the first index where the standard deviation is more than twice the minimum. This line will return an array that holds these differences.

53: Next, we need to know the indices that correspond to the contiguous range where the standard deviation of the second derivative is nearly zero. We do that here using a list comprehension. We zip up the array of valid indices and the array that tells us whether the range is contiguous and return a new list that holds only the indices for the contiguous range.

56 & 57: Now we can simply fit the contiguous range we identified.

60: Use the best fit slope and intercept to compute an array holding the capacitive current at each value of the potential.

61: Return the capacitive component of the current to the main code.

64–74: This is all the same as the refactored version of the code in Chapter 12, which itself is based on the code discussed in that chapter, so see the discussion of that code for the details.

77 & 78: Using our function, compute the capacitive currents for the reduction and oxidation waves.

81–83: Print the potentials and currents that we found.

86–109: Make a plot that shows the peaks found and the capacitive extrapolations so that we can do a sanity check.

CHAPTER 13 EXERCISES

TARGETED EXERCISES

Exercises for Section 13.1.0

0. Refactor the final code from Chapter 1. Create separate functions to handle the printing output, the `calcPlateVols()` functionality, and defining the geometry and concentrations of the well plate. The latter should take the following as arguments: the number of rows, the number of columns, and the starting and ending concentrations and ionic strengths.

1. Refactor the final code from Chapter 9. Produce a function that will automatically background subtract data passed to it.

Exercises for Section 13.1.2

2. For each of the following functions, determine the accuracy of a numerical derivative. Create data with an x-axis from 0 to 2π with 10 data points. Make a figure with two subplots, where the top plot shows the function, its analytical derivative, and its numerical derivative using the forward backward method, and the bottom plot shows the difference between the numerical and analytical solutions:

a) sine

b) Gaussian

c) Lorentzian

3. Starting with your code from Exercise 2 Create a plot of the root-mean squared deviation between the numerical and analytical solutions versus number of data points for data with 10, 20, 30, 40, and 50 points. What do you notice?

4. Redo Exercise 3 using only forward derivatives and compare the accuracy of the two methods.

Exercises for Section 13.1.1

5. Return to the final code for Chapter 7 and replace the block that imported code from an external script with code that uses `importFromPy()`.

6. Using `importFromPy()` get the x- and y-data from the final code in Chapter 2. Then use `importFromPy()` to get the final extinction coefficient from the code in Chapter 5. Then make a plot of the data, but rescale the y-values to be in terms of concentration, rather than absorbance. Label the axes appropriately.

COMPREHENSIVE EXERCISES

7. Derivatives tend to amplify noise. Using the X-ray powder diffraction data from the website for this book, compare the results when computing the derivative first, then performing Savitzky–Golay filtering to the opposite. Which one produces more noise? You should do this in two steps, separately computing derivatives and smoothing, but note (for future use) that `scipy.savgol_filter()` can give you derivatives automatically with the `deriv` keyword.

8. Write your own peak finding algorithm using derivatives. Using the file 'HCl-IR.csv' from the website for this book, which contains the FTIR spectrum of HCl, create a code that finds peaks without using `scipy.find_peaks()`. Remember that at a peak, the first derivative is zero and the second derivative is negative:

   - Read the file and plot the data, its first derivative, and its second derivative. Note that the first and second derivatives may be much smaller or larger than the data, so you may want to apply a scaling factor to each such that they are all about the same height as the data. Zoom in on a smaller peak to confirm that the derivatives look as expected.
   - A peak should be a maximum in the signal, a negative-going zero crossing in the first derivative, and a minimum for the second derivative. Because noise looks like peaks to a computer, you will also need to have a "threshold" or a minimum intensity in the spectrum that distinguishes a peak from noise. Any zero crossing at a point where the signal is less than the threshold should be ignored.

   The first derivative is unlikely to be precisely zero anywhere. It is much easier to find the point just before and just after the zero crossing of the first derivative and pick as the peak whichever of the two points is higher in the signal. Plot the peaks you picked on top of the data using a scatter plot. Zoom in to a few peaks to confirm that your code works correctly.
   We have also provided three other files, another FTIR spectrum "CO2-IR.csv", NMR spectrum "NMR.csv", and a powder diffraction pattern "KBr.csv". Using this same code (changing only the thresholds and scaling factors for the derivatives), find the peaks. Note that the diffraction data is noisy—comment on how does this affect your peak picker. What could you do about it?

   - A good trick can be to 'weight' the derivatives to make it easier to tell the real peaks from the noise ones. This can be accomplished by multiplying the derivative by the signal, so that the amplitude of the derivative at each peak is enhanced. Implement this and see how this impacts your results.

- You could get a more accurate estimate of the peak by explicitly determining the zero crossing. For the two points in your derivative data that are just above and just below zero, compute the x-value at which the derivative crosses zero. You can do this by computing the slope, $m = (y_2 - y_1)/(x_2 - x_1)$, with point 2 as the point just below zero and point 1 the point just above zero. Then, solve the same equation for x_2 (your zero crossing) where $y_2 = 0$ and m is the slope you just computed. Implement this and comment on how it impacts your results.

9. Sometimes spectroscopists find it easier to determine the number of peaks in a complex, overlapping band, by taking the derivative of the spectrum. To figure out why, and apply this to generate initial guesses for the code at the end of Chapter 6. Hints: You can answer the fundamental question by making your own overlapping peaks and taking their derivative. Then, ask yourself: what are the different ways to find peaks!

PART FOUR

Reporting and Sharing Results

CHAPTER 14

Producing Structured Data Files

- ☐ **New concepts**
  - — *and their prerequisites*

- ☐ **Writing text files with arbitrary formatting using `with open()`**
  - — *Section 2.1.5: Describing the location of files or folders on your computer using Pathlib*
  - — *Section 5.1.4: Inserting calculation results into strings using f-strings*
- ☐ **Making Microsoft word tables using docx**
  - — *Section 2.1.1: Importing subsets of libraries using `from`*
  - — *Section 5.1.5: Expressing data as readable text using strings*

14.0 PROBLEM

It is a normal day in the lab, which means there is good news and bad news. The good news is that your experiments have gone well and you are starting to draft a paper based on them. The bad news is that you need to gather all your results to report for this paper. Specifically, you need to prepare two files for the supplementary information (SI) of the paper: in one file, your advisor wishes to collect all the results into a csv file so that these files can be used by anyone on any type of computer and does not require commercial software to read. However, your advisor also wants to report just the parameters you obtained by your fitting in a table, which will be part of a Microsoft Word document that explains the fitting procedure and results. Thus, you need to make a Word table of containing these parameters.

Before learning to code, the above would have meant that you would spend an afternoon copying and pasting the exact same data between at least two programs, then formatting these files by hand.

You now know that such formulaic processes can be handled well by coding. You ask your labmate about it, and they smile. "Absolutely! We can automatically produce complex csv and Microsoft Word files. Let's start with the csv files and then move on to Word."

14.1 WRITING A csv FILE

14.1.0 PLANNING

"There are lots of reasons to use the **csv** format," your labmate says. "It is a **plain text** file, so you don't need specialized software to open it. It is human readable, so you can just open the file in any text editor and read the data. Both of these points means it is a reasonably transferable and future proof format. This makes it a nice choice when you are sharing it with other people that may have software or workflows you don't know about. Like in the supplementary information of a paper."

> **Binary Data Files versus csv files**
> Although plain text files like csv files are convenient, the representation of data as text typically requires more storage space per unit of information than the equivalent representation in a **binary format**. Many programs and instruments will, by default, store data in binary-formatted files to reduce the storage space required for the computers that operate them. The disadvantage of binary files is that it is very difficult to retrieve the information stored in them without a program designed to do it, or at least knowing the exact formatting used to generate the file. Because of this, many programs will allow you to export a plain text file, often csv, if you want it.

"Also," they continue, "the structure is simple: it is just the data you want, as text, separated by commas. We saw this the first time we worked with a spectrum.[1] So, you basically already know everything you need about the structure. We only need to introduce a few new ideas in order to make the code to produce it."

[1] See Chapter 2.

14.1.1 WRITING TEXT FILES WITH ARBITRARY FORMATTING USING `with ... open()`

"There are going to be times that we want to save text files to the computer. Numpy does have some functions for writing **Numpy arrays** to files, like csv files, but you will want the formatting of your file to be different than what those function can easily produce. One case is when all of the data that you want to write is not in arrays that are the same length, or especially if you want to write some arrays and some single values. So, it is beneficial to learn the general case for writing text files."

"We'll start by learning how to open a file—before you can write to it, you have to open it. This is done using the `open()` function, and providing a **path** as the first **positional argument**. Additionally, we can supply another **argument** stating if we need to read ('`r`'), write ('`w`'), or append ('`a`') the file. Read is used when you don't want to change the contents of the file, write is used to create an entirely new file, and append is used to *add* to the end of an existing file. Here, we will open in read mode."

"If we simply use the `open()` command we will also need to close the file, for instance, using the `close()` method. If we don't close the file, it stays open, which can use up **memory** and possibly lead to problems with the file down the line. But it is sort of silly to have to both manually open and close files. Thus, there is another way."

"**This other way automatically closes the file for you**, working under the assumption that we normally open a file, do something with it, and once we are done, we wish to close it. If this is true, we can use a new **control structure**, `with`, combined with `open()`. The construct `with open():` starts a new **block** using indented lines. When we end the codeblock (i.e., we un-indent the code) the file is automatically closed because we have exited the control structure associated with the `with open():` **statement**. We strongly recommend using this `with open():` structure, and we also recommend providing an **alias** for the file using `with open(<path to file>) as <alias>:`, similar to how this is done when **import**ing a Python **library**."

Once the file is open, it is opened as a 'file object,' which we can then write to. We can do this using the `write()` method of the file object and providing the string we wish to write as the first positional argument of this function.

14.2 CODE TO PRODUCE A csv FILE

You think through the process that you need to go through to write this code, realizing that you will need a loop to iterate through each line to write, and assemble the line from the data. You also know that the header line will have to be treated differently than the

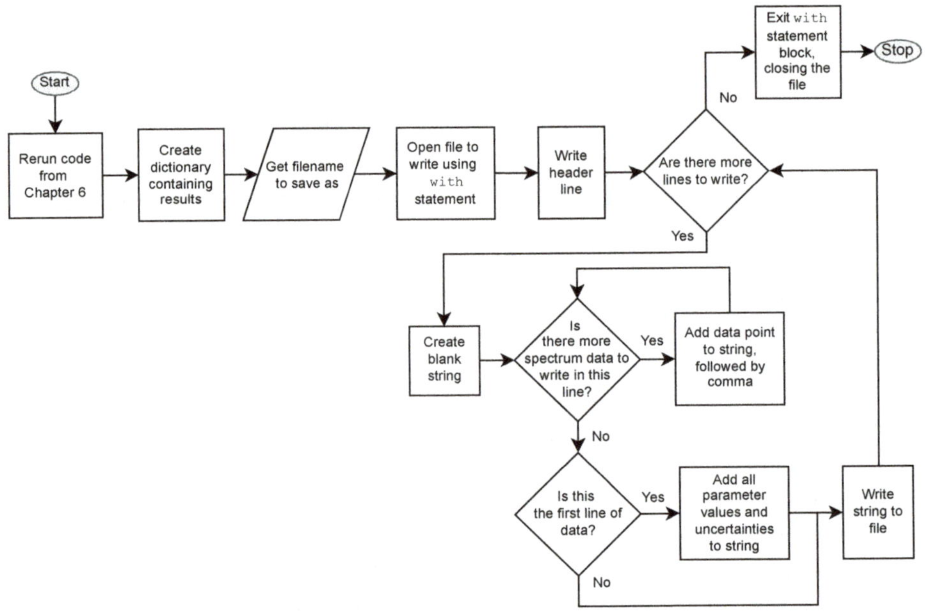

FIGURE 14.1. Flowchart for code that will take the results from Chapter 6, and save it to a csv file.

data lines. Also, the first line will contain both the spectral data and the fit results, but the rest of the lines will only have spectral data. You come up with the plan shown in Figure 14.1.

Working from this plan, you produce a code:

```
'''
A script to produce .csv files with data from my titration paper
    Requires: DecomposeSpectrum.py
    Written by: Ben Lear and Chris Johnson (authors@codingforchemistsbook.com)
    v1.0.0 - 250502 - initial version
'''
from codechembook.quickTools import importFromPy, quickSaveFilename, quickPopupMessage
importFromPy('DecomposeSpectrum.py', 'trimmed_wavelength', 'trimmed_absorbance', 'result')

# Gather the fit parameters and spectral data into a dictionary
spectral_data = dict(wavelength = trimmed_wavelength, absorption = trimmed_absorbance)

# Load the individual components of the fit into the dictionary
comps = result.eval_components(x = trimmed_wavelength)
for c in comps:
    spectral_data[f'{c[:-1]}'] = comps[c]

# Get the path for the file that we want to save
quickPopupMessage('Choose a CSV file to save the data.')
file = quickSaveFilename(filetypes='CSV files, *.csv',
    initialpath='./report.csv',
    title='Save your file')

```

```
24 # Write the spectral data and fit parameters to a CSV file
25 with open(file) as f: # open a file to write, in write mode
26     # Write the header line, starting with the data
27     line = 'wavelength, exp, fit, gaussian1, gaussian2, gaussian3, '
28 
29     # Now add the headers for the columns with gaussian fits
30     for j in range(3):
31         line = line + f'g{j+1} amplitude, g{j+1} amplitude uncertainty, g{j+1} center, g{j+1} center uncertainty, g{j+1} sigma, g{j+1} sigma uncertainty, '
32 
33     # Now add the linear part
34     line = line + 'slope, slope uncertainty, intercept, intercept uncertainty\n'
35     f.write(line)
36 
37     # Construct the complex CSV file.  We need to write out N_data_data_points rows
38     # Some rows at the top will have extra columns for the fit data
39     for i in range(len(trimmed_wavelength)):    # loop through the rows we need
40         line = '' # start with a blank line on each iteration
41         for data in spectral_data:    # iterate through keys in the spectral data dictionary
42             line = line + f'{spectral_data[data][i]:E}, ' # add data values to line in engineering format
43 
44         # Treat rows 1-3 differently to print the best fitting parameters for the gaussian components
45         if i == 0: # we can add in parameter value information
46             for j in range(3):
47                 line = line + f'{result.params[f"g{j+1}_amplitude"].value}, '
48                 line = line + f'{result.params[f"g{j+1}_amplitude"].stderr}, '
49                 line = line + f'{result.params[f"g{j+1}_center"].value}, '
50                 line = line + f'{result.params[f"g{j+1}_center"].stderr}, '
51                 line = line + f'{result.params[f"g{j+1}_sigma"].value}, '
52                 line = line + f'{result.params[f"g{j+1}_sigma"].stderr}, '
53             line = line + f'{result.params["lin_slope"].value:E}, '
54             line = line + f'{result.params["lin_slope"].stderr:E},'
55             line = line + f'{result.params["lin_intercept"].value:E}, '
56             line = line + f'{result.params["lin_intercept"].stderr:E}'
57 
58         line = line + '\n' # add a newline character to the end of our line
59         f.write(line) # write the line
```

You run the script, select the code that you wrote to decompose the UV-vis peak into three Gaussians (DecomposeSpectrum.py, in the Ch06 folder of your Code directory) for importing, select the file 0.001.csv from your Catalyst/Titration/UVvis directory, and then in a moment, the script produces a new csv file. If you open this in a text editor, it will be a bit hard to read, but if you open this in Microsoft Excel, you will see the structure shown in Figure 14.2.

14.2.0 LINE-BY-LINE EXPLANATION

7: Import the libraries we need.

8: We use `importFromPy()` to specify a Python file we wish to read and names of objects we want to import from it. This function will attempt to open the file, and

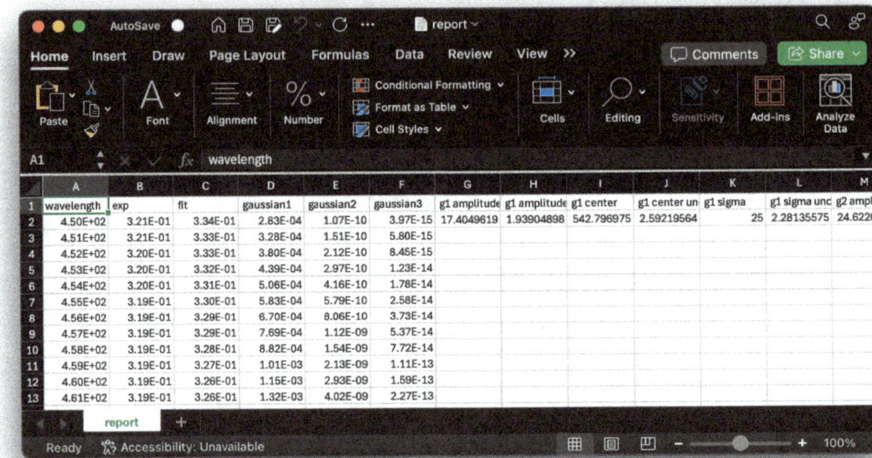

	A	B	C	D	E	F	G	H	I	J	K	L	M
1	wavelength	exp	fit	gaussian1	gaussian2	gaussian3	g1 amplitude	g1 amplitude	g1 center	g1 center un	g1 sigma	g1 sigma unc	g2 amplit
2	4.50E+02	3.21E-01	3.34E-01	2.83E-04	1.07E-10	3.97E-15	17.4049619	1.93904898	542.796975	2.59219564	25	2.28135575	24.6226
3	4.51E+02	3.21E-01	3.33E-01	3.28E-04	1.51E-10	5.80E-15							
4	4.52E+02	3.20E-01	3.33E-01	3.80E-04	2.12E-10	8.45E-15							
5	4.53E+02	3.20E-01	3.32E-01	4.39E-04	2.97E-10	1.23E-14							
6	4.54E+02	3.20E-01	3.31E-01	5.06E-04	4.16E-10	1.78E-14							
7	4.55E+02	3.19E-01	3.30E-01	5.83E-04	5.79E-10	2.58E-14							
8	4.56E+02	3.19E-01	3.29E-01	6.70E-04	8.06E-10	3.73E-14							
9	4.57E+02	3.19E-01	3.29E-01	7.69E-04	1.12E-09	5.37E-14							
10	4.58E+02	3.19E-01	3.28E-01	8.82E-04	1.54E-09	7.72E-14							
11	4.59E+02	3.19E-01	3.27E-01	1.01E-03	2.13E-09	1.11E-13							
12	4.60E+02	3.19E-01	3.26E-01	1.15E-03	2.93E-09	1.59E-13							
13	4.61E+02	3.19E-01	3.26E-01	1.32E-03	4.02E-09	2.27E-13							

FIGURE 14.2. Screenshot of the csv file produced by the above code, opened up in Microsoft Excel.

if it is not in the current working directory, then it will prompt the user to identify where it is, using the system's native file explorer. After the function executes, we will have direct access to the objects `trimmed_wavelength`, `trimmed_absorbance`, and `result`, exactly as they were in the script "DecomposeSpectrum.py." This means that the script in "DecomposeSpectrum.py" will be run, and you will need to follow the prompts from that script. At the end, the variable `result` will hold the lmfit result object.

11: We record the variables holding the x- and y-data in a dictionary. We will add more data to the dictionary later on, and the reason for using a dictionary will become a bit more clear then (lines 15 and 16).

14: We get the components of the fit, using the `eval_components()` method of the fit object, and specify the x-values over which they should be evaluated. Because our model consisted of a linear model and three different Gaussian models, the result is a dictionary with four items: the keys are the names of the components and the y-values associated with each of these components.

15 & 16: We use a **`for loop`** to take each component in turn and then add them to the `spectral_data` dictionary, using the name of the component as the key. However we strip out the last character, which was a _. If you need to recall this, look back at the code produced in Chapter 6.

19–22: We will want to save the csv file we are going to generate. So, we pop up a message telling the user this is what is going on, then we use `quickSaveFilename()` to allow them to specify the file name they wish to save the report as and the folder in which to save it. This is done using the system's native file explorer. It returns a pathlib Path object, which is assigned to a variable for later use.

25–59: Write the file we want. We start this by using the `with open() as <alias>:` construction, which starts a new code block in which our open file object is stored in the variable `f`.

27: We write a string that explicitly specifies the first row, which serves as the header of the file. We are giving the header for each column, separated by commas.

30 & 31: We will now add the column headers for the fitting parameters. Since there are three gaussians, there are three sets of fitting parameters, and we will use a for loop so as not to type repetitively. We will just keep adding this text to the end of our line, with each entry separated by a comma.

34: Finally, we add the header entries for the linear part of the fit and end with a newline character (`\n`).

35: The line is complete, so we write it to the file.

39–59: Now we want to actually write the data. There will be as many more rows as we have data points in the `trimmed wavelength` array, so we will just use that as our array to iterate over.

40: We are going to construct a string to hold the text for the line that we then write. So, we start with a blank string.

41 & 42: We then iterate through our spectral data, one at a time in a for loop, where we access the value of the data and concatenate it with our growing string.

45–56: We check if we are working on the first line of data. If so, that means we are we are going to write out all the parameter data as well, so we include that. Here, each line is a value (or uncertainty) for the parameters, corresponding to the order specified in the header. Some of these values are formatted using **e notation**. If you need a refresher on how to access the results of fitting, you can read Chapter 6.

58: We add a newline character to our string to end it.

59: We write our final assembled string to our file. The script ends here, because as soon as we are done with the `for` loop, we will read the end of the `with open` block and the file will be closed, and then the script will end.

14.3 WRITING A MICROSOFT WORD FILE

You realize that you are unlikely to want to produce the *entire* Microsoft Word document in Python. However, you find it a real pain to produce tables in Microsoft Word. It would be really nice if you could produce a document that had a section title and a table—in other words, the *section* of the supporting information you will need. Then you can either copy and paste it into the final document, or write around this part of the file.

"It is a bit frustrating to work with Microsoft Word in Python," your labmate says. "The finicky formatting can be a bit hard to handle. However, it is worth seeing how to create Microsoft Word documents, and the elements within them and *for sure* creating docx tables in Python is much simpler than using the graphical user interface (GUI) in Word."

"Really the only new thing here is to understand how docx files are structured and then how to write them. There is a lot of specialty stuff for how to write them, and so the best way to illustrate this is just to create one. However, before we do this, we should consider the general structure of a docx file."

14.3.0 MAKING MICROSOFT WORD TABLES USING DOCX

"When you open a Microsoft Word document using the Microsoft Word program, it *looks* as though it is only text—albeit formatted text. But there is much more going on behind the scenes. Microsoft has all sorts of additional information that is not shown, concerning things like the size of the paper, the font that should be used, the margin size, and so on. Even things that appear to be just text often are not."

"Microsoft Word provides the ability to designate different types of text—for instance, we can select text and designate it a 'heading' type. Although a heading (like a title) may just look like it is a bolder, larger set of text, Microsoft Word considers it to be a different kind of object. This is why you can change the formatting for a heading type in Word, and then all the text that is in headings will change throughout the document at once. Thus, there is all sorts of 'metadata' involved in a Word document. Often **binary encoding** is the most efficient way to keep track of such information. If you try to open a docx file in a text editor, then you will just get a series of numbers—the binary encoded data."

"The main reason to point all this out is so that you realize that writing a Microsoft Word document is *never* really just writing text. We have to write text and make sure it is given the proper text type, as well as any other formatting information we may want it to have. This also means that things that seem simple when using the GUI interface within Word actually can be quite complex constructions. For this reason, you are going to see that we import many different **module**s for working with docx files. Additionally, there is going to be a significant amount of terminology associated with making these files. For instance, the most basic kind of text is called a 'run,' and we will need to use an `add_run()` **function** to write even the most basic text."

"But rather than try to front-load all of this, let us just jump in and make the file of interest. Once we have the code, we can explain how it works, and hopefully give you enough understanding so that you can read up on how to make any other changes you like."

14.4 CODE TO PRODUCE A docx FILE

The flowchart for the solution is given in Figure 14.3.

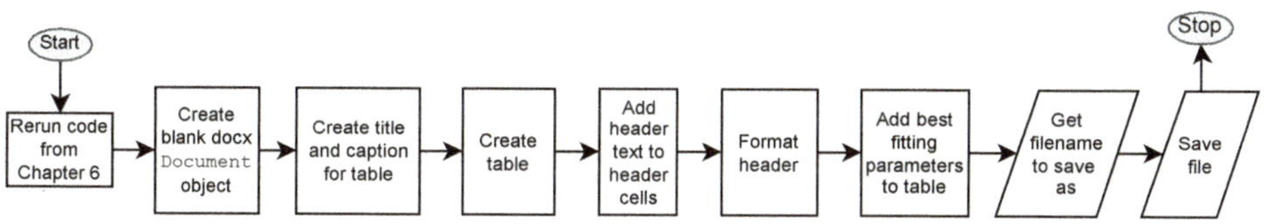

FIGURE 14.3. Flowchart for code that will take the results from Chapter 6 and write them to table in a Microsoft Word file.

```
'''
Produce a Word document with a table for the titration paper
    Requires: DecomposeSpectrum.py
    Written by: Ben Lear and Chris Johnson (authors@codingforchemistsbook.com)
    v1.0.0 - 250502 - initial version
'''
from docx import Document
from docx.oxml import OxmlElement, parse_xml
from docx.oxml.ns import qn, nsdecls
from docx.enum.text import WD_PARAGRAPH_ALIGNMENT
from codechembook.quickTools import importFromPy, quickSaveFilename, quickPopupMessage
importFromPy('DecomposeSpectrum.py', 'result')

# Start a new Word document
doc = Document()

# Add a section title - we have to handle the subscript as a separate paragraph
title = doc.add_heading(level=1) # Use the preset heading font
title_run = title.add_run('Composition of the Ni(pz)') # Text for the start of the heading; pz = pyrazine
subscript_run = title.add_run('2') # Text that is supposed to be subscripted
subscript_run.font.subscript = True # Change the text formatting to subscript
title.add_run(' MLCT band') # Change back to normal font and keep going

# Add the introductory paragraph with subscript
paragraph = doc.add_paragraph('Parameters extracted from the spectral decomposition of the MLCT band for Ni(pz)')
paragraph_run = paragraph.add_run('2')
paragraph_run.font.subscript = True
paragraph.add_run(' into Gaussians contributions. ')
```

```
29
30 # Create a table
31 parameters = ['amplitude', 'center', 'sigma']
32 ncomponents = 3 # the number of gaussians
33 ncols = len(parameters)*2 # each gaussian has three parameters, each with uncertainty
34 table = doc.add_table(rows=ncomponents + 1, cols=ncols + 1) # create a blank table
35 table.style = 'Table Grid' # style the table
36
37 # Add header row with the names of the fit parameters
38 hdr_cells = table.rows[0].cells
39 hdr_cells[0].text = 'Component'
40 for i, name in enumerate(parameters):
41     hdr_cells[2*i+1].text = f'{name}'
42     hdr_cells[2*i+2].text = f'{name} uncertainty'
43
44 # Make text in header cells bold and shaded gray, and with a light gray background
45 for cell in hdr_cells:
46     # Get the instructions for Word to apply the text shading and giving it to the cell
47     shading_elm = parse_xml(r'<w:shd {} w:fill="D9D9D9"/>'.format(nsdecls('w'))) # must
       create each time
48     cell._tc.get_or_add_tcPr().append(shading_elm)
49     # Loop through each run in each paragraph in the cell
50     for paragraph in cell.paragraphs:
51         for run in paragraph.runs:
52             run.font.bold = True
53             shading_elm = OxmlElement('w:shd') # create a shading element
54             shading_elm.set(qn('w:fill'), 'C0C0C0') # set the text color
55             run._element.append(shading_elm) # add it back in
56
57 # Loop through each gaussian component and add the parameter values to the table
58 for i, model in enumerate(['g1', 'g2', 'g3']):
59     data_cells = table.rows[i+1].cells # get a collection of new cells
60     data_cells[0].text = f'{model}' # set the row label
61     # Loop through the parameters and print their values in the corresponding cell
62     for j, parameter in enumerate(parameters):
63         data_cells[2*j+1].text = f'{result.params[f"{model}_{parameter}"].value:.2f}'
64         data_cells[2*j+2].text = f'{result.params[f"{model}_{parameter}"].stderr:.2f}'
65         data_cells[2*j+1].paragraphs[0].alignment = WD_PARAGRAPH_ALIGNMENT.RIGHT
66         data_cells[2*j+2].paragraphs[0].alignment = WD_PARAGRAPH_ALIGNMENT.RIGHT
67
68 # Open a file dialog to pick the file to save to
69 quickPopupMessage(message = 'Choose a filename for the Word document.')
70 file = quickSaveFilename(title = 'Please choose a filename to save as.',
71                          filetypes = 'Word Documents, *.docx')
72
73 # Save the document
74 doc.save(file)
```

14.4.0 LINE-BY-LINE EXPLANATION

7–11: Import the libraries we need. All but the last of these import statements are specifically for working with docx files.

7: This will allow to create and write the document object.

8: "oxml" stands for "Office Open XML". In turn, "XML" stands for "eXtensible Markup Language," which is a way to supply formatting and type tags to data. Essentially this line lets us handle the internal structure of the .docx.

9: This line allows us to handle numeric data in .docx format.

10: This line allows us to control alignment with in the .docx.

12: Because the point of the `importFromPy()` function is to import arbitrary Python code, we add its usage right up against the import block. We are importing the script from Chapter 6. We are importing the same script as used in the other report builder in this chapter, so refer to the line-by-line comments there for more information.

15: We create a document object, and assign it to the variable doc.

18: We add a level 1 (i.e., highest level) heading object and assign it to the variable title.

19: We create a run object that will hold the text for the title. We create this up to the first special formatting we encounter, which will be a subscript 2.

20: We create a separate run object, that holds just the numeral 2. Although this is added to the title run, the fact that it is also a separate object means we can format it separately.

21: We now take the run that has the 2 and tell it to be subscripted.

22: We then finish up the title.

25: We now start a paragraph with normal text, assigning it a variable.

26 & 27: We go through the same process of adding in the subscript 2 that we used in lines 20 and 21.

28: We finish up the paragraph.

31: We create a list of three strings that are the words we will use for the heading of our table.

32: It will prove convenient to know how many components we want to report, and so we set a variable for this. Although we often try to avoid hard coding of variables, here it is okay because we are only using this to generate a single specific table. We set this to 3 because we only want to report the parameters for the Gaussian components, and there are three of them. You can refer to Chapter 6 for the details of the model we used.

33: We also set the number of columns we will need, which is two times the number of parameters. We multiply by two because we want to have a column for each parameter value, and then a column for the uncertainty in each parameter value.

34: We create a table with the number of rows and columns we want. We have one extra row and column because we also need headings for these.

35: We set the style of the table to a style pre-defined in the standard for .docx documents.

38: We select the cells of the first row of the table.

39: We add text to the first cell in this row, and we add the text "Component" to the cell.

40–42: We start a **`for` loop** that will iterate through our parameter names and add them to the cells in the header row of our table. We use the `enumerate()` function to count which row we are working on, and we have to add one because the first column is reserved for component names. We add the parameter name, and then the parameter name + "uncertainty."

45–55: We are going to apply formatting to the header row cells now. We want the header row to be filled with gray and the rest to be white. Also, the text in the top row should be bold. We are not going to explain the inner workings of this here, but you can refer to the documentation for the docx library. Or you can have a nice conversation with an AI model, like ChatGPT.

58–66: We now write the rest of the data. We are going to employ an approach like that used in lines 40–42 for the header, just looping through more values to write. We add one to the row number to account for the header row (row 0) and either one or two to the column number to account for the Component column (for value columns) or the Component and value columns (for uncertainty columns).

69: We use a popup to tell the user that we want to save this word document. We then open up native file explorer (using `quickSaveFilename()`) and have them identify the name and folder to save in. We restrict the filenames to those that will end in .docx—we want this to be a file that Microsoft Word will automatically recognize.

74: We write the docx file.

Composition of the Ni(pz)2 MLCT band

Parameters extracted from the spectral decomposition of the MLCT band for Ni(pz)2 into Gaussians contributions.

Component	amplitude	amplitude uncertainty	center	center uncertainty	sigma	sigma uncertainty
g1	17.38	1.95	542.77	2.61	25.00	2.42
g2	24.69	2.58	579.52	0.51	19.42	0.66
g3	27.54	0.89	621.17	0.64	21.21	0.29

FIGURE 14.4. Screenshot of the content of the Microsoft Word document produced by the above code.

You run this code the same way as the CSV code you just made, and see a new .docx file appear in your folder. When you open it in Microsoft Word,[2] you see a document like that shown in Figure 14.4.

You simply copy and paste this table into your SI document. Now, you can just rerun this code for each file and avoid having to transcribe all of these fit results into tables by hand!

[2] At the time of writing, Google Docs and Apple Pages successfully open the file, too.

CHAPTER 14 EXERCISES

TARGETED EXERCISES

Exercises for Section 14.1.1

0. Produce a narrative fit report that explains the model, the fit approach, and the results for the linear fit produced by the final code in Chapter 5.

Exercises for Section 14.3.0

1. Write out the slope and intercept values from the final code in Chapter 5 as a table in Microsoft word.

COMPREHENSIVE EXERCISES

2. We have just learned we can open files and write them using the `with open()` construction. We could have done this to read files as well. For well-structured files, functions like `genfromtxt()` are quicker and more efficient. However, there are times that you will need to use the `with open()` to process a file line-by-line. This is done using:

```
with open ( < file > , 'r ') as f : # the 'r ' indicates the file is open in read mode
    for line in f : # this reads the file line by line
   < do what you want with each line >
```

With this knowledge, go back and re-create the plot we made at the end of Chapter 2, but import the data from the file using the `with open()` construction.

3. Starting with the code written at the end of this chapter, modify it so that it also includes the image of the fit result that is produced by the code. Include a helpful caption. You can use the Spyder plots window to save the image to a file.

CHAPTER 15

Reading, Editing, and Writing Excel Files

- ☐ **New concepts**
  - — *and their prerequisites*

- ☐ **Reading and writing excel files using openpyxl**
  - — *Section 0.1.6: Expressing data as readable text using strings*
  - — *Section 2.1.1: Importing subsets of libraries using* `from`
  - — *Section 5.1.5: Expressing data as readable text using strings*
- ☐ **Specifying colors using the hexadecimal standard**
  - — *Section 5.1.5: Expressing data as readable text using strings*
  - — *Section 4.1.5: Specifying colors using the rgb standard*

15.0 PROBLEM

Things continue to go well! You have taking a large amount of good data, and your advisor is excited about the results.

There is a non-scientific problem now though... Your advisor is asking for results formatted in very specific ways. For instance, this morning, you received an email from your advisor with a Microsoft Excel file containing the data from your titration and a plot of it (Figure 15.1).

They are then asking that you add in the results of your fitting to this in Column C, as well as the results of that fitting to the plot. They also want the extracted parameter value in the purple boxes, and they want you to preserve the colors they have, as well as color the new cells a reddish color.

"This is so frustrating! Now I have to spend an hour or so just moving information around. Like, do I copy it from the variable explorer in the Spyder IDE? I thought I was trying to avoid all this point-and-click manipulation."

Your labmate cracks a small smile and says, "Do you really think I spend my afternoons copying things into Excel? Instead, I let Python do that. I can read data from Excel, create or modify Excel files, and save all the info I want directly into Excel sheets. I can use formulas and include plots. It is actually really quite simple."

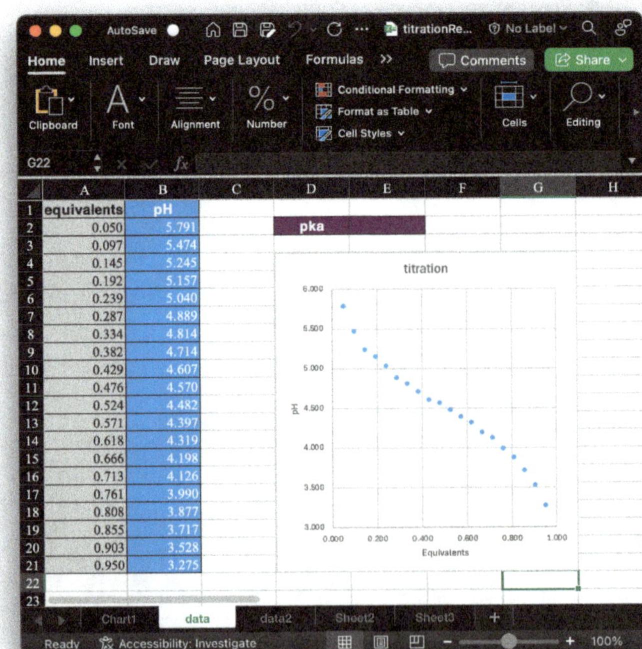

FIGURE 15.1. Screenshot of the Excel file that you wish to modify.

You sit there, thinking of all the time you have spent simply copying, pasting, and formatting data into Excel. You get a glimpse of the things you could be doing instead. Planning experiments, hanging out with friends, cooking...

With some relish, you ask your labmate to see their script and have them walk you through a solution to your problem.

> **A word on Excel**
> Allow us to get on our soapbox for a moment and discuss the use of Microsoft Excel. This program currently dominates the quantitative workflows of experimental chemistry. There are good reasons for this. It presents a nice grid-based appearance of data. It can be used to create plots, and juxtapose the plots with the data. It can be used to perform fairly sophisticated analysis (indeed, at the time of this writing, you can include Python in Excel on Windows machines... so it can do *anything* we show in this book). Certainly, it has enhanced many a working chemist's analysis.
>
> At the same time, Excel uses a proprietary format, and the standard interface (Microsoft Excel) costs a significant amount of money. It usually requires intensive manual user input, which can significantly slow many workflows and increase the chance of errors. Working with very large data sets can be especially slow. At the moment, the web-based version of Excel lacks **many** features of the one that you can install on your computer. And, it takes work to make Excel charts appear up to the standards of scientific publications. For these reason, we believe it is worth considering moving beyond Excel, especially when you need to archive your data. But, Excel remains a standard and is useful enough that it is worth learning how to work with it programmatically.

15.1 SOLUTION

15.1.0 PLANNING

"Okay," your labmate says. "Here is what you are going to want to do:"

1. Understand the structure of the Excel file you start with. This way, you can know where to read data from and where you can write data to. In this case, you have data in Columns A and B.

2. Identify the structure you want for your final Excel file. Basically, what do you want to report and how do you want to structure it? This includes any formatting you might like to apply, and you decide that:

   - You want to write the results of your fitting into Column C. You want to color these cells red and outline them to match the other cells.

- You also need to add in the extracted pK_a and its uncertainty into the purple cells. You will need to ensure the text is white in order to see it against the purple.

- Finally, you will want to modify the existing plot to add in the fit as a line that is color-coded to the fill of the cells.

3. You want to identify how you will get all this information. Because you are working with a file that has data, you can extract the data from the file and then fit it using the fitting routine you already wrote.

4. Once you have the data and the fit, you can modify the Excel file.

"Of course, there are a few new things to learn."

15.1.1 READING AND WRITING EXCEL FILES USING OPENPYXL

We are going to want to write an Excel file, but this is not as simple as writing a text file. The reason is that there are all sorts of useful things that Excel files can do that simple text files cannot. For instance, we can fill cells, add colors, add formulas, add data visualizations, and so on. This all means that the data stored in an Excel file is encoded in a way that is not easily understood by humans.

Fortunately, there are Python **libraries** that can handle interfacing with Excel files. We are going to use openpyxl, and probably the simplest way to illustrate its usage is just to write the code we need for this chapter. As a note, we are going to assume you have some familiarity with Excel, specifically with entering data, entering formulas, and creating charts. If you have not yet learned to do this in Excel, then you may benefit from reading or watching some basic Excel tutorials. Also, it is worth a quick discussion of the structure of an Excel file, to make sure we are on the same page.

If you open up a Microsoft Excel file, you will notice a few things:

- Across the top, the columns are identified by capital letters: A, B, C, ..., AA, AB, and so on.

- Across the left side, the rows are identified by integers, starting with the number one.

- On the bottom there is a tab for the current sheet, and it contains the name of the sheet. By default this is "Sheet1", but you can change it.

- There can be many different sheets in the same file. By default, there is only one, but we can add more. The collection of sheets is known as the Excel workbook.

Mirroring this structure, the Excel .xlsx file (and thus OpenPyXL's internal structure) is organized in a Workbook→Worksheet→Cell hierarchy. An overview of this hierarchy is shown in Figure 15.2.

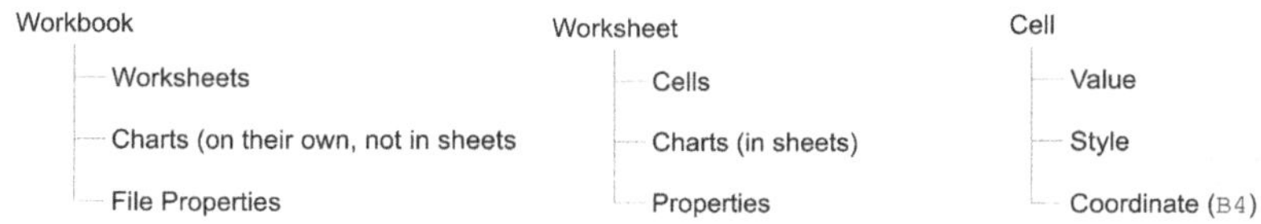

FIGURE 15.2. A brief overview of the organization of an Excel file.

"Excel holds values, formulas, and so on within individual cells, and these cells have addresses that are given by identifying the workbook they are in, the sheet they are in, as well as the column and row, in that order. You can think of this a bit like a directory structure, where the address of any cell is given by: <workbook><sheet><COLUMN-row#>. Each individual cell also can have formatting applied to it. We can add a fill color, change the color of the text, add borders, and a range of other things. All of these can be thought of as attributes of the cell. Additionally, we can also add 'charts,' which is Excel's name for data visualizations. These are used to plot data referenced in the workbook."

15.1.2 SPECIFYING COLORS USING THE HEXADECIMAL STANDARD

When you were making colored chemical drawings, you saw that we can specify colors using the rbg or rbga standards.[1] This is not the only standard, and here we are going to have to use **hexadecimal** representations of colors. These are actually easier to understand if we first go back and consider the rgb standard. In the rgb standard, we provides values for the red, green, and blue channels. These span values that often run from either 0 to 1 or from 0 to 255.

[1] See Section 4.1.5.

For people new to coding, 255 might seem like an arbitrary number, but it is not. Computers work in base 2. That is, computers only have two digits to work with: zero and one. In this base, the number three is represented as 11. If you have two number places, you can represent up to four numbers: 00, 01, 10, and 11. If you had three number places, you could have eight numbers. In general, you can represent 2^n numbers, when you have n number places. Thus, powers of 2 have special significance, and it turns out that there are 256 values of 0 to 255, or 2^8 numbers.

If there are 2^8 numbers, then there are also 16^2 numbers. You can see this by noting that $2^8 = 2^{4+4} = 2^4 \times 2^4 = 16 \times 16 = 16^2$. If we made a square grid that was 16 numbers on a side, then we could put all the values from 0 to 255 in them. If we then assigned each row and column an identifier, we could represent each of the 256 values using just two identifiers—which might prove quite convenient.

Easy identifiers would be numbers, and because there are 16 rows and 16 columns to identify, we can use the hexadecimal system, which contains the digits 0, 1, 2, 3, 4, 5, 6, 7, 8, 9, a, b, c, d, e, and f. And now, we can represent the full range of 256 values using only two numbers. In this approach, we can run from a value of 0 (`00`) to a value of 255 (`ff`).

Remembering that we need values for red, green, and blue, we just repeat this thought process for each, meaning that we can simply put together three pairs of such identifiers to get the full color designation. This will mean that each color in the rgb

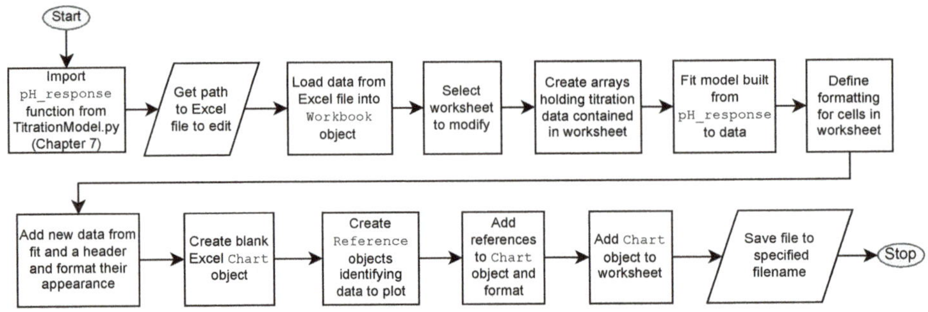

FIGURE 15.3. Flowchart for code that will read in data from an Excel file and then modify this file to include the results from Chapter 6.

standard can also be specified using six digits of hexadecimal number, or three pairs of digits. The first pair is the red value, the second the green value, and the third the blue value. It is convention to indicate this is a hexadecimal number by placing the # sign in front. So, #000000 is black (no intensity of any color), #ff0000 is the brightest red (full intensity red, no intensity of any other color), #ff00ff is magenta (full intensity of red and blue), #ffffff is white (all three primary colors at their full intensity), and so on. There are online converters to go between rgb and hexadecimal colors, or you can use color choosing tools in Microsoft Office or other programs. It is also worth noting that hexadecimal colors are sometimes called HTML because they are widely used in designing web pages.

15.2 CODE

"And now we are ready to produce the Excel file for our advisor! Of course, you can write the file one bit at a time," your labmate says. "If you know how to write the information for the pH titration data, you really know how to write the data for the UV-vis data as well. So why don't I show you how to write the titration sheet and then you can try to do the rest on your own?"

You work together to produce a flowchart (Figure 15.3) and the associated script.

```
'''
Fit data in an excel file and update the file with the results
    Requires: existing titration data excel file
    Written by: Ben Lear and Chris Johnson (authors@codingforchemistsbook.com)
    v1.0.0 - 250508 - initial version
'''
import numpy as np
from lmfit import Model
from openpyxl import load_workbook
from openpyxl.chart import Reference, Series
from openpyxl.styles import PatternFill, Border, Side, Font
```

```
from codechembook.quickTools import importFromPy, quickOpenFilename, quickSaveFilename,
    quickPopupMessage
importFromPy('TitrationModel.py', 'pH_response')

# Read the excel file and go to the right worksheet
quickPopupMessage(message = 'Select the Excel workbook containing the data.')
filename = quickOpenFilename(filetypes = 'Excel Workbooks, *.xlsx')
wb = load_workbook(filename) # Load the workbook into Python
ws = wb['data'] # Select a worksheet

# Get the data from the columns of the worksheet row by row
exp_eqs, exp_pHs = [], []
for eq, pH in zip(ws['A'], ws['B']):
    try:
        exp_eqs.append(float(eq.value))
        exp_pHs.append(float(pH.value))
    except:
        pass

# Set up a new model with the pH_response function and 'eqs' as the indpedent variable
pH_model = Model(pH_response, independent_vars=['eqs'])

# Set up the fit parameter with the average pH as the initial guess
pH_params = pH_model.make_params()
pH_params.add('pKa', value = np.mean(exp_pHs)) # specifications for the parameter

# Fit the model to the data
pH_fit = pH_model.fit(data = exp_pHs, eqs = exp_eqs, params = pH_params,  base_i = exp_eqs
    [0])

# Specify the colors and borders to use in the worksheet
font_color = Font(color='FFFFFF')
pink_fill = PatternFill(start_color='FF9194', end_color='FF9194', fill_type='solid')
cell_border = Border(left=Side(style='thin'),
                     right=Side(style='thin'),
                     top=Side(style='thin'),
                     bottom=Side(style='thin'))

# Label the header cells and format
ws['C1'] = 'fit' # column heading
ws['C1'].fill = pink_fill
ws['C1'].border = cell_border
ws['D1'] = 'parameter'
ws['E1'] = 'value'
ws['F1'] = 'uncertainty'

# Loop through each value in the best fitting line and add it to the corresponding cell
for i, value in enumerate(pH_fit.best_fit):
    ws[f'C{i+2}'] = round(value, 3) # set the value
    ws[f'C{i+2}'].number_format = '0.000' # ensure we have three decimal places showing
    ws[f'C{i+2}'].fill = pink_fill
    ws[f'C{i+2}'].border = cell_border

# Add the best fitting parameters to the corresponding cells and format
ws['E2'] = f'{pH_fit.params["pKa"].value:.3f}'
ws['F2'] = f'{pH_fit.params["pKa"].stderr:.3f}'
```

```
66 ws['F2'].font = font_color
67 ws['F2'].fill = PatternFill(start_color='51154A', end_color='51154A', fill_type='solid')
68
69 # Get a variable for the chart in the worksheet
70 chart = ws._charts[0]
71
72 # Create reference to the data to be plotted in the chart
73 x_values = Reference(ws, min_col=1, min_row=2, max_col=1, max_row=len(pH_fit.best_fit)+1)
74 y_values = Reference(ws, min_col=3, min_row=2, max_col=3, max_row=len(pH_fit.best_fit)+1)
75
76 # Create a data series for the data
77 fit_data = Series(y_values, x_values, title_from_data=False)
78
79 # Set the data series line color and width
80 fit_data.graphicalProperties.line.solidFill = 'FF9194'  # Red color
81 fit_data.graphicalProperties.line.width = 28575  # Width in EMUs (1 pt = 12700 EMUs)
82
83 # Add the fit line to the chart
84 chart.series.append(fit_data)
85
86 # Save the changes back to the same file
87 quickPopupMessage(message = 'Choose a filename to save the new Excel workbook.')
88 save_filename = quickSaveFilename(filetypes = 'Excel Workbooks, *.xlsx')
89 wb.save(save_filename)
```

You run this code, point it to where to find TitrationModel.py, select the file "titrationReport.xlsx" in the Catalyst/Titraton/pH folder in your computer, and it outputs an Excel .xlsx file. Opening the file, you see a document like that shown in Figure 15.4.

"I guess this is how you get so much done," you say to you labmate. "You don't spend your time on little annoying things like this!"

15.2.0 LINE-BY-LINE EXPLANATION

7–12: Import the libraries that we will need. In addition to the familiar `numpy` and `lmfit` libraries, we also have three imports for dealing with Microsoft Excel. The first is a function we use so we can load workbooks. The second imports components of the module that deals with charts, while the last imports components that allow us to style the cells in the workbook. We will also use `importFromPy()`, `quickOpenFilename()`, `quickSaveFilename()`, and `quickPopupMessage()` from `codechembook.quickTools`.

16–17: As the user to find the Excel workbook containing the data and store the path to that file.

18: We use the `load_workbook()` function and provide it the path to the Excel workbook we wish to use.

19: We then select the sheet within the workbook we want to use. We can reference this using a key-type formulation, referencing the name of a sheet within

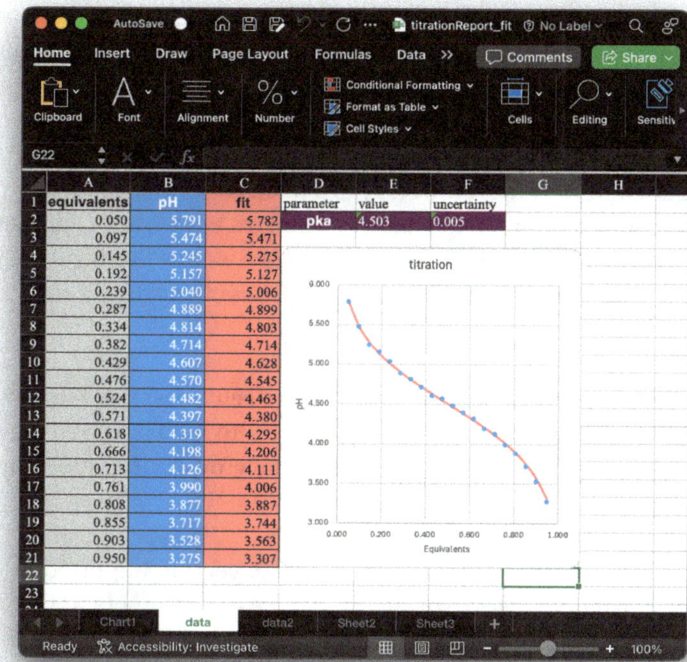

FIGURE 15.4. Screenshot of an Excel file after running the code above. This added results from Chapter 6 and formatted the Excel file, all using Python.

the workbook. If this were a default Excel workbook, then the name would be "Sheet 1." However, the sheet had been renamed by your advisor, so we reference it using this name.

22: It is time to get the data from the Excel workbook. We want two empty lists that we can add data to as we read it cell by cell. We will eventually fit the titration model that we developed in Chapter 7 to this data. We create these lists by having two empty lists that we unpack to two different variable names.

23–28: We want to pull the data from the Excel file, and it is held in Column A and Column B of the worksheet. We can reference these columns using them as a key. The default behavior is to return an **iterable** collection of all the *cells* that have values in them. Because Column A and Column B have the same number of entries, they will return collections of the same length. Thus, we can synchronize our iteration through these collections using `zip()`. We assign the values we unpack to two variables: `eq` and `pH`.

24: We are going to want to add the data as **floats** to our lists. The first entry is not a number (it is a heading), and so we are not going to be able to convert it to a float. One way to avoid this would be to slice out the first value in the collection (i.e., `ws['A'][1:]`). However, we would need to change the slice if the structure of the worksheet changed. A more robust way of handling this would be to use the `try... except...` **keyword**. Inside the `try` block, we attempt to convert the entry to a float. If this works, then we will end up appending a number. If it fails (as when we have strings and not numbers), then we will enter the `except` block, where we use the keyword `pass` to continue with the loop. In this way, we can handle the header without needing to slice.

22–23: For these lines, we get the value in the cell using the `.value` **attribute** of our cell. We then attempt to convert this to a float. If we can convert to a float, then we end up appending it to the growing list holding our values of interest.

31: Using the `Model()` function from `lmfit`, we convert our imported `pH_response()` function to a model object, specifying that the `eqs` parameter is the independent variable for the model.

34–35: We create a parameters object and then add the fit parameter to it. This is the same as was done in Chapter 7.

38: We perform the fit of the `pH_response()` function to the data we took from the Excel file. The usage is the basically the same as was shown in Chapter 7, except here, we obtain the initial amount of base from the first entry in our equivalents list and use it directly as an argument in the function call rather than initializing a new variable for it. At this point, we have the fit result object and are ready to add it to our Microsoft Excel worksheet.

41: Using the `Font()` function, we supply a color we wish to use for the font of the cell's text, and then assign this to a variable. The color is white and the `Font()` function represents this in a way that Excel can use internally. In this way, each time we want to use this color, we can simply use the variable.

42: Using the `PatternFill()` function, we create a pink-ish color for the best-fitting line and the cell area for the fit data.

43–46: Using the `Border()` function, we generate information about the border we want and assign that to a variable. This way, we can use the border construction simply by using the variable name. In this case, we are using a series of **keyword arguments** to specify the formatting of the border on the four sides. Each of these keyword arguments is also a function that accepts a keyword argument about the style of the border. In this case, we make all borders thin. The reason to include these borders is that when we color in the fill of a cell, we will no longer see the default borders that Excel shows. So we have to add in our own.

49: Using the address of the cell as a key for our worksheet, we assign the string `'fit'` to the cell. This will serve as a header for the fit values in column C.

50: We set the `fill` attribute of the cell to our fill color variable. This will fill the cell with that color.

51: We now use the `border` attribute of the cell and set it equal to our previously created border variable.

52–54: The cells that hold our parameter value and uncertainties did not have headings. We use these three lines to add them.

57–61: Use `for`—`enumerate()` to go through each item in `pH_fit.best_fit`, which is the best-fit value for each dependent variable data point (i.e., equivalents added). We wish to add each variable to a new cell, moving down the row. By using `enumerate()` we will get indices for each value in `pH_fit.best_fit`. We can use this to assign values to the correct cells in the column. We will unpack the index and the value into two variables: `i` and `value`.

58: We add the value to the cell. We use **f-string**s to convert the value of the index to a number that we can use in the key. We add two to the value of the index because index counting starts at zero, and we wish to add the first value to the second row (counting in Excel starts at one).

59: We set the format for rounding the number in the cell using the Excel format string, which gives a pattern to follow using only zeros. Here, we want three decimal places, so we use "0.000" to pattern it.

60: We assign the fill color to the cell.

61: We assign the border to the cell.

64: We next assign the value of the best-fitting parameter to the cell E2. We do this inside of an f-string, using the `value` attribute of the Parameter object representing pK_a in the `params` dictionary. We also use the formatting `:.3f` to make the value a float that has three decimals of precision.

65: We repeat the above, but for the uncertainty in the parameter, using the `stderr` attribute.

66–67: The cell F2 was not yet purple, and the font was not yet white. So, we make sure the font color will be white (so we can see it against the dark purple) and that the cell is dark purple. For the pattern, we directly use the `PatternFill()` function, without assigning it to a variable.

70: Before we can add anything to the chart, we have to identify where it is. This is done by accessing the collection of charts and pulling out the one we are interested in. In our sheet, there is only a single chart, and so we just get the first item in the list of charts. For convenience, we assign this to a variable.

73–74: In order to add data to the chart, we need to identify what that data is. We do this by setting up a reference to the data, using the `Reference()` function. This function accepts a **positional argument** that is the worksheet where the data resides, followed by four keyword arguments that specify the minimum and maximum row and column addresses for the data. This defines a 2D grid of data that will be used as a reference. For the x-values and y-values, we only need a column, and we start at the first row and end at a row that is equal to the length of our fit data + 1. The +1 is again because Excel starts counting rows at one.

77: We create a series, which is an object that can be added to a chart in Excel. We use the `Series()` function for this, supplying two positional arguments (y-data and then x-data) as well as if we want to include a title (i.e., a legend entry). In this case, because we are color-coding the data and the line in the chart, we elect not to include the title in the legend.

80–81: We format the series we have created. In line 80, we specify that we want a line with a solid fill, and then provide the hex code for the color. In line 81, we specify the width of the line in EMUs, which is an internal measurement used in Microsoft Office.

84: We add the series to the chart by appending it to the series attribute of the chart object.

87–89: We save the workbook. First, we ask the user to pick a file name, then we use the `save()` method of the `Workbook` object to save a new file. The supplied positional argument is the path to the file to be saved. If you used the same path as was used when you opened the file, then it will be overwritten (i.e., normal behavior when saving an Excel file).

CHAPTER 15 EXERCISES

TARGETED EXERCISES

Exercises for Section 15.1.1

0. Go get the "Stress-Strain.xlsm" file from our website (https://www.codingforchemistsbook.com/). Then

   a) Extract the name of the all the sheets in this file.

   b) Get the value of the cell B3 for the first sheet.

   c) Transpose the results in the first sheet and save the results in a new sheet.

Exercises for Section 15.1.2

1. Find the colors associated with the logo of a school you have attended or a organization you have worked at and express them in hexadecimal.

2. Change the final code of Chapter 3 to specify line colors using hexadecimal values rather than rgb color designations. Verify the code still works.

COMPREHENSIVE EXERCISES

3. In this chapter, we saw how to handle results similar to that produced by the final code of Chapter 7. Go back to Chapters 5 and 6. Create a Microsoft Excel file, that has a sheet associated with each chapter. In each sheet, include the data used for the fit (i.e., the experimental y-values), the fit y-values, and the extracted parameters and their uncertainties. You will likely need to read the openpyxl documentation in order to understand how to create an Excel workbook and add sheets to the workbook.

4. Some journals will only accept supplementary information in PDFs or Word documents. This means you cannot simply upload a Microsoft Excel file or a .csv with your data. Write a program that will take the pH titration values from the Excel file used in this chapter, and then convert them to a Microsoft word table.

CHAPTER 16

Automating Basic Image Processing

☐ **New concepts**
— *and their prerequisites*

☐ **Understanding how computers represent images**
— *Section 1.1.3: Operating on collections of data efficiently with Numpy arrays*

☐ **Importing and manipulating images using the Pillow library**
— *Section 2.1.1: Importing subsets of libraries using* `from`

16.0 PROBLEM

The day you realize that word has gotten around about your coding skills is the day that a friend of a friend walks into your lab looking for your help. He works for a professor over in the Materials Science department and, although his project involves some chemistry, they are often dealing with images rather than spectra.

"Ummm..." he starts, "So Kim tells me that you have been learning coding. I was wondering if you could help me out."

He goes on to explain that he has been taking some thermal images of polymeric materials heated up by shining light on them. Although the images show what they want, the software on their thermal imaging camera does not provide the coloring options they want. He asks if you know how to recolor images in Python.

"Not yet!" you say. "But why don't you show me what you want, and I can do a bit of reading online and get back to you."

So, he opens up his laptop and shows you a few images (Figure 16.1).

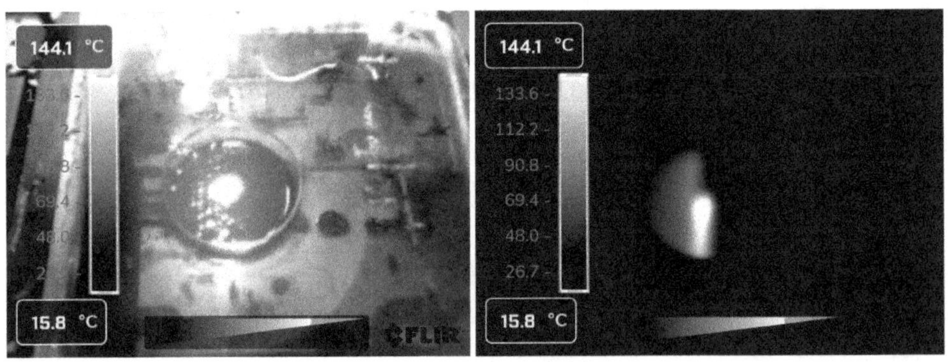

FIGURE 16.1. Images that the thermal camera outputs. On the left is the "visual" image—sort of a "normal" picture. On the right is the thermal image.

"Basically, we have a greyscale photo of our setup, and a greyscale thermal image during the experiment. The camera takes both and can overlay them. But the issue is that we don't want just greyscale colors, and the color options the camera provides are not great. For instance, it can provide a colorscale based on a rainbow, but we don't to use that colorscale. It would be great if we could use a colorscale based around reds, yellows, or oranges because people typically associate those colors with heat. Also, it would be nice if we could crop the images to remove the colorscale from the image. We want just the part that is hot to be shown." He looks at you, clearly a bit nervous to be taking up your time. "Ummm... and yeah, I can do all of this in Photoshop, but it takes so much time for each image. I have *hundreds* of these images, and I don't want to lose an entire weekend to this. I know its not your problem, but if you could help out, then I could pay you back in beer or pizza or something."

"Well, let me do a bit of Googling and see what I come up with. Don't worry about it too much, I am interested in learning to process images. I'll let you know if you owe me anything after I see how long this takes."

16.1 SOLUTION

Even though you don't quite know how to do this in Python, you can still come up with a general plan of what you will need to do:

1. Import the image into Python so you can work with it.
2. Change the image data to be the colors you want.
3. Crop the image.
4. Save the image.

So, really you only need to figure out how to do these four steps, and you are done.

> **Where to next?**
> Here, we use Pillow to work with images because it is both simple and fast. If you find yourself wanting to accomplish things that Pillow cannot, we recommend moving to OpenCV next. This is a very powerful library, but does have a more complex interface and can be slower than Pillow. Thus, for simple tasks, the authors of this book turn to Pillow first.

With just a small amount of searching on the internet,[1] you find that there are several different libraries that are dedicated to working with images in Python. These include Pillow, OpenCV, and scikit-image. Of these, it seems that Pillow is the simplest to use and should be adequate for the tasks you want to accomplish. Also, it seems that Pillow directly handles steps 1, 3, and 4 above.

[1] As AI chatbots like ChatGPT emerge, they will likely be a much better choice for finding this information.

Before you dive in, you need to understand how images are structured, and also introduce a few features of the Pillow library.

16.1.0 UNDERSTANDING HOW COMPUTERS REPRESENT IMAGES

The very first thing you learn about images is that there are two general types: **scalable vector graphics** and **bitmaps**.

Vector graphics break apart images into components and then represent each component via an equation that is solved in order to draw the image. For instance, if we wanted to draw a smiley face, we could include the equation for a large circle, two small circles, and an arc. Then we just place these in the appropriate relative positions, and we have the image we want. The advantage of this approach is that our image can be scaled to any size and the equations are the same, so that our image always looks crisp and sharp. The disadvantage is that not all images can be readily broken apart into such simple components. For instance, consider a photograph of a person. There is simply too many different shapes, colors, and so on to easily represent as a series of equations.

Bitmaps represent images by breaking them down into small bits, called pixels, and then specifying the color of each of these pixels in turn. For instance, imagine we had only black and white. We could represent an image as a series of zeros (black) and ones (white). This approach provides explicit directions on how to draw our image. Thus, if we wanted to draw a smiley face (Figure 16.2), we could represent a series of black-and-white pixels in the shape of this face. Already, you can see a disadvantage of this approach: it is hard to represent curves well. In real implementation, we typically use much larger densities of pixels, and grays and the edges to make a curve look smooth. Additionally, if we resize a bitmap image after making it, we cannot know how to color any new pixels that might occur, and so we will always preserve the blocky nature of our image. The advantage, however, is that we can represent very complex images in a straightforward manner.

Pillow is designed to work with bitmaps.[2] We are using Pillow in this chapter, because we are working with images taken with a camera, which are almost exclusively saved as bitmaps. This is important to realize because knowing the type of image we have dictates the way we wish to work with it. In particular, because bitmaps are just a series of pixels, we can represent the information in these pixels as a 2D array of numbers—in other words, Numpy arrays can be useful to use.

[2] If you want to work with vector graphics in Python, consider using svgwrite (for creating and modifying svg files) or cairosvg (for converting from svg to bitmap formats).

Understanding that you are working with a bitmap that will ultimately be represented as a series of arrays is just the first step. Next is to decide what bitmap format

```
[
[1,1,1,1,1,1,0,0,0,0,1,1,1,1,1,1],
[1,1,1,1,0,0,1,1,1,1,0,0,1,1,1,1],
[1,1,1,0,1,1,1,1,1,1,1,1,0,1,1,1],
[1,1,0,1,1,1,1,1,1,1,1,1,1,0,1,1],
[1,0,1,1,1,1,1,1,1,1,1,1,1,1,0,1],
[1,0,1,1,0,0,1,1,1,1,0,0,1,1,0,1],
[0,1,1,1,0,0,1,1,1,1,0,0,1,1,1,0],
[0,1,1,1,1,1,1,1,1,1,1,1,1,1,1,0],
[0,1,1,1,1,1,1,1,1,1,1,1,1,1,1,0],
[0,1,1,1,1,1,1,1,1,1,1,1,1,1,1,0],
[1,0,1,1,0,1,1,1,1,1,1,0,1,1,0,1],
[1,0,1,1,1,0,1,1,1,1,0,1,1,1,0,1],
[1,1,0,1,1,1,0,0,0,0,1,1,1,0,1,1],
[1,1,1,0,1,1,1,1,1,1,1,1,0,1,1,1],
[1,1,1,1,0,0,1,1,1,1,0,0,1,1,1,1],
[1,1,1,1,1,1,0,0,0,0,1,1,1,1,1,1]
]
```

```
[
[6,4,6],
[4,2,4,2,4],
[3,1,8,1,3],
[2,1,8,1,2],
[1,1,b,1,1],
[1,1,2,2,4,2,2,1],
[0,1,3,2,4,2,3,1]],
[0,1,d,1],
[0,1,d,1],
[0,1,d,1],
[1,1,2,1,6,1,2,1],
[1,1,3,1,4,1,3,1,1,],
[2,1,3,4,3,1,2],
[3,1,8,1,3],
[4,2,4,2,4],
[6,4,6]
]
```

FIGURE 16.2. A black-and-white bitmap image on the left, with a representation of this using arrays of ones and zeros in the middle, and the same data under a simple compression scheme on the right. To some extent, all bitmap images employ similar encoding.

to use. There are a large number of options, but the four most common are tiff, bmp, png, and jpg (or jpeg). Of course, this begs another question: if bitmaps are just lists of numbers specifying the color of pixels, why do we have so many different formats?

The short answer is that most bitmaps are not simple lists of pixel colors, because this can be quite wasteful of memory. Consider our bitmap of our smiley face (Figure 16.2), and imagine we were reading this row by row. There are several places where we have multiple black spots that follow one another. The top row starts with six white pixels followed by four black ones. One way to represent this is 1,1,1,1,1,1,0,0,0,0. But another way might be [6,1,4,0], where the first number tells us how many times to write the second number. Or you save even more room by knowing that there are only two options for color (zero and one) so you could just list the numbers of each number, also starting with the number of ones you need. In this case, we would write this sequence as 6,4. Looking at the image, we can see that there are instances where there are 12 or 14 numbers in a row, so using **hexadecimal**, you can replace these with b and d, respectively. For our smiley face, if we apply this final scheme, we get the result shown on the right of Figure 16.2. This process is known as encoding. When chosen properly, we can use a smaller amount of data to represent the same information—a concept known as compression.

Our above encoding led to lossless compression—we produce the exact same image. However, there are ways to encode that might lead to loss. For instance, what if we took a 3×3 grid and represented it by a single number: the dominant color? That would lead to a smaller image, but one that is not accurate. We call this compression "lossy."

The different image formats use different compression encodings, and can have different amount of information loss. tiff images are lossless, but are very large as a result. jpg images are small, but often lose too much information—especially in a scientific context.[3] However, png files provide reasonable preservation of data while making smallish files. Thus, we prefer png files. Of course, as is the case below, sometimes instruments or cameras produce jpg files, and so that will be your only option. Also, we note that tiff is nice to use when you really want to preserve everything, and this has long been the standard for serious graphic designers. However, no matter the format you use, we can always end up representing an image as a series of numbers.

[3] Back when the internet was young, it was also very slow, and so passing graphical data could take a long time. Thus, there was a premium on compression and people were more willing to accept loss of data. This is why jpg images often look "old."

16.1.1 Importing and Manipulating Images Using the Pillow Library

There are many modules in the Pillow library, but we can accomplish everything using the `Image` module. In particular, it provides five functions that we will use:

- `Image.open()` accepts a positional argument that is the path to an image file. It returns an image object that can be manipulated in Python. Thus, this function is the way that we gain access to the image. Once we have access to this image, we can do many things to it, like crop it, combine it with other images, or even get information on each pixel and convert that to Numpy arrays for further manipulation.

- `Image.fromarray( )` accepts a positional argument that is an array of numbers, and then converts these numbers into a Image object that can be further manipulated in Pillow. Combined with the `Image.open( )` function, we can see a convenient way to recolor images. We can open the image object, get the pixel data (which is just color information) as a Numpy array, change the values of the entries in the Numpy array, and then convert this array into a image object. This will effectively recolor our image.

- `Image.show( )` is used to export an image object. The `Image.show( )` **method** displays the image using the default output. It is a simple way to directly see the image you have created. Also, since it will open using a standard image viewer on your computer, you can save it from there. If you wish to save directly, there is an `Image.write( )` function, although we will not use it below.

- `Image.blend( )` is used to combine information from two different images. It accepts two positional arguments (the two image objects to be combined) as well as several keyword arguments. Here, we will use just one, `alpha`, which controls the opacity of the second image.

- `crop( )`, which is a method of an image object. This allows us to control how much we wish to remove from the left, top, right, and bottom parts of an image. We can use this to remove the parts that are not the focus of our image.

And with this information, you realize you have everything you need to recolor an image in Python!

16.2 CODE

You first sketch out a flowchart for the process (Figure 16.3) and then produce the script.

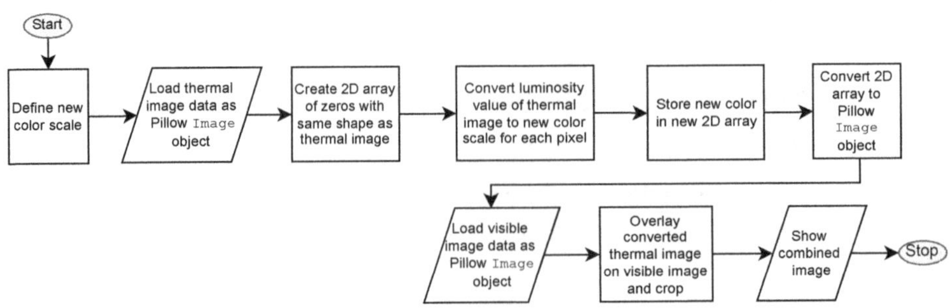

FIGURE 16.3. Flowchart for code that reads in the content of bitmap images, recolors the image, and then saves the recolored images.

```
'''
Recolor thermal images and add them on top of photos or other visual images
    Requires: two images in .jpg format
    Written by: Ben Lear and Chris Johnson (authors@codingforchemistsbook.com)
    v1.0.0 - 250527 - initial version
'''
import numpy as np
from PIL import Image
from codechembook.quickTools import quickOpenFilename, quickPopupMessage

# Create a new color scale for recoloring the pixels
new_scale = []
for i in range(3): # R, G, and B
    for j in range(256): # 8 bits each
        color = [0, 0, 0] # start with black
        for k in range(i):
            color[k] = 255
        color[i] = j
        new_scale.append(color)

# Open the thermal image and convert it to a 2d array
quickPopupMessage(message = 'Select the thermal image file.')
original_thermal_image = Image.open(quickOpenFilename(filetypes = 'Thermal Image file, *_thermal.jpg'))
original_thermal_data = np.array(original_thermal_image)

# Make a new 2d array of zeros to hold the recolored image
new_thermal_data = np.zeros_like(original_thermal_data)

# Determine the shape of the image in pixels x pixels
width, height = original_thermal_data.shape[0:2]

# Loop over pixels and set color of new one based on luminosity of thermal
for i in range(width):
    for j in range(height):
        current_color = original_thermal_data[i, j][0] # get the luminosity of the pixel
        new_thermal_data[i, j] = new_scale[current_color*3] # find corresponding color in new color scale

# Convert new data to image
new_thermal_image = Image.fromarray(new_thermal_data.astype('uint8'))

# Combine new thermal image with visual image and crop
quickPopupMessage(message = 'Select the visual image file.')
old_visual_image = Image.open(quickOpenFilename(filetypes = 'Visual Image file, *_visual.jpg')) #open visual image
result = Image.blend(new_thermal_image, old_visual_image, alpha=0.3) # Blend the images together
cropped_image = result.crop([0.4 * width,           # left
                             0.2*height,            # top
                             width - 0.1*width,     # right
                             height - 0.42*height])   # bottom

# Show final combined and cropped image
cropped_image.show()
```

You run this code, select the thermal file "inb_thermal.png" in your Thermal Imaging folder, select the file "inb_visual.png" in the same folder, and the image (Figure 16.4) appears on your screen:

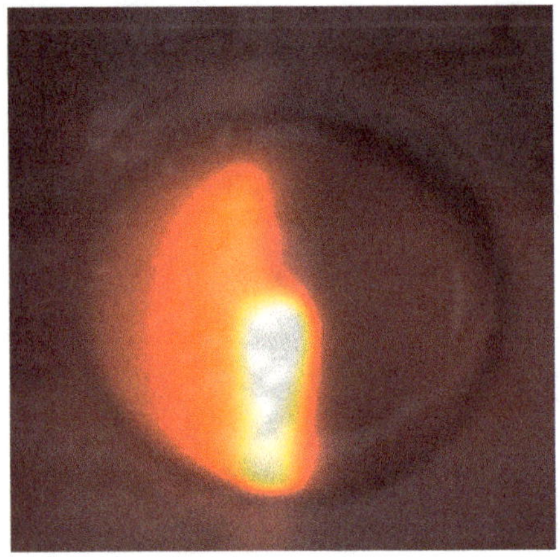

FIGURE 16.4. Results of running the code in this chapter.

You see that you can easily build the capabilities to select several different files and perform the same action on all of them. So, at this point, the problem is solved.

You show this solution to your friend, who says, "Wow, that is amazing! I really wish I knew how to do this."

"Tell you what," you say, thinking of the favor your labmate did for you, "I am free over lunch, so if you wanted to take an hour, I could give you a quick introduction to coding and get you to the point that we can start solving real problems for you."

16.2.0 LINE-BY-LINE EXPLANATION

7–9: Libraries we will need for this code.

12: We are going to need to build a new color scale, held in a list. So we create an empty list that will contain specifications for the colors we wish to map the greyscale onto when re-coloring.

13–19: We are going to use three nested **`for` loops** to build our color scale. Each entry in our color scale has three values, for red, green, and blue. The plan is to start at black [0, 0, 0], then go through to pure red [255, 0, 0], then through to pure yellow [255, 255, 0], and then to white [255, 255, 255]. This will give the impression of smoothly increasing the brightness.

13: We start with a for loop that will iterate over the numbers 0, 1, and 2. This will let us understand which index in our color specification we are currently increasing.

14: We now immediately start the next loop and iterate through the numbers 0–255. These are the numbers we are increasing.

15: We start with black [0, 0, 0].

16 and 17: We start the final loop. Here we iterate from zero to the index *prior* to the one that we are currently increasing. Within this loop, we simply set all the values we are not currently increasing to their maximum (255). The reason is that we have already increased them to their maximum.

18: We then change the value for the index we are increasing to the value we are currently on.

19: We then append this color to our list of colors. Once we complete all these loops, we will have a list 3×256 items long, which specifies smoothly changing colors from red through yellow to white.

22: We prompt the user with a popup message that tells them we want them to identify a thermal image (not a visual one).

23: We then use `quickOpenFilename()` to let the user point to the file we will use. We use the `filetypes` keyword to specify that the filename *must* end in "_thermal.jpg" which will ensure they do not open a visual image—provided the user is following the naming convention we expect. We then use the path object returned by this function and open it, and pass the resulting Image object to a variable.

24: We convert the Image object into a Numpy array. This will create a 2D array that contains a number of interior arrays equal to the number of pixels tall the image is, and each interior array is as long as the number of pixels wide the image is.

27: We create a 2D Numpy array that is exactly the same size as the one we made in line 24; however, we set all the values initially to zero. We assign this array to a variable. This is the array that we will recolor.

30: We use the `shape()` method of Numpy arrays to get the numbers of inner arrays, and the number of elements in each inner array of the 2D array. This corresponds to the number of pixels tall and the number of pixels wide the image will be.

We have to use a **slice** from zero to two because each pixel is represented by an element of the inner arrays, which is itself a three-item array (for rgb values). Thus, it is only the first two shape values that we want (just the number of inner arrays and the number of elements in them).

33–36: We start the process of recoloring the image by iterating over numbers associated with the width of the image.

34: We then start another for loop where we iterate over numbers associated with the height of the image.

35: Using the indices, we check the color for a pixel in the original image. The greyscale colors are specified by a list [r, g, b], but in which r=g=b, so we only need one of these values. We take the first one.

36: We then specify the color for the new thermal image at the same pixel. We set the color equal to the value in our `scale_bar` list. We simply pull the value at the index associated with 3× the greyscale value. The reason is that the `scale_bar` list is three times longer than the larger than a list associated with greyscale colors. Thus, we are simply re-scaling the greyscale onto our color scale. Once we complete the list, we will have a Numpy array that carries all the colors we want for our new thermal image.

39: We use the `fromarray( )` function from the `Image` module of Pillow to create a image object from the 2D Numpy array we just constructed above.

42: We now prompt the user to select the visual image that we wish to combine with the recolored thermal image.

43: We then use the same approach as on line 23 to get the path to the visual image, this time restricting the files so that they must end in "_visual.jpg" and then pass the path to a function that opens the image and pass the Image object to a variable.

44: We now overlay the two images, using the `blend` function from the `Image` module of Pillow. This accepts two positional arguments, which are the images to be blended, and we supply a keyword argument that specifies how opaque the second image should be. Here we set it to 30% opaque (70% transparent). The result is as if the second image was laid on top of the first, at that opacity.

45–48: We need to crop our image, using the `crop` method of the image object. This accepts four positional arguments that give the amount (in pixels) to crop the left, top, right, and bottom sides, respectively. It can be hard to know exactly how much to crop any given image in pixels, so we specify the fraction of the width and height we wish to crop each side, which we think is easier to estimate when looking at an image.

51: Finally, we show the image. This should bring it up in a new window on your computer. After it is shown, we can save it, or do anything else with it.

CHAPTER 16 EXERCISES

TARGETED EXERCISES

Exercises for Section 16.1.0

0. Create an array of ones and zeros where the arrangement represents a picture of a frog.

1. Modify the example code shown for Section 4.1.3 to save the molecular drawing as both svg and png images. This can be done by append the following to the end of the code:

```
Draw.MolToFile(acetate_mol, "acetate.png", size = (100, 100)) # saves a png

svg_data = Draw.MolToSVG(acetate_mol, size = (100, 100)) # generates svg specifications as a string

with open("acetate.svg") as a:
  a.write(svg_data)

```

Once you have the images, bring both of them into a Microsoft Word document or Powerpoint presentation (or similar software). Make both 3 inches wide. Change the colors of the bonds (if you can). Do you notice any advantages of svg or png?

Exercises for Section 16.1.1

2. Take a selfie. Then:

a) Turn the selfie into black and white.

b) Color all black parts red, and all white parts blue.

3. Using Pillow, combined with Numpy arrays, create a png image that is a smiley face.

4. Take a selfie and crop out your body, much like you would do to create a headshot.

COMPREHENSIVE EXERCISES

5. One of the old lab members in your group made a .png figure of some spectra. You want to include this image in an analysis of your current data, and you want to preserve the color scheme you are using in your publication. You don't have access to their data anymore. Write a script that will go through any such images and replace one color with another one. You can test with the file "old-spectra.png" found on the companion website (https://www.codingforchemistsbook.com/).

6. You also think it could be good to extract the values from the old spectra discussed in the previous exercise. You are lucky that the extremes of the x- and y-axes are labeled, so that you can simply crop the image to contain only the plot area and know the data values of the edges. This also means that the coordinates of the pixels in the cropped image are readily mappable to the values on the x and y axes, providing a way to extract the data values. This would let you re-plot this in any Plotly figure you want. Write a code that will crop this image, and then extract locations of the data and assign them the correct values. Save the extracted values to a .csv or Excel workbook. Then plot the resulting data in Plotly. Comment on how well this worked for extracting the data. Does the plot look similar?

PART FIVE

Moving Beyond This Book

CHAPTER 17

THINKING ABOUT YOUR NEXT STEPS

It is a bittersweet moment in the lab. Your labmate has successfully defended their thesis, which you are very happy about. They are also moving on to their dream job, which you are ecstatic about. But this also means that they will no longer be a desk away, and so asking them questions about Python any time you like is about to come to an end.

On the walk back from a celebratory lunch, your labmate makes sure to find you. "Don't worry too much. You have already learned so much about programming! I even heard you're solving other people's problems now."

You agree, but you also feel there is still so much to learn and you aren't even sure what might be worth learning next. You already know how to read data, plot data, fit data, simulate data, record data, and work with common file formats. You know how to integrate, smooth, and interpolate data, as well as process images. Thinking through your normal workflow in lab, it isn't clear what else you might like to learn.

"Oh, I can appreciate that! But just keep your eyes open. Every time you find yourself doing a task on a computer that you wish *you* didn't have to do, think about if you could automate it with Python. Every time you think of something you wish you could do with your data, but don't know how, look online for other people doing that thing using Python. And if the solutions you are finding aren't quite clear immediately, that's okay. Think back to learning Python for the first time. You were confused, but then were able to figure things out. Just keep doing that. And we can always stay in touch! If I find something I think you might like to know about, I can give you a call or send you an email."

The next Monday, you come into lab, and your labmate is gone, but there is an email from them in your inbox. It reads:

"Hey! I am just getting settled in, but on the drive to my new place, I thought of a few things you might like to consider learning about. Here they are!"

You smile, thinking about how fortunate you were to have your labmate as a friend and mentor, and start reading the email...

17.1 MANAGING RESEARCH PROJECTS

17.1.1 IMPLEMENTING A REASONABLE DATA MANAGEMENT PLAN

This is not strictly a Python problem, but it is worth thinking about, because it will impact your Python workflows. Data management refers to how you will store your data, organize it, and ensure that you know where it came from and how it has been modified.

One way to do this is to establish a consistent file name convention, which might specify the date data was taken, the instrument that recorded it, and other useful information (e.g., temperature). The advantage of this approach is that the name is linked to the file. The disadvantage is that you end up with very long file names, which can be hard to read and work with.

If your data is held in **.csv** files or Excel files, you can add in **headers** to these files that will explain the nature of the data. This is nice because you can use shorter file names. But a disadvantage is that it is likely that not all the data you collect will come in as .csv or Excel files, so you will need to convert your data to a format where you can add in arbitrary headers.

Yet another option is to keep data in folders, and then include a **plain text** file in the folder that explains all the data. This kind of documentation can work with any sort of data file. However, it is relatively easy to separate the file with the descriptions from the data files, and you have to remember to update the description when you add or change data.

Finally, you can use software that is designed to help maintain data in a manner where you know what it is, and even if it was changed. Two relatively popular means to do so right now is to use GitHub or create a database. These are discussed next, in turn.

17.1.2 MANAGING CODING PROJECTS WITH GITHUB

If you change the data in a file, what do you do about it? The best practice is to make a note of the changes, so you can back track, if need be. However, if you work with lots of data or perform many different transformations, this can become cumbersome. One way that people default to is to make multiple copies of the same file, giving it successively longer and longer names to keep track of the changes. Anyone how has received a file that ends in "...final final FINAL4_final" has seen the results of such practices. This is perhaps the *worst* way to document changes. A much better practice is to keep a single file, but keep track of the changes you have made to it. Perhaps by growing a longer and longer header...

There is a better approach, however: Git. The most popular and accessible implementation of this is GitHub: an online repository that can be used to store and share code and data, that also manages **version control**. That is, each time you make a change, GitHub asks you to describe what you did, and then remembers what was changed. This means that you, and anyone else that you give access too, has a complete record of the history of your code. This is a wonderful way to manage code and data—especially when working across multiple devices or with multiple people.

GitHub is just one implementation of Git, and you can create your own, but GitHub has a relatively easy interface to learn, and there are lots of tutorials out there on how to use it. See the book's website (https://www.codingforchemistsbook.com/) for links to resources on learning Git.

17.1.3 ORGANIZING YOUR DATA IN DATABASES

Even if you are a pretty organized person, it can be easy to end up with so much data that you lose track of it. And if you wanted to keep track of data from other people in the lab, this can be extremely challenging.

There is a tool invented for managing large amounts of information: the database. You can store objects in this database with information about who took it, what project it was for, what instrument it was taken on, experimental conditions, and so on, and all of this can be searchable. Thus, this is a nice way to organize data.

Python provides many ways to build, maintain, and interact with these databases, so this may be of interest. Some of the more popular libraries to explore are sql-lite and duck-db. It will take some thought to consider how you want to set up a database, but you might find it worth your time.

17.1.4 ENSURING YOUR DATA MEET FAIR DATA STANDARDS

FAIR data standards are a set of principles designed to ensure that scientific data is Findable, Accessible, Interoperable, and Reusable. Developed by a broad community, FAIR standards are guidelines to help researchers manage data effectively, enabling collaboration, reproducibility, and long-term usability by other researchers. Because you are now in control of the manner of production and the format of your data, you have an opportunity to increase its impact by following these guidelines. Briefly, *findability* ensures that data is well-described with metadata and assigned unique identifiers. *Accessibility* promotes open sharing while respecting security and legal constraints. *Interoperability* ensures compatibility across different software and formats through standardized vocabularies and data structures. *Reusability* mandates clear licensing, provenance tracking, and adherence to community standards, allowing data to be repurposed for new research. You can find out more by visiting the FAIR website, https://www.go-fair.org/.

17.2 INCORPORATING ADVANCED ANALYSIS

17.2.1 IMPLEMENTING MORE ADVANCED DATA ANALYSIS TECHNIQUES

We have touched on some of the standard approaches to data analysis and manipulation in our book—things like integration, peak finding, fitting, and so on. But there is so much more that could be done. Once you have gathered a significant amount of data, you can search for patterns and meaning in it using approaches like ANOVA (analysis of variance), principle component analysis, and so on. Python has many libraries to help accomplish these tasks. There are too many approaches to list here, but a good approach to starting might be to read a basic college-level statistics book, and then search for how to do the analysis presented there in Python. Or you might consider reading a book on this topic, such as *Practical Statistics for Data Scientists: 50+ Essential Concepts Using R and Python,* by Peter Bruce, Andrew Bruce, and Peter Gedeck. Certainly you now have sufficient understanding of Python to read that book and ones like it.

17.2.2 Making Your Own Libraries

Rather than having to add a local folder paths to your **PythonPATH** or use the functionality from the codechembook library to incorporate external code, you can take code you wish to reuse, package it up, and then add it to places like conda-forge or PyPi. If you do this, then it becomes installable exactly the way you have been installing other **libraries**, which means that it is easier for you to access in other scripts and other people can use it to. If you have encountered a problem that benefited from making a code solution, maybe other people will find it useful as well. That is, in part, how the codechembook library started.

Although it might seem daunting to create your own libraries, they are really just .py files with code, and then a few other **plain text** files that specify details on how Python should use the library. If you have gone through this book, you certainly have the ability to learn how to make your own library. A good resource to get started is https://realpython.com/pypi-publish-python-package/.

17.2.3 Incorporating Machine Learning in Your Workflow

A huge amount of the emerging machine learning software uses Python as the interface for giving commands. Thus, if you want to learn how to use machine learning, you are now in a position to start using these interfaces. The landscape is evolving so quickly that it is not worth trying to provide a specific material, but some web searches should get you going in the correct direction. We will try to keep a list of resources we like at the book's website: https://www.codingforchemistsbook.com/.

17.2.4 Interfacing with Hardware and Software Through Vendor-Provided APIs

API is an acronym for Application-Programmer Interface. They are libraries, methods, and so on that allow for a programmer to interface with other software. In this book, we used APIs for external software like Microsoft Excel. But there are many other API libraries out there, so if you have software you want to use and are wondering if you can use it via Python, search the web! You can find APIs for many scientific instruments and software, e.g., for making PDF files and interacting with other spreadsheet software like Google Sheets or Microsoft Powerpoint.

You will also find that Python is so widely used that many other programs are written in it, or natively provide APIs for the program. For instance, one of the best 3D rendering programs, Blender, allows interfacing to it using Python. Thus, if you learn this API, you can programmatically assemble a scene for creating pretty renderings of your molecules, all from Python.

17.2.5 COLLECTING DATA FROM THE INTERNET VIA SCRAPING

There is a huge amount of information compiled into pages and databases on the web. There is pricing on chemicals, authors of papers you like, stock market information, and so on. This can be obtained by "visiting" the website using Python and pulling the information back into your code to be parsed. This is known as web scraping, and Python provides several libraries for such things. There are also always multiple tutorials being written on this subject on the web, so you can also just search for this topic, and find the most up-to-date approaches. At the time of this writing, looking for tutorials on using the library "beautiful soup" would provide nice resources.

17.2.6 IMPLEMENTING CHEMINFORMATICS WORKFLOWS

One area not addressed in this book was ChemInformatics. But there are many libraries for such practices. See the excellent ACS InFocus book *Python for Chemists* for details on applications of Python to ChemInformatics problems. These libraries provide tools for processing data about molecules and chemical reactions, and this book provides a great place to get started.

17.3 EXPANDING YOUR CODING TOOLSET

17.3.1 TAKING ADVANTAGE OF LARGE LANGUAGE MODEL–BASED ASSISTANTS LIKE CHATGPT

When considering of machine learning, there are also tools that can help you code and learn to write code. Currently, ChatGPT is the most well-known, but others exist, like Gemini, Microsoft (or GitHub) Copilot, and Anthropic's Claude, among others. Most of these models do an admirable job of producing code, although they are not yet reliable or capable of producing complex solutions without supervision so you really do need to have basic coding skills to efficiently use these tools. Fortunately for you, you now have these skills. When using AI coding tools, don't forget to not only ask for code examples, but to ask questions about how these code examples work. Used this way, these tools have an amazing potential to hasten your learning and speed up code development.

17.3.2 UNDERSTANDING THE STRENGTHS AND WEAKNESSES OF OTHER IDEs

Although this book focused on the Spyder IDE, which we think is likely the best choice for basic chemical workflows, there are large number of other IDEs out there, each offering some advantages. Probably the best-known is VScode, which is a free IDE maintained by Microsoft. This IDE can be used for many other languages, meaning that if you are doing coding in multiple languages, this is a one-stop shop. Additionally, it can connect to and control external computers and clusters using SSH (secure shell) and interface natively with GitHub (with Spyder, you will use external software to achieve the same results), so that you can have a seamless interaction with resources off your personal computer. Another popular choice is Jupyter Labs, which provides a notebook-style interface, which can be nice when documenting how you work through a problem (see below). Google colab offers an all-online IDE that enables real-time collaboration between you and any research partners. An emerging IDE that the authors of this look like is Positron, which is sort of a mix between Spyder and VScode, and Marimo notebooks (see next section). It is worth emphasizing that the Positron IDE is still in the early stages of its development, but it is already a nice option.

17.3.3 CREATING A RECORD OF DATA WORKUP IN PYTHON NOTEBOOKS

Speaking of notebooks, it is worth talking about this approach to an IDE a bit more. The idea behind a notebook is that you can document a workflow for working up data. The notebook provides "cells" (areas that you can write and run code) separated by regions where the outputs of running the code is displayed. Thus, you could write code that creates a plot of data and show this plot. Then the next cell could manipulate this data somehow—maybe normalizing it, and you can output that plot. In this manner you have a log of what you have done. In many Python notebooks, you can also place other things in the cells—such as Markdown text (which provides a means by which to create fancy text output). For instance, you can render nice looking chemical and mathematical equations using Markdown. Altogether, these can be a great resource for keeping track of how you have analyzed your data.

At the time of writing this book, Jupyter notebooks was by far the most popular implementation of notebooks for Python. However, there are two very large caveats to keep in mind when you consider using Jupyter notebooks:

1. Jupyter notebooks are not simple Python files. Thus, you cannot really read this file easily using another Python IDE. This means that if you are writing code in a Jupyter notebook that you would like to use as a module, it will be challenging to do so.

2. Jupyter notebooks also have a serious problem whereby cells can become "stale." That is, you can run one cell and change data, and then another cell can display

the state of the data prior to you changing it. This means that you can have multiple outputs that show different values for the same **object!** This is a serious problem because when you look at your notebook you cannot be assured that you seeing the current state of all your data.

For the reasons above, we don't use Jupyter notebooks much. However, there is an emerging solution: Marimo notebooks. This notebook addresses both of the shortcomings of Jupyter notebooks highlighted above. The Marimo notebook is just a Python file, so you can access code written in it as if it were a module. Additionally, Marimo keeps track of the current state of your variables, and if you change one, it will run *all* cells that depend on that variable—the result being that your cells do not become "stale" and you are assured that the notebook you are looking at is an accurate reflection of the state of your data. Marimo is still being developed, but already it is the preferred notebook for the authors of this book.

17.4 SHARING YOUR RESEARCH PRODUCTS

17.4.1 SIMPLIFYING THE USE OF YOUR CODE USING GRAPHICAL USER INTERFACES OR WEBASSEMBLY

In this book, we made some scripts that solved problems for chemical workflows. But nearly all of the interfaces with the scripts were via the command line. The only GUI interface we invoked was for identifying and saving files. The codechembook library also features a number of simple dialog boxes to give and receive information from users. However, writing full-on GUI interfaces for programs is not hard to do. We include some tutorials on our website (https://www.codingforchemistsbook.com/). In this way, you can create professional looking and customized GUI interfaces for your scripts.

Another option is to use webassembly, which creates a Python environment inside a web browser, allowing you to interact with your scripts using the familiar text boxes, buttons, and so on from websites. One benefit of the Marimo notebook mentioned above is that it includes this capability naturally. At the time of this writing, the notebooks included a simple button to generate the web assembly needed. Another nice thing about Marimo notebooks is that they include ready-made interactivity, such as creation of slider bars, check boxes, and so on, which can be included in the web-assembled version of the notebook. Thus, even though Marimo may not produce the absolute most tidy looking interactive programs, they produce very nice and powerful ones quickly. We definitely recommend looking into this functionality of Marimo.

17.4.2 FORMATTING SCIENTIFIC COMMUNICATIONS WITH LATEX

If you decide to get deeper into notebooks, you will start using the Markdown language. This provides the means to format text quickly... for instance creating itemized lists, check boxes, drawing mathematical formula, representing chemical formula well, and so on.

However, there are even more powerful means to generate beautifully formatted bits of text. Perhaps the most widely used tool in scientific writing is $\LaTeX$ (Latex, pronounced lay-tech). This is a typesetting program built around a paradigm called "what you *say* is what you get." The idea is to provide a series of instructions to the Latex interpreter, which will then render your text was you want. You just need to learn the **syntax** for Latex. For instance, writing the formula $k = \nu_N e^{-\frac{E_A}{k_B T}}$ is done using `k = \nu_N e^{-\frac{E_A}{k_BT}}`. If you have gone through this book, then you should be quite familiar with providing instructions to your computer, and so you should feel confident learning the syntax for Latex.

As you learn Latex, you will find that it can handle many of the frustrations of scientific writing. It provides a native means to add and place figures, add citations, and more. You can organize large amounts of data across a project using Latex. Indeed, the entirety of this book, including the table of contents, formulas, representations of code, figures, and the cross-linked glossary was written entirely in Latex.

If you want to start your Latex journey, then we think the best resource for this is to just start using Latex at www.overleaf.com. At the time of this writing, you can sign up for a free account, which provide online access to a Latex interpreter and a very nice interface to the program. Overleaf handles most of the details of making Latex work, and it also has some nice tutorials and excellent documentation. So, don't be afraid to jump on in!

17.5 FINDING INSTRUCTION NOT COVERED IN THIS BOOK

Finally, this book is, by no means, intended to be the end all of resources for learning Python. Instead, our goal was to get you solving chemical problems, while preparing you to be able to learn from more advanced resources. Here is a partial list of these advanced resources:

Chemistry-specific

- Christian Hill's *Python for Chemists*, Cambridge University Press (Physical Chemistry Focus)
- Charles Weiss's *Scientific Computing for Chemists with Python*, via https://weisscharlesj.github.io/SciCompforChemists/
- Tanemura, Sierra-Costa, and Merz's *Python for Chemists*, ACS InFocus (ChemInformatics Focus)

General Python

- Al Sweigert's *Automate the Boring Stuff with Python*, No Starch Press
- RealPython—https://realpython.com/.

Numerical methods

- Kong, Siauw, and Bayen's *Python Programming and Numerical Methods: A Guide for Engineers and Scientists*, Academic Press, or via https://pythonnumericalmethods.studentorg.berkeley.edu/notebooks/Index.html.
- Sullivan, *Numerical Methods: An Inquiry-Based Approach With Python*, via https://numericalmethodssullivan.github.io/.

We will try to keep more exhaustive and up-to-date list, in addition to descriptions of rapidly evolving Python capabilities, on the companion website for this book, https://www.codingforchemistsbook.com/.

CHAPTER 17 EXERCISES

0. 1. There are no exercises for this chapter. Congratulations on completing the book!

Here is some bonus code for you!

```
import numpy as np
import plotly.graph_objects as go

theta = np.linspace(0, 2*np.pi, 200)
f_x = np.cos(theta)
f_y = np.sin(theta)

left = (-0.3, 0.4)
right = (0.3, 0.4)

m_theta = np.linspace(0, np.pi, 100)
m_x = 0.5 * np.cos(m_theta)
m_y = -0.4 + 0.2 * -np.sin(m_theta)
m_x = 0.66*np.cos(theta)
m_y = -abs(0.66*np.sin(theta))

fig = go.Figure()

fig.add_scatter(x=f_x, y=f_y,
                line_color='black',
                )

```

```
fig.add_scatter(x=[left[0]], y=[left[1]], mode='markers',
                marker=dict(size=20, color='black'))

fig.add_scatter(x=[right[0]], y=[right[1]], mode='markers',
                marker=dict(size=20, color='black'))

fig.add_scatter(x=m_x, y=m_y, mode='lines', line=dict(width=4, color='black'))

fig.update_layout(
    title="Congratulations!",
    template = "simple_white",
    paper_bgcolor='yellow',
    plot_bgcolor='yellow',
    xaxis=dict(showgrid=False, zeroline=False, visible=False),
    yaxis=dict(showgrid=False, zeroline=False, visible=False),
    width=500,
    height=500,
    margin = dict(l = 40, r = 40, t = 40, b = 40),
    showlegend=False
)

fig.update_yaxes(scaleanchor="x", scaleratio=1)
fig.show("png")
```

PART SIX

TECHNICAL DETAILS

CHAPTER A

COMPUTERS

A.0 A COMPUTER IS NOT JUST A LAPTOP OR A RACK OF SERVERS

You are probably used to thinking of computers as the laptop or desktop that you use to write lab reports, analyze data, operate instruments, or look up papers. Although this is certainly true, computers are all around us, most of which run software created using very similar tools to those taught in this book. The first "programmable computers" (in other words, computers that can perform different computations depending on instructions from a user) were in fact mechanical, but still functioned overall similarly to those of today. Many of the objects with which you interact today are computers: microwaves, graphing calculators, and computers all operate fundamentally on the same principles. So what is a computer?

A.1 COMPUTERS CONVERT INPUT DATA INTO OUTPUT DATA

At their most basic level, computers can be thought of as the converting of data, as outlined in Figure A.1. This data can be in the form of inputs (data that the computer needs to solve the problem) and outputs (data that forms a representation of the solution). This conversion happens in a series of computations, in which new data is created via a series of processes fully determined by the input data. As users, we can both give the computer commands (e.g., from a keyboard or mouse, by speaking, via email, or any number of other avenues) that instruct it on what we wish it to do, and perceive the results (e.g., on a display, through a printer or speaker, and so on). The process of receiving commands from the user and translating results into a form that can be perceived in the real world often makes up the bulk of programming.

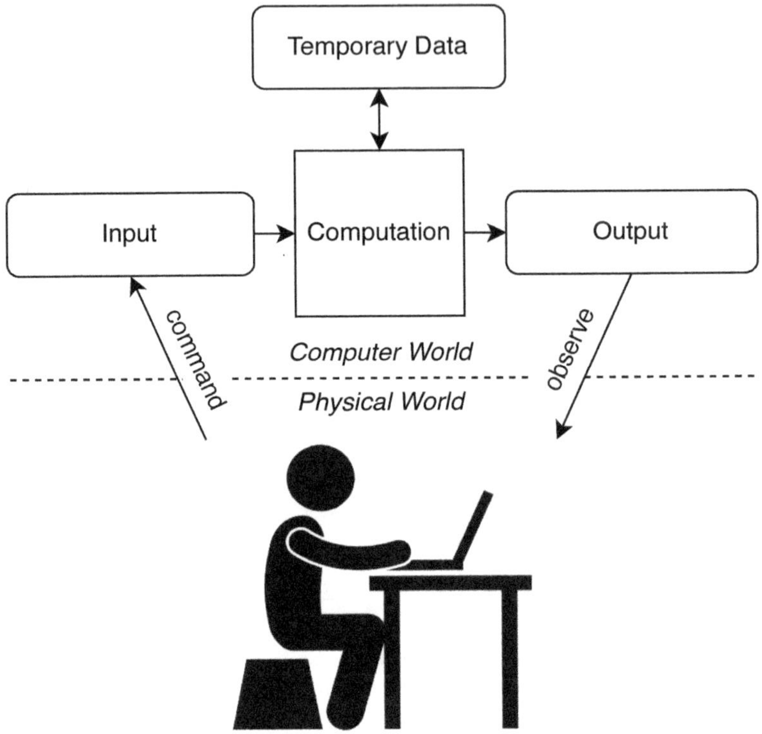

FIGURE A.1. The flow of data in a computer. Input data of many different types are converted into output data via a process known as "computation."

Computation is the basic problem solving steps you would use to perform any task, like the evaluation of mathematical expressions or the organization of information into categories and groups. The combination of many of these individual computational steps forms an **algorithm**, a self-contained set of instructions to perform a specific task. Thus, data conversion boils down to the application of a series of algorithms to data to transform it in the desired way. The power of using a computer to do this is largely four fold:

1. The computer can execute them many orders of magnitude faster than you can.

2. The computer can work with far more data at a single moment than you can.

3. The computer does not get bored doing the same thing over and over, and very rarely makes "mistakes."

4. It can perform the computation while you are doing something else, freeing up you time for tasks that are not (yet) well-suited to a computer.

But if computation is the process of transforming data, this raises the question: what is data?

A.1.0 WHAT IS DATA?

You are probably used to thinking of data as an NMR spectrum, a table of experimental results, or the thing you sometimes run out of if you've been streaming video to your phone. And, in all cases, you are correct. But data is much more general. Data is a representation of information. More philosophical texts might discuss the underlying link between data and information theory, but here we will stick to the pragmatic (to get you solving chemistry problems as quickly as possible). For the purpose of this book, data is information that is transferred to and/or from the computer and/or stored in the computer. The results of your experiments, which you are used to calling "data," is thus one type of data with which we must deal. However, you will see that from the perspective of a computer program, data is a much broader concept.

In a computer, every everything is data. Let's take a simple hand-held calculator as an example. You might think that the data is the numbers you press, and the computation is the plus or equals buttons that you press. However, from the calculator's perspective, there is much more data than this. We can see that this is the case, by conceptualizing the process of data conversion in the calculator as follows:

1. The primary data is all of the buttons the user pressed, whether they represent numbers or not. At first, the calculator does not need to know what those buttons mean, just which ones were pressed in which order.

2. The first computation required is to convert which button is pressed (the real world) into data in the form of numbers or commands. Here, pressing "2" "+" "2" "=" will be converted into two pieces of data representing the two different number 2s, a command that the numbers should be added, and a command to execute the calculation.

3. Once the command to execute the calculation is given, the program must perform the arithmetic and remember the result, in this case (hopefully) 4.

4. The user cannot know that 4 is the result unless the calculator presents it appropriately. This means, for a simple calculator with a display made up of a series of seven-segment LCD devices, that the calculator must determine which segments should be darkened to create a shape that is interpreted by the user as the numeral "4."

Note here that pressing the plus button or the equals button is not computation, that is still input (data). This data tells the calculator that you have requested it to perform a specific arithmetic operation. But it does not "do" that operation. In addition to all of this, there is a significant amount of data that is not obvious in our example. For instance, some data is instructions on *how* to darken the parts of the LCD screen that will represent "4" to the user, but changing voltages in circuits and so on. Most of the data involved in the calculator is focused on user input and output, not the calculation to be done!

A.1.1 How Computers Store and Represent Data

Given this diverse array of data that we must consider as programmers, we next have to discuss how the computer actually *remembers* the data. For reasons that we won't go into detail here, computers operate fundamentally using only two values: often represented as `0` and `1`. This limited range of expression of information has necessitated the development of a complex set of systems for representing all other types of information as a series of these two numbers. Each digit in the series is called a "bit," and eight bits make a "byte." Generally, the longer the series of ones and zeros, the more expressive power the system has to represent larger or more complex types of data. However, because zero and one are called "binary," the numbers that are expressed using a series of zeros and ones are called "binary numbers" and the scheme by which they are translated to and from "regular numbers" is called **binary encoding**. These systems are described in arbitrarily large detail by a number of sources, so we won't reproduce those efforts here, but we will highlight two systems as basic examples.

A.1.1.0 Memory Structure

In the abstract, you can think of main memory as simply a sea of bits that the computer can use as it sees fit. This sea is divided into one-byte chunks (1 byte = 8 bits), where each byte has a corresponding "address" that the computer uses to access the data in a given chunk or overwrite data that is already there. Addresses are also binary numbers, typically (but not always) the number of bits used by your computer processor (64-bit is the most common as of this writing). This number of bits sets a theoretical upper limit on the size of your memory because you will run out of binary numbers to use as an address after 2^{64}. Practically, for most of us, this is not a concern because that works out to more than 16 million TB! However, even relatively recently, 32-bit processors were common, with a theoretical limit of 4 GB.

Memory addresses are commonly expressed as **hexadecimal** numbers, where each digit represents four bytes. So, a 64-bit memory address can be written as `0x8f397a395dbe06b6`, where `0x` signifies that we should expect a hexadecimal number and the memory address appears as a 16-digit, 64-bit hexadecimal number. You may see such numbers flash on your screen when your computer encounters a particularly nasty error. In principle, these memory address numbers in the error message carry information that could be used to debug the error; in practice, they carry only sadness.

A.1.1.1 Storing Data in Memory

It might seem that a lot of work will have to go into remembering what data is stored in what memory address. Fortunately, this can be abstracted using by using **variables**, which are essentially named placeholders for the memory addresses. Don't confuse variables in coding with variables in math; there are some important differences. In coding, variables are bits of text that (hopefully) succinctly describes the nature of the information that is stored in a specific part of the memory.

In an abstract sense, people think of variables as "holding" data for you, even if that is not really what is happening behind the scenes. Slightly more accurately, a variable points to a location in **memory** where data is stored. A variable must always have some value, and as a programmer you can use that value in a computation, create a new variable with some value, or replace the current value in a pre-existing variable with a new one (in most cases).

The two main uses of variables are in **expressions**, which are just mathematical instructions, and **assignment operations**, which produce a result that you want the computer to store at least temporarily. In Chapter 0, you can see how to use the values in a variable, by writing a mathematical expression in which the variable's name is used rather than explicitly writing its value. To create or change the value stored in a variable, you must use as assignment operation, in which the variable name into which you would like to store the result is always on the left hand side of the assignment operator (equals sign), and the expression you would like to evaluate is on the right hand side. When executing this line of code, the computer will first evaluate the expression on the right-hand side, and then store the resulting value in the memory, pointed to by the variable on the left-hand side. Behind the scenes, what Python is doing is adding the variable name, and the place in memory that it points to, into the so-called "namespace"—the collection of **object**s that Python has access to.

Different languages have different conventions, but in Python, variable names must start with a letter (A–Z or a–z) or an underscore (_), but cannot begin with a number. They can contain letters, numbers, and underscores, but no other special characters or spaces. Variable names are case-sensitive (myVar and myvar are different), and they cannot be the same as Python's reserved keywords (e.g., `round`, `abs`, `def`). For readability, it is recommended to use descriptive names and follow your favorite conventions for diambiguating between variable names, **function** names, and **class** names.

A.1.1.2 Type of Data Stored in a Variable

In some programming languages, you have to decide what type of data (i.e., text, numeric values, and so on; more on that in Chapter B) will be stored in the variable when you first create it, a concept called **static typing**. However, in Python, the type of data that can be stored in a variable can change. You can assign the number 1.507 to the variable `sample_mass` and then later change that value to the text `"puppies"`s and Python will not complain. This can be both a good and a bad thing, but it is something you should be aware of. A complete discussion of these different **data types** can be found in Chapter B.

A.1.2 How Computers Manipulate Data

Now that we've discussed some of the fundamentals how computers store data, we can talk in broad strokes about how computers perform computation on the data. Computation in computers is the conversion of data based on instructions for how to convert the data. Essentially every step of computation will create new data from existing data. Computers have a limited subset of conversions that they can do, essentially arithmetic, moving data around, and some basic shortcuts that make common operations go faster. Every computation thus involves many steps of these basic conversions, much like a complex chemical reaction involves many elementary reactions.

Data conversion in computers is performed by a "controller" or, in traditional computers, a "microprocessor." We will leave a detailed explanation of how these components work to other resources, but give a brief overview here. At its most basic, a processor is a series of logic gates that perform boolean operations on binary numbers. These logic gates are made of transistors, and each operation that a processor can do is designed by hard wiring a series of these transistors together such that the result of the operation depends deterministically upon the combination of binary numbers being operated on. For example, to add two numbers, the binary representation of them must first be loaded into specific registers, locations in the processor that store data only as long as necessary to complete an operation. The operation is then triggered, and an output register is populated with a series of bits corresponding to the result of the operation. The data in this output register must then be moved back to memory to allow the next operation to run.

Luckily, modern software developers do not have to think at this "low level" about how the computer works. Extremely powerful programming languages, using systems much more abstract and easier for humans to think about, have been developed, and this is why it is possible for you to be an excellent chemist and a serviceable programmer simultaneously.

A.2 YOU ARE USED TO WORKING WITH A "SOFTWARE STACK"

When you are doing work on a computer, you are typically thinking about the specific program or application that you are running. These programs are very rarely self-contained, but rather depend on a number of other pieces of software, all of which work together to provide the functionality you need.

A.2.0 THE OPERATING SYSTEM

The operating system (OS) takes care of most of the housekeeping of your computer: talking to input and output devices (e.g., mouse and displays), managing files and folders, managing installed software, and a number of other important but often behind-the-scenes processes. Common OSes you may encounter include Microsoft Windows, Apple MacOS (formerly OS X), various Linux distributions, iOS, and Android—although many others exist as well. You can think of the OS as the "host" of everything that runs on your computer, and most software must be written somewhat differently for different specific

OSes. The "widgets" such as buttons, scroll bars, or menus that you are used to interacting with when using applications are one example of this OS-dependence, requiring user interfaces to be rewritten for different OSes. Luckily, cross-platform (i.e., works the same across different OSes) user interface packages exist to make this easier, although they sacrifice the look of a "native" application.

A.2.1 LIBRARIES

If you, as a programmer, had to figure out how to make a scroll bar, or access data on a website, yourself, it would take a long time to produce new software! Luckily, nearly all of these problems have already been solved, and those solutions are often made available to you in a modular form known as **libraries**. Libraries allow you to borrow other programmers' solutions to common problems, so that you don't have to figure out the solution yourself. This saves you time and makes it more likely that your code will have fewer bugs. Python features an extensive set of standard libraries, and its users have created a huge array of other libraries that you can use with just a small amount of effort. Many chemistry-specific libraries are available, such as RDKit (introduced in Chapter 4).

Libraries (should) come with documentation (that can be of varying quality) that explain to you how to use the library without having to know the implementation details of the code behind the scenes. Note that many of these libraries are free and open-source, which means that you are depending on volunteers to produce the code and documentation and to maintain them. The pitfall here is that some libraries are often abandoned or have minimal technical support available, and it is up to you to try to evaluate the longevity of the library you are using and validate that they work as intended.

A.2.1.0 Software Development Kits

An important subset of libraries are Software Development Kits (SDKs), which (in the context of chemistry) are sometimes available as companions to applications that control instruments or perform specific chemistry tasks. SDKs are libraries that allow you to control, or make use of capabilities of, applications for which you do not have source code. This could be useful, for example, if you need to write a code that automates experiments involving several pieces of equipment that need to be synchronized in some way.

A.2.2 APPLICATIONS

Applications are the software that you mostly use on a day-to-day basis: web browsers, instrument control software, spreadsheets, and more. These are the programs that you are likely most familiar with. Applications come in many shapes and sizes, from complex programs like Microsoft Powerpoint to simple widgets like calculators. Applications are, in some ways, self-contained. They may require libraries on which they depend to be installed on your computer, but they typically do not require some other software other than the OS to be running in order to function.

A.2.3 SCRIPTS

Scripts are pieces of software that *do* require some other software to be running in order to function, and thus are not self-contained in the way that applications are. The programs produced in the narrative part of this book are scripts, just because it is easier to not have to worry about the coding overhead to make a self-contained application. Scripts may impose a performance penalty over applications, but typically this performance penalty is negligible—especially for the workflows discussed in this book—and well worth the trade-off for ease of development.

CHAPTER B

DATA TYPES

As noted in Chapters 0 and A, *everything* that a computer operates on can ultimately be framed as numbers. At the same time, most people make a distinction between multiple categories of data, like words versus numbers, integers versus numbers with decimals, and between single objects and groups of objects. We do this because it helps us set expectations for how data will behave and what we can do with it. Most people have no difficulty carrying out multiplication between integers ($2 \times 4 = 8$) or between integers and numbers with decimals ($2 \times 3.2 = 6.4$), but will find multiplication between words to be non-sensical (apathy $\times$ magenta=???).

One way to frame this treatment is to say that there are different "types" of data, and that these types have different possible behaviors. Modern programming languages try to capture this instinctive view of data by explicitly defining types of data, and then using these types to determine behavior. In this chapter, we will discuss the built-in data types in Python, some of the data types commonly added from libraries, and their basic behaviors. The goal is not to be comprehensive, but to provide enough detail to understand the examples presented throughout this book, avoid common errors that are encountered when data type behavior is ignored, and also prepare you to be able to read further on this topic.

B.0 TYPES THAT DEAL WITH NUMBERS

Because everything in your computer is a number, it is probably no surprise that there are several **data types** that deal with numbers. Here, we consider each in turn.

B.0.0 Numeric Types Are Used for "Normal" Math

As scientists, we are fairly familiar with integers, numbers with decimals, and numbers that include multiples of the square root of -1. Python has built in types for all three of these: `int`, `float`, and `complex`.

B.0.0.0 `int` and `float`

The Python type for integers is `int`. This holds the "whole" numbers, both positive and negative. These are numbers that are most useful for counting. We naturally say things like: "I own 5 shirts." So, a quick rule of thumb is that if you are counting things, you should be using the `int` type.

The other type of basic number that we use are those that contain decimals. These are not the counting numbers: we would never say, "I have 5.0 shirts." Instead, we tend to use these in calculations that are not simply counting. In Python, we call these numbers floats,[1] and the type is `float`.

[1] The terminology comes from the term "floating point."

There are a few ways to create each of these data types. We can directly assign them to **variables**, using the assignment operator (=).

```
In:  i = 2
In:  f = 3.8
```

This will result in the creation of two variables, one that points to an `int` data type with a value of 2 and one that points to a `float` data type with a value of 3.8.

You will note that we didn't tell Python what type to use. Instead, it made a guess, based on the form of the data we gave it. We can check to make sure that the correct type was assigned using the `type( )` **function**.

```
In:  print(type(i))
Out: int
```

```
In:  print(type(f))
Out: float
```

Rather than leaving Python to guess at the data type we want, we can explicitly tell Python what data type we want, using a function.[2] The function to create an integer is `int( )`, and the function to create a float is `float( )`. We can give these at the time of creating the data, or we can also use them to make one data type another. In the language of programming, this is known as *coercion*. Here are examples of coercing our `int` to a `float`, and vice versa:

[2] When used to create the initial instance of a type, the function is also called a "constructor."

```
In: int_from_float = int(f)
In: float_from_int = float(i)
```

We can get these values as follows:

```
In:  print(int_from_float)
Out: 3
```

```
In: print(float_from_int)
Out: 2.0
```

A few things to note here: first, when we use `int()` on a float, the value of the float is truncated, rather than rounded. If we want to round instead, we would need to use the `round()` function, which returns an integer (when no **keyword arguments** are supplied), after rounding.

```
In: int_from_float_2 = round(f)
In: print(int_from_float_2)
Out: 4
```

Second, floats *always* display at least one decimal place. This is one way that Python indicates that an object is a `float`, which is why Python represents the second value as 2.0.

B.0.0.1 `complex`

Python can also handle complex numbers. These are numbers that contain some factor of the square root of -1. Although we will not use them in this book, depending on the nature of your research, you might. The type for these numbers is `complex`.

We can create complex numbers in similar ways to above. For instance, we can directly write the value we want, and leave Python to automatically assign the type:

```
In: c1 = 2.1 + 3j
```

Here, the inclusion of the square root of -1 (i.e., j) is how Python knows this is a complex number.[3] However we can also use the function `complex()` to create a complex number. In this case, the function accepts two values (called **arguments**), the first is assigned to the real part of the complex number, and the second to the imaginary part of the complex number.

[3] Many people are more familiar with the use of i for the square root of -1, but j is also common.

```
In: c2 = complex(2.1, 3)
```

And c1 and c2 will have equivalent values. It is important to realize that we can also provide the `complex()` function a single argument, in which case it is assigned to the real part, and the complex part is assigned a value of zero.

```
In: c3 = complex(7)
In: print(c3)
Out: 7+0j
```

We can try to coerce a number to a complex using the same function, providing either one or two values.

```
In: c4 = complex(i, f)
In: print(c4)
Out: 2+3.8j
```

There are two more things to note about complex types. First, we can access the real and imaginary parts separately. We do this by appending .real or .imaginary to the number. In other words, they are **attribute**s of the `complex` data type.

```
In: real = c4.real
In: imaginary = c4.imaginary
```

The second thing to note is that the numbers stored in the complex are floats. You can see this if we print the values of for real component:

```
In: print(real)
Out: 2.0
```

Thus, even though Python had not printed our complex number with a decimal place, it still views this number as a float, when pulled away from the complex number.

B.0.0.2 Using Numeric Types

We will conclude this discussion of the "common" math types by noting that they can be used with the familiar mathematical operations. Although there are *many* possible mathematical operations, Python includes only 12 (though more can be added with **libraries**. Of these 12, we will only use 7 in this book: addition, subtraction, multiplication, division, exponentiation, absolute value, round, and modulo.[4] You will note that this list *does not* include logarithms. In order to use logarithms, we will need to import a library. Many libraries include this operation, but the one we will use most often is Numpy. Table B.1 presents these operations, the symbol or function used to implement it in Python, and an example of its use.

[4] If you are unfamiliar with modulo, it returns the remainder of a division problem, as explained here: https://simple.wikipedia.org/wiki/Modulo_operation.

And this concludes our introduction to the "normal" math types. However, as we are about to see, Python has others.

TABLE B.1 Operations that can be performed on numerical data in Python. Those above the double horizontal line are built-in to Python.

operation	symbol / function	example
addition	+	$2 + 3.8 = 5.8$
subtraction	−	$2 - 3.8 = -1.8$
multiplication	*	$2 \cdot 3.8 = 7.6$
division	/	$2 / 3.8 = 0.5263157894736842$
exponentiation	**	$2^{3.8} = 13.928809012737984$
absolute value	abs()	$\text{abs}(-1.8) = 1.8$
round	round()	$\text{round}(-1.8) = -2$
modulo (remainder)	%	$7\%5 = 2$
ln	numpy.log()	$\log(10^7) = 2.5649493574615367$
$\log_{10}$	numpy.log10()	$\log_{10}(10^7) = 7$

B.0.1 THE BOOLEAN TYPE IS USED WHEN EVALUATING TRUTH STATEMENTS

Another type of "number" that we commonly encounter is the boolean, which in Python is called `bool` and which hold "truth values." In Python, a `bool` can have one of two values: `True` or `False`. Please note that capitalization is critical here: TRUE, true, FALSE, and false are **not** proper values for the `bool` type.

We can create a boolean directly.

```
In: T = True
In: F = False
```

But we can also attain booleans by performing **operation**s or **function**s on data. Booleans are often also associated with certain "boolean" operations that test if a particular situation is true or not.

For instance, what if we wanted to know if 4 was less than 5? We can use the boolean operator "<" within a **comparison operation** to test this.

```
In: 4<5
Out: True
```

Here we say that the operation "**returned**" a value of `False`.

There are other questions we can ask, and they are summarized in Table B.2. When looking at this table, there are two kinds of operations. Those that are above the double line are typically applied to non-booleans, but result in a boolean. That is, we will often check if one number is larger than another, or if the value of one variable is equal to the value of another.

However, below the double line, we find boolean operators are mostly used to operate on booleans (also returning booleans), for instance checking to see if both in a pair of booleans are True (`and`), if at least one in a pair of booleans is True (`or`), or changing the value of a boolean (`not`).

```
In: 3 < 5 and 5 < 3
Out: False
```

```
In: 3 < 5 or 5 < 3
Out: True
```

```
In: not 3 < 5
Out: False
```

Python also includes the xor operator, which is different from how we use "or" in everyday discourse. When we say "or" we mean that either one, or the other, or *both* can be true (i.e., at least one value is true). This is the way that the operator `or` works. However, in logic, there is an "exclusive or (xor)", which is only True if one or the other value is True, but *not both*. The operator for this in Python is ^ (the caret character). Its use can be illustrated as follows:

```
In: not 3 < 5 ^ 7 < 2
Out: True

In: not 8 < 1 ^ 7 < 3
Out: False

In: not 3 < 5 ^ 1 < 9
Out: False
```

Booleans will be most often encountered in **control structures**, like `if` and `while` loops (See Chapter F). Outside of that, they can be hidden in the operations of other functions, like `in` and `numpy.loadtext()`. These will be flagged as needed in the text.

TABLE B.2 Operations that can be performed using Boolean logic.

operation	symbol	example	returns
equality	==	3 == 3.0	True
inequality	!=	2 != 2	False
greater than			
greater or equal to			
less than			
less or equal to			
and	and	True and False	False
or	or	True and False	True
xor	^	True ^ True	False
not	not	not True	False

B.0.2 THE BINARY TYPES ALLOW DIRECT ACCESS OF COMPUTER MEMORY

Although the basic unit of computer number is the bit (often represented by zero or one) there is no data type in Python that corresponds directly to zero and one.[5] Instead, the smallest element that Python provides access to is the byte. Python provides the means to access, store, and manipulate these bytes using three types: `byte`, `bytearray`, and `memoryview`. Although these exist, it is unlikely that you will use them; instead, you will be working with data types that map onto more familiar numbers. For this reason, we will not expound on these further here, mentioning them only for completeness.

[5] Although one could try to conceive that the booleans discussed above are zero and one, they are not actually zero and one, in the way computers perceive them. For instance, a single zero or one should occupy one bit of **memory**. In Python `True` and `False` commonly occupy 24 *bytes* of memory.

B.0.3 THE NoneType LETS US REPRESENT THE VALUE OF NOTHING

The number zero was not invented until around 500 CE. It must have seemed like a strange idea: how to represent not having anything? Why even bother? Nevertheless, it was a useful invention: allowing us to express somewhat strange ideas like, "having" zero shirts.

The utility of indicating nothing carries over to programming and, in Python, we represent nothing with the `NoneType`. This type has only one value: `None`. Note the capitalization: neither NONE nor none are proper values. This type is a way for Python to note a difference between the number zero and there being actually...nothing. As a more concrete example: `None` also provides the means to discriminate between a collection of items that contains 0 objects (i.e., the collection exists, but is empty) and one that does not exist (so the value is `None`). This is akin to discriminating between a bank account that has no money in it, and one that does not yet exist—clearly a useful thing to be able to do.

We can create NoneType objects directly:

```
In: n = None
```

This approach can be used to create variables that are placeholders (they will later hold data), or maybe to indicate that there is no value to yet assign.

Another means to generate them by assigning a variable to the results of a **function** or **operation** that does not **return** a value. For instance, the `print()` function in Python does not return any value, so, if we try to assign the results of a print function to a variable, it will result in `None`.

```
In: p = print("HI!")
In: print(p)
Out: None
```

We realize this discussion is somewhat abstract, but it is easier to understand the need for `None` within more complex code examples. We will point out the occurrences as they happen in the book. For now, we note that this concludes the tour of built in Python data types that hold only one value at a time. Now, we turn to those data types that can hold multiple values simultaneously.

B.1 COLLECTIONS LET US HOLD MULTIPLE ELEMENTS AND HAVE NEW BEHAVIORS

It is rare in life that we only deal with single objects. Instead, it is often quite convenient to collect multiple values together—sometimes even grouping different types of data. We might make a grocery list in which we have multiple ingredient names, combined with numbers indicating how much to buy. Next we explore how Python enables such behavior. We will first discuss on type of collection (the list) in order to illustrate some behaviors, and then proceed to consider all the collections that are in the base version of Python (more can be added by libraries).

B.1.0 THE LIST, AND BEHAVIORS OF COLLECTIONS

There are many types of data in Python that allow us to hold several values ("elements"). However, before we discuss them in general, we need to introduce some new behaviors. To do so, we will introduce the Python `list` type.

A Python list is a data type that can hold any number of elements, of any type. A list can hold integers, floats, booleans, or even other lists. Any data type we discuss here can be an entry in a Python list. Lists can be created directly by placing zero or more values, separated by commas, between square brackets.

```
In: l1 = [None, 4, True]
In: print(l1)
Out: [None, 4, True]
```

Alternatively, we can build long lists from short lists, using the repetition operator (*). If we were using numeric types, this would be multiplication, but for lists, we simply repeat the list the specified number of times.

```
In: l2 = l1 *2
In: print(l2)
Out: [None, 4, True, None, 4, True]
```

Where you can see that we have a new list that now repeats the first list twice. This approach can be nice when you want to have a list that contains a large number of the same entry. For instance, imagine we wanted to have a list that held the integer 1, repeated 100 times. You could simply use:

```
In: l1 = [1]*100
```

The fact that the list is a data type with multiple elements within it introduces several new behaviors. There are three that we feel are critical to understand, and we discuss next.

B.1.1 ITERABLE

Some (but not all) data types that hold multiple elements are iterable (see Table B.3). When a data type is iterable, this means that we can run through the elements in the data type one at a time. This operation is known as "iteration." There are several common built-in Python functions that rely on this property. Again, there are three that we think are critical to understand. They are:

- len()
  This function returns the length of the iterable, which is equal to the number of elements it holds.

```
In: len(l1)
Out: 3
```

If you give this function a data type that is not iterable (for instance, a float), then it will return an exception.

```
In: len(2.9)
Out: object of type 'float' has no len()
```

- in
  This operator checks to see if an element is already present within the object. This works by iterating over every element and evaluating the boolean operation "==". If the element is present, a value of `True` is returned, otherwise a value of `False` is returned.

```
In: 4 in l1
Out: True
```

If you try to use `in` on a non-iterable data type, Python generates an exception:

```
In: 2 in 2.9
Out: argument of type 'float' is not iterable
```

- for
  This is most commonly used as part of a **control structure** called a "`for` **loop**"[6] and is used in conjunction with a variable, the `in` operator, and an iterable object. The behavior is that we iterate through the list, each element in the iterable is returned and assigned to the variable. We can then use that variable however we like before we move on to the next element. For instance, we might print the value of the element.

[6] For more information, you can read Chapter F.

```
In: for i in l1:
In:     print(i)
Out: None
Out: 4
Out: True
```

You can see that this `for` loop has taken each element in our iterable (`l1`) in turn, passed it to the variable `i`, and then printed this variable, before moving on to the next element. Once we have done this for every element, the loop stops.
There are some interesting formatting in this example, such as the colon at the end of the first line, and the indentation of the second line. For a detailed description of this, please see Chapter F. For now, we will just consider the output.
If we try to use this on a non-iterable object, Python generates an error.

```
In: for i in 6.0:
In:     print(i)
Out: 'float' object is not iterable
```

- * and ** These operators are known as "unpacking" operators. What they do is return all the elements within the iterable object. This is different than getting the value of the iterable object:

```
In: print(l1)
Out: [None, 4, True]
```

```
In: print(*l1)
Out: None 4 True
```

Note that the first input results in printing the actual list (an object containing three elements), while the second returns the actual values of these elements. In a sense, this is a bit like running a `for` loop and getting the return elements all at once, though it is certainly used differently than in `for` loops.

One important point to make here is that the * operator is used with *almost* every iterable object. However, there is one iterable type where it does not work: the dictionary. We will discuss dictionaries below, but for here we simply note that the correct unpacking operator for dictionaries is **.

B.1.2 SLICABLE

Some (but not all) data types that hold multiple elements are slicable (see Table B.3). When a data type is slicable, it means that the elements held within it are assigned an integer index[7] that can be used to refer to that element *and* which can be used to create

[7] This index is also known as a "subscript." However, subscript is the more general term for how we can refer to elements in an object. There are non-numeric subscripts. For example, the subscripts for dictionaries are strings and are called "keys." In this book, the behavior of data types with numeric subscripts (a.k.a., indices) is useful enough that we focus on this aspect.

TABLE B.3 Date types and their properties.

type	page	iterable	slicable	mutable
		built-in types		
`nonetype`				
`bool`				
`int`				
`float`				
`complex`				
`byte`		True		
`bytearray`		True	True	True
`list`		True	True	True
`tuple`		True	True	
`str`		True	True	
`dict`		True		True
`set`		True		True
`frozenset`		True		
		types that can be added using libraries		
`numpy array`		True	True	True
`Pillow Image`		True	True	
`Plotly Figure`		True		

new sub-collections of the items.[8] For both historical reasons and reasons of computational efficiency, index counting starts at zero. In Python, we can also use negative numbers to count from the end backwards. Because zero is already used for the first index, we start from -1. This is illustrated in Figure B.1 and the following examples.

```
indices    0       1     2
        [ "red" ,  4  , True ]
indices   -3      -2    -1
```

FIGURE B.1. Schematic showing how index numbering is assigned to a list.

```
In: l1[0]
Out: None
```

```
In: l1[-2]
Out: 4
```

We can also use these numbers to extract sub-collections. If we supply two numbers separated by a colon, then Python will return the elements that have the first index (inclusive) through the last index (exclusive). That is, we get the element corresponding to the first index, but do not get the element corresponding to the last number, rather we get the element corresponding to the last number minus one.

```
In: l1[1:-1]
Out: [4, True]
```

This is known as "slicing." If you wish to **slice** from some initial index to the end of the list, you can simply use a colon, but leave the final number out:

```
In: l1[1:]
Out: [4, True]
```

Similarly, if you want to start at the beginning of the list, you can also omit the initial number before the colon:

```
In: l1[:-2]
Out: [None]
```

If you try to use indexes on an object that is not slicable (either to get a single object or a slice), Python will generate an exception.

```
In: 6.0[1]
Out: float' object is not subscriptable
```

Here, Python is letting you know that you cannot use the subscript (index) you supplied. It is also using "subscritble" as a stand-in for slicable. We prefer to use the term slicable because there are data types in Python that are subscriptable, but not slicable (i.e., `bytearrays`), so we think the terminology we have selected is more accurate.

[8] There are a few objects that do have integer subscripts, but cannot be sliced, for instance, the `byte` data type. However, because we do not use the byte type in this book, we are not dwelling on this distinction.

B.1.3 MUTABLE

Some (but not all) data types that hold multiple elements are mutable (see Table B.3). If an object can be modified *after* it is created, then it is called *mutable*. Otherwise, it is *immutable*. To understand what this means, and how it works, we need to consider how Python keeps track of variable and values.

When you make an assignment to a variable in Python, what is happening is that Python is writing that variable somewhere in its memory, *then* it points the variable to that *place in memory*. If we think about assigning the value of 2.0 to the variable f1

```
In: f1 = 2.0
```

we can envision it something like this:

f1 ⟶ 2.0

where the grey box is symbolizing a place in memory.

If you then assign a new value to f1,

```
In: f1 = "a"
```

Python creates a *new* value in a different place in the memory, and then points f1 to that instead.

2.0
f1 ⟶ "a"

This is the behavior of a data type that is *immutable*. That is, the value of the actual object does not change. Instead, completely new values are created and the variable now points to that new object.

Mutable objects work differently. If an object is mutable, then we can change the value in the same place of memory, without changing where our variable points.

The list we have been working with is mutable. When we assign a list to Python,

```
In: l1 = ["red", 4, True]
```

we assign the overall list a value in memory, and point the variable to it.

l1 ⟶ ["red" , 4 , True]

But this points to the entire *list*, not the things inside it. So, if we change the elements inside it, maybe using indexes to choose which one:

```
In: l[1] = "a"
```

the variable still points to the same place in memory where the list is. In fact, even the items in the list are locations in memory that are kept track of internally in the list object.

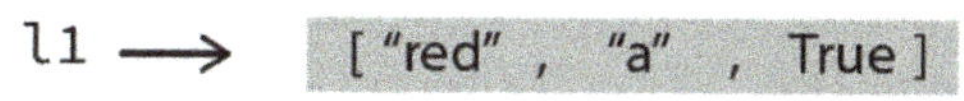

And printing the list gives this new value at position 1.

```
In: print(l1)
Out: ["red", "a", True]
```

Because we can change the elements in the list, we say the list is mutable.

Now, the reason to make this distinction between mutable and immutable is because it changes some behaviors in interesting ways. Let us first consider a case for immutable objects.

```
In: f1 = 2.0
In: f2 = f1
In: f1 = "a"
In: print(f1)
Out: "a"
```

```
In: print(f2)
Out: 2.0
```

What has happened here is that we created the value 2.0, and pointed `f1` at it. Then we pointed `f2` at the same value as `f1`, meaning that *we pointed* `f2` *at the same object as* `f1`.

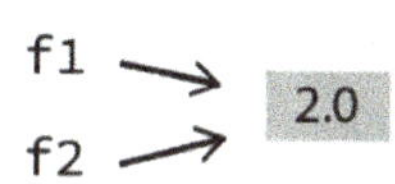

Then we made a new object "a" and pointed f1 at it. Now we have f1 and f2 pointing at different objects, with different values.

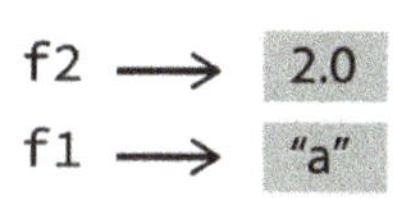

This is the behavior for *immutable* objects. If we look at the same set of operations for a mutable object, we will get a different behavior.

```
In: l1 = ["red", 4, True]
In: l2 = l1
```

Now we have created a list object, and pointed `l1` at it. Then we tell `l2` to point at the same object as `l1`. Now we once again have two variables pointing at the same object in memory.

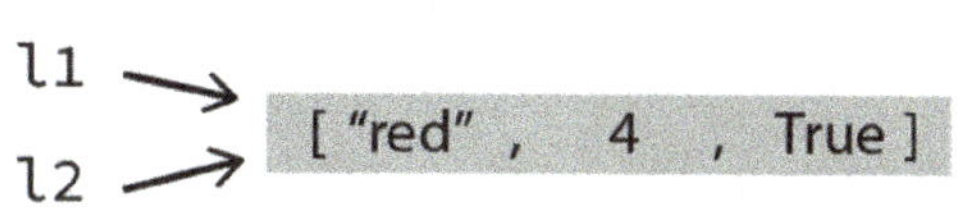

The difference occurs when we change the elements of the mutable list. In this case, changing the second element.

```
In: l1[1] = "a"
In: print(l1)
Out: ["red", "a", True]
```

```
In: print(l2)
Out: ["red", "a", True]
```

The values for *both* `l1` and `l2` have changed! Actually, that is not quite accurate. What has happened is that we changed the element within a single *mutable* object in memory, and we simply had both `l1` and `l2` pointed at this same object.

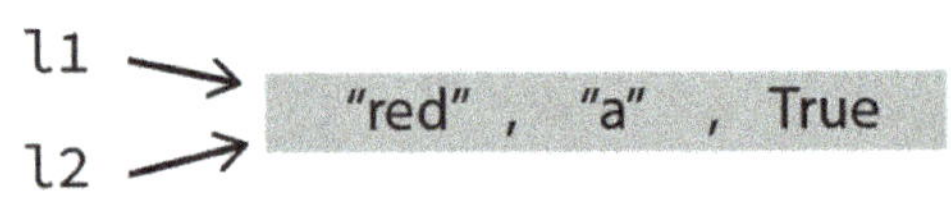

A common problem people new to Python encounter is assigning two variables to the same mutable objects, when they think they have assigned the variable to two different objects. Then, they change the first object, and unwittingly also change the second.

What if we don't want to point two things at the same mutable object in memory? Well, when we make the second assignment, we need to copy the list to a new memory position, so that we have two list objects. This is done using the copy module and `copy()` or `copy.deepcopy()`. The difference between these is that copy makes a copy of only the outermost list (known as a "shallow copy"), whereas deepcopy makes a copy of the outerlist and all sublists that might exist (also known as recursive copying). We believe that the most common thing that you will want to do is deepcopy (even when not strictly necessary), so we will just recommend you use that, unless you have a compelling reason not to.

```
In: l1 = ["red", 4, True]
In: import copy
In: l2 = copy.deepcopy(l1)
In: l1[1] = "a"
In: print(l1)
Out: ["red", "a", True]
```

```
In: print(l2)
Out: ["red", 4, True]
```

Here we can see that, even though we altered `l1`, we didn't alter `l2`, this is because the `copy.deepcopy()` function created a completely new list, and then we assigned this list to `l2`.

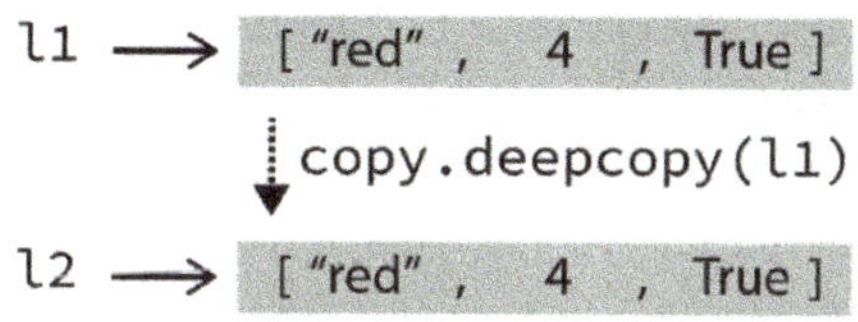

Then we only modified the object assigned to `l1`.

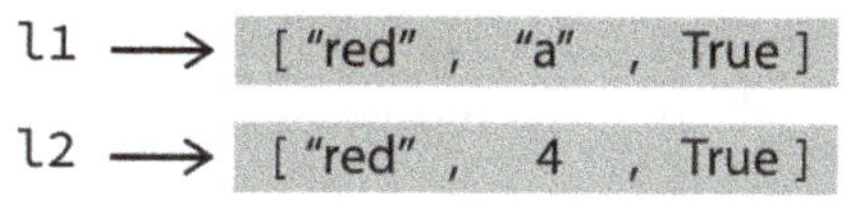

A final interesting behavior we consider here is with regard to the Python `method` `.sort()` and the `function` `sorted()`. These attempt to change the order of objects within a iterable object so that they follow a particular order (i.e., alphabetical). The difference is that `.sort()` does this "in place", meaning that it tries to reorder the items within the collection. However, this is a behavior that is only possible for mutable objects. Because immutable objects cannot be altered after their creation, you would need to make a copy of this object that is sorted, and this is what `sorted()` does. In this book, in order to avoid confusion, we are going to always use `sorted()`. Although this is slower and less memory efficient than `sort()`, this will not matter for the types of programs we are writing. An example of its usage is:

```
In:  l3 = [3, 5, 1]
In:  l4 = sorted(l3)
In:  print(p4)
Out: [1, 3, 5]
```

We can understand this pictorially as well:

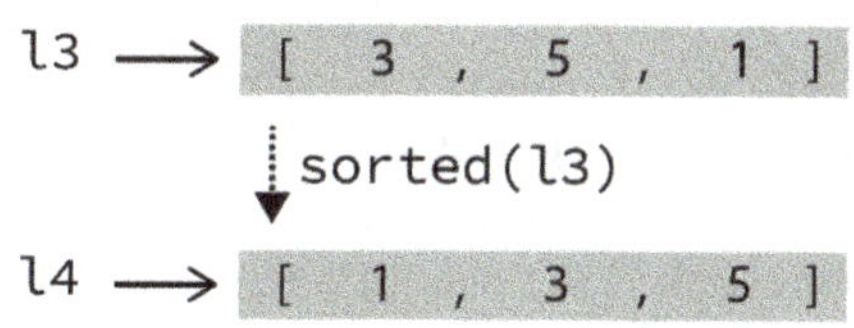

We note that the `sorted()` function accepts other **arguments**, the most useful of which is the keyword `reverse`, when set to **True**, the function returns the collection sorted in the reverse order.

B.1.3.0 Growing and Shrinking Lists

Because lists are mutable, we can adjust them dynamically. We have already seen how we can change single elements of a list, but we can also add elements to the list, or remove elements.

If we want to grow a list, we have a few options.

`.append()` is a method that adds a single element to the *end* of a list. If you supply a list, then you add that entire list as a single element.

```
In: l1 = ['red', 4, True]
In: l1.append('hi')
In: print(l1)
Out: ["red", 4, True, 'hi']
```

You can see that the element was added to the list, and we didn't need to assign the result to a new list. This behavior is described as modifying the list "in place."

`.insert()` is a method with two **positional arguments**. This method adds the supplied element at the specified index. All elements that were at this index and later are shifted one index greater. As with `append()`, we can supply only a single new element and the modification happens to the list in place.

```
In: l1 = ['red', 4, True]
In: l1.insert('yo', 1)
In: print(l1)
Out: ["red", 'yo',  4, True]
```

Here, the element "yo" was added at the index 1.

`.extend()` is a method that allows us to append multiple elements to the end of a list. In this case, we supply a list (or other iterable object) that will be added to the end.

```
In: l1 = ['red', 4, True]
In: l1.extend(["oh", "hi", 9])
In: print(l1)
Out: ["red", 'yo',  4, True, "oh", "hi", 9]
```

`+` is an operator that allows us to join to together two lists (called "concatenation"). The second list is added to the end of the first list. We can string together multiple lists this way, but concatenation results in the production of a new list, meaning that we have to assign the results to a variable, if we want to use it.

```
In: l1 = ['red', 4, True]
In: l2 = l1 + ["this", "is", "list", 2] + ["third", "list"]
In: print(l2)
Out: ["red", 'yo',  4, True, "this", "is", "list", 2, "third", "list"]
```

where we concatenated three lists together.

There are also ways to remove elements from a list.

`.remove()`. This method iterates through the list and removes the first instance of the supplied value. This is done on the list in place.

```
In: l2 = ["red", 'yo', 4, True, "this", "is", "list", 2, "third", "list"]
In: l2.remove("list")
In: print(l2)
Out: l2 = ["red", 'yo', 4, True, "this", "is", 2, "third", "list"]
```

`.pop()`. This removes an item from a list at a specified index. This is done in place and the method also returns the value at that index. This is useful, because you can then assign this to a variable.

```
In: l1 = ["red", 'yo', 4, True]
In: popped = l1.pop(2)
In: print(l1)
Out: ["red", 'yo', True]
```

```
In: print(popped)
Out: 4
```

B.1.4 Moving on to Other Collection Data Types

With an understanding of iterable, slicable, and mutable, we are ready to explore the rest of our data types. If you wish to see a table that lists all the types, and if they have these properties, please refer to Table B.3.

B.1.5 Sequence Types Hold Things in Specific Orders (List, Tuple, Range)

B.1.5.0 `list`

The Python `list` was already discussed in detail above; we only mention it here to note that it is a member of the sequence family.

B.1.5.1 `tuple`

The `tuple` is closely related to the `list`. We create tuples by placing the items you wish, separated by commas, between parentheses.

```
In: t1 = (8.0, None, "love")
In: print(t1)
Out: (8.0, None, "love")
```

When printing, Python indicates that the data type is a list by placing the values, separated by commas, between parentheses.

The tuple can hold values of any type, and it is iterable and slicable. Where it differs from the list is that a tuple is immutable. So, once we create it, we cannot change it or any of the elements within it. If we try to do so, Python will generate an error.

```
In:  t1[1] = "all you need"
Out: 'tuple' object does not support item assignment
```

B.1.5.2 range

The last sequence type is range. This data type is iterable, slicable, and immutable. Unlike list and tuple, range only holds one data type: the range type. We realize this seems circular, but... to some extent it really is. Python was defined so that there is a specific kind of data, known as a range. If we try to dig into it, we might conceive of the range type as representing an immutable sequence of integers, and so it might seem a bit like a tuple. However, it is *not* a tuple. It is a sequence of numbers of the range type.

This distinction is perhaps best illustrated by discussing how range types are created and used. We create a range type by using the function range( ), and then providing it between one and three integer **arguments**. The most explicit usage is with three values. The first value specifies the starting point of our number list, the second specifies the stopping point, and the third specifies the step size. One thing to note is that the start value is inclusive and the stop value is exclusive. We can also count in the positive or negative direction, depending on if the step size is positive or negative.

```
In:  r1 = range (1, 6, 2)
```

This will return a range object that has the numbers 1, 3, 5.

```
In:  r2 = range (2, -9, -3)
```

This will return an object that has the numbers 2, −1, −4, −7. However, these numbers are not directly represented when we call the full object. Instead, if you print the value, you get:

```
In:  print(r1)
Out: range(1, 6, 2)
```

So, Python does not directly produce the sequence of integers. Instead, the range **object** contains instructions on how to produce the numbers we specified. These instructions are followed and the numbers are then generated when needed. For this reason, range is also called a “generator.”

Because the instructions do allow the range to produce numbers on demand, we can check to see if a number is in the range object:

```
In:  3 in r1
Out: True
```

Additionally, we can use indices to access the numbers, by using the **index** we would use if the numbers where in a list or tuple data type. Because the range places numbers in order, we can see that the second number in our sequence should be three.

```
In:  r1[1]
Out: 3
```

Returning to creating a `range` object, if we use only two numbers, they are used as the start and stop values, and the step size is assumed to be one. Thus, we can only count forwards by one.

```
In: r3 = range(-1, 3)
```

This will produce an object with the numbers −1, 0, 1, 2. Also, if we print a range object where the increment is one, then it will not print this value.

```
In: print(r3)
Out: range(-1, 3)
```

If we use only a single number, then that number is used as the stop value, the start value is assumed to be zero, and the step size is assumed to be one.

```
In: r4 = range(4)
```

This will produce an object with the numbers 0, 1, 2, 3. If we print the range object, we will get the first value (0) in the print.

```
In: print(r4)
Out: range(0, 4)
```

Finally, if you specify conditions for which no values are found, then an empty object will be returned.

```
In: r5 = range(1, -3, 1)
```

This returns an empty value, because there are no numbers between 1 and −3 that can be arrived at using a positive stepsize. Printing the range will still tell us what we asked for:

```
In: print(r5)
Out: range(1, -3)
```

But we can verify this is empty by asking for the length of the range object:

```
In: len(r5)
Out: 0
```

And the length is zero because the object is empty.

Now, at first glance, it might seem strange to have a range object; why not just have a list with the numbers we want? There are three answers to this question. First, it is less verbose to use range than to make a list. For instance, getting the numbers 1 to 100 using range is simple: range(1, 101, 1). If you created a list with these numbers, it will take more typing. Now you might ask: why does Python return a range object, rather than just create a list or tuple from the range command? The answer is that the range data type can provide faster processing and use less memory, although it should be noted that *none* of the programs we write in this book would noticeably benefit from such considerations. If you go on to write much more complex programs, however, this consideration might become important. Now you might *still* think that you would want these elements as a list or tuple instead of a range type, which brings us to the third answer, and the next part of this section: Python doesn't need to have range be a `list` or `tuple`, because we can easily convert it to one. In fact, we can easily convert between any of our sequence types.

B.1.5.3 Converting Between Sequence Types

Converting to a list or to a tuple is straightforward: we simply use `list()` or `tuple()` **functions** on the sequence data type we wish to convert:

```
In: t1 = tuple(['red', 4, True])
In: t2 = tuple(range(1,10,2))
In: l1 = list((8.0, None, "love"))
In: l2 = list(range(3, -9, -3))
```

The first two these creates a tuple, first from list and then from a range. The last two create lists, first from a tuple and then from a range.

Although in some special cases, it is possible to create a range object from a list of tuple, we are not going to encounter the need to do so. Thus, in the interest of both clarity and space, we are not going to discuss how to do this.

B.1.6 Set Types Are Used to Ensure We Don't Duplicate Items

Imagine that you and a friend are going to host a party and you are going to meet up to shop. However, you both independently make a grocery list beforehand. You have some different ideas regarding what you want to buy:

you: beer, chips, salsa, paper plates

friend: soda, paper plates, ice cream

Now, when you meet up to shop, what will you do? You probably don't need double the amount of paper plates, so you will just combine the lists, keeping only the unique elements:

two buy: beer, soda, ships, salsa, paper plates, ice cream.

In mathematical parlance, you have just created a new set by forming the union of your individual sets.

This is the fundamental behavior of sets: they contain only unique elements (elements do not repeat) and one can perform set operations on them, like union.

In Python, there are two built-in types with these behaviors: sets and frozen sets.

The simplest way to make objects of this data type is to make a list and convert it to them using `set()` of `frozenset()` functions.[9] Because a set cannot repeat values, any duplicate values will be discarded.

[9] The type `set` can also be created using curly brackets, although we will not use this approach in our book.

```
In: s = set(["one", 2.0, 3, 2.0])
In: print(s)
Out: {"one", 2.0, 3}
```

```
In: fs = frozenset([1, 2, 3.0, 2.0])
```

```
In:  print(fs)
Out: {1, 2, 3}
```

In the example of the `frozenset` you will have *either* 2 or 2.0, but not both. The set treats numbers with the same value as the same, even if they have a different type. However, you cannot know which data type will be retained because sets are unordered: the order items are held and added is not preserved. This also means that if you are implementing these examples on your own computer, you might get a slightly different ordering of elements. That is `print(fs)` could have also returned `{2, 1, 3}`, or any other order of the elements.

One once we have these sets, we can perform set operations on them. In Python, there are four built-in operators to use between sets: union, intersection, difference, and symmetric difference:

- **union (`|`):**[10] This operation returns a set that contains all the unique elements across both sets. This can be thought of as combining the two sets, and the removing duplicate copies of any elements:

[10] If you are not used to coding, you might struggle to find out how to make this "vertical bar." If you have a US keyboard, then it is on the same key as the \ character.

```
In:  s | fs
Out: {1, 2, 3.0, 'one'}
```

You can see we have only the elements that were in common, and only one copy of each.

- **intersection (`&`):** This operation returns a set that contains all the elements present in *both* lists:

```
In:  s \& fs
Out: {2.0, 3}
```

- **difference (`-`):** This operation returns those elements that are present in the first set, but not the second:

```
In:  s - fs
Out: {"one"}
```

```
In:  fs - s
Out: frozenset{1}
```

Here, you can also see that the type of set returned is dictated by the type of set for the first set used.

- **symmetric difference (`^`):** This operation returns a set that contains all the elements that are present in only a single set. Thus, it can be thought of as performing the union of both possible differences: (s - fs) | (s - fs)

```
In: s ^ fs
Out: {1, "one"}
```

All of the above are useful operations, but do come at the cost of some of the other useful behaviors we have seen so far. Namely, sets are not slicable. So we cannot refer to elements by their index; if we try, we will get an error.

```
In: s[1]
Out: 'set' object is not subscriptable
```

```
In: fs[0]
Out: 'frozenset' object is not subscriptable
```

However, both sets and frozensets are iterable, and so we can use functions like `in` and `len( )`.

The difference between (`set`) and (`frozenset`) is that the first is mutable and the second immutable.

We will not explicitly encounter sets in the narrative portion of the book, although you might in the real world.

B.1.7 The Text Type Lets Us Represent Language

One type of data that we have used, but not discussed are **strings**. In Python, the name of this type is `str`, and it is created by placing text between two `'` or `'''` characters or between a series of three of these (i.e., `'''` or `"""`). The reason for having either single characters or triple characters is that the triple characters allow for declaration of strings that span multiple lines.

```
In: s1 = 'hi'
In: s2 = "hello"
In: s3 = '''Hello there.
In: Nice to meet you'''
In: s4 = """Howdy there partner.
In: You from these parts?"""
```

All of these create valid strings, which we can see by printing them.

```
In: s1
Out: hi
```

```
In: print(s2)
Out: hello
```

```
In: print(s3)
Out: Hello there. Nice to meet you.
```

```
In: print(s4)
Out: Howdy there partner. You from these parts?
```

Here, you can see that the line breaks used when we used the triple characters arc preserved in the output.

We can also create a string from another type of object, using the `str()` function.

```
In: s5 = str(2.0)
In: s5
Out: `2.0'
```

Here, the fact that the 2.0 is held between single quotations indicates it is a string, not a float. We can see this by trying to perform an operation that is only valid for numerical types, such as exponentiation.

```
In: s5**2
Out: unsupported operand type(s) for ** or pow(): 'str' and 'int'
```

In some sense, strings can be thought of as a tuple of characters. We say this because the string type is iterable and slicable, but not mutable—just as are tuples. Because they are iterable, one can use operators like `in` and `len()` on them.

```
In: len(s4)
Out: 29
```

Because they are slicable, we can get information by index.

```
In: s4[2]
Out: 'l'
```

We have to remember that indexing starts at zero.

And because they are immutable, we cannot change them once they are created.

```
In: s4[2] = "e"
Out: str' object does not support item assignment
```

Also, because strings act like they are sequences, we can use the concatenation (+) and repetition (*) operators.[11]

```
In: s5 = s1 + s2
In: s5
Out: 'hi hello'
```

```
In: s6 = s1 * 6
In: s6
Out: 'hi hi hi hi hi hi'
```

[11] For a discussion of these, please see the entry on lists on page B.1.3.0.

At this point, you might be wondering exactly how strings differ from tuples. Perhaps the most important difference is that tuples can hold any data type (including strings), but strings can only hold an ordered collection of characters. By characters, we mean the objects used to represent text, such as those used by ASCII. The default encoding for Python is UTF-8. If we try to explicitly pass some other object within a string, the entire sequence simply gets interpreted as a string.

```
In: "I have int(2) apples."
Out: 'I have int(2) apples.'
```

But this brings us to another important difference for strings: strings do some interpreting of the characters that they are given. That is, what you type into a string is not always what you get back. This (and generally working with text) is going to be so important that we have an entire chapter on it (Chapter C). So, if you want to know how to work with strings more, please go there.

B.1.8 The Dictionary Type Allows Us to Map, or Connect, Values

Our final built-in data type is the dictionary; called `dict` in Python. The dictionary mutable. It is iterable, but has some unique behaviors. It is also non-slicable, but can be made to respond to indexes. Both of these somewhat unusual behaviors will be explained below.

A dictionary is made up of a collection of **key:value** pairs. Perhaps the most useful way to conceive of this is as two paired lists. There is a list of keys and a list of values. Each entry in our list of keys is associated with exactly one entry in our list of values. We say that the values are "mapped" onto the keys.

The list of values is quite similar to a list in that each item can be any data type, including another dictionary. However, the list of keys can only hold string type data. Thus, the keys can be thought of as names, in some sense.

There are several ways to make a dictionary. The one most often used is to place key:value pairs, separated by commas, between curly brackets. When doing so, the keys need to be placed between quotations to indicate they are strings:

```
In:     d1 = {"integer":7, "float":3.14, "list":["hi", "there", True]}
```

This dictionary contains three entries: an `int`, a `float`, and a `list`.

One can also make a dictionary using the `dict()` **function** in two different ways. First, one gives a list of keys and values separated by the = operator.

```
In: d2 = dict(name='John', age=25, city='New York')
```

In this case, we do not need to explicitly make the key a string—the thing we type will be automatically converted to a string by Python.

In yet another approach to making a dictionary, one provides a list of two-item lists or tuples. Python converts the two-item lists into key-value pairs.

```
In: d3 = dict([('name', 'Ben'), ('age', 45), ('city', 'State College')])
```

There is one last way, which is to convert two lists into key:value pairs using the `zip()` function, and converting the zip object to a dictionary:

```
In: z4 = zip(["greeting", "year", "month"], ["hi", 2023, "June"])
In: d4 = dict(z4)
```

Although this goes through an intermediate stage of using `zip()`,[12] it is an efficient way to take two long lists and convert them into a dictionary.

[12] Perhaps the best explanation of how `zip()` works can be found in Section4.1.12 of the narrative section of this book.

We also wish to note that you can merge dictionaries using the unpacking operator (**). The unpacking operator will return key:value pairs for a dictionary, and so we can use this behavior to combine dictionaries:

```
In: merged_dict = {**d1, **d2}
In: print(merged_dict)
Out: {'integer':7, 'float':3.14, 'list':['hi', 'there', True], 'name':'John', ↩
     ↪'age':25, 'city':'New York'}
```

We can see that the two dictionaries are now combined into a single dictionary, and we can see how Python prints dictionaries.

At this point, it might seem like dictionaries are just pointlessly complex lists. However, as we start to describe how information in accessed from dictionaries, we hope their utility will become apparent.

B.1.8.0 Accessing Information from Dictionaries

If you have a dictionary, then you can get lists of both the key and the values using the `keys()` and `values()` **methods**, respectively.

```
In: d2.keys()
Out: ['name', 'age', 'city']
```

```
In: d2.values()
Out: ['John', 25, 'New York']
```

The most standard means to use dictionaries is to use the key to access the value:

```
In: d1["float"]
Out: 3.14
```

And we can combine these two to access values by iterating through keys...

```
In: for k in d2.keys():
In:     print(d2[k])
```

```
Out: John
Out: 25
Out: New York
```

There are two things to note here. First, the default behavior for interacting through a dictionary is to return the keys, so you don't actually need to specify that you want the keys. However, this does help with readability. Second, one could just iterate through the list returned by the `.values()` method; however, doing so side-steps the utility of the mapping supplied by the dictionary. If you just want a list of values, you might as well use a list, which is sliceable as well.

But this implies that there is some advantage to the mapping. And there is. The nice thing about a dictionary is that you have one set of things that you know the name for (keys) and they can point to things that you don't know the values for (values).

In science, imagine that you were fitting data to a line, and you get a slope and intercept. We could hold these in a dictionary that has keys of "slope" and "intercept." Now we have a single object that holds all the information for the line, rather than two variables. Also, even though the values of slope and intercept may change, we will always want to know what these values are. Thus, in some sense you can think of dictionaries of collections of variables that will point to values you want to use later.

We will see more clear examples in this book, but you hopefully start to see the power and utility of dictionaries.

B.2 COMMON TYPES ADDED BY LIBRARIES

The above discussion was limited to those types that are built-in to Python. However, it is often the case that we will be using types that are not built in. We will keep each of the following extremely brief. However, it is worth a brief word on why we introduce each of these.

B.3 NUMPY ARRAYS ALLOW FOR VECTORIZATION OF DATA

The Numpy array looks a bit like a list, in that it is iterable, slicable, and mutable. However, there are a few key differences between it and the list:

- *It is not straightforward for numpy arrays hold any type of data. They can only hold a single data type at a time.* Although mixing data types can be done, we will not show examples of how to do this because we think the power of arrays is most obvious when they hold a single type of data.

- *You can apply an operation to every element of a numpy array without using a for loop.* Thus, if you had an array (`a1`) that held the values 1, 2, 3 and you wanted to multiply all of them by 5, you could simple use `a1*5`.

- *It is much slower to add an element to a Numpy array than to a Python list.* This is important to understand because there are times you want to grow a collection and so if you ultimately want the final collection to be a numpy array, it is going to be significantly more efficient to grow the collection as a list, and then convert it to a numpy array.

B.4 THE PILLOW IMAGE TYPE LETS YOU STORE, MANIPULATE, AND DISPLAY IMAGES

There will be times you want to manipulate images or show them in results (i.e., as an insert in a plot). Pillow Image type can hold the data from an image and allow you to do these operations. It handles bitmaps and so is storing information pixel by pixel. It is iterable and slicable, but not mutable.

B.5 THE PLOTLY FIGURE LETS YOU BUILD AND DISPLAY PLOTS OF DATA

I doubt we have to convince you why plotting data is a good idea, so we will just point out that we are going to teach you to use the Plotly library, and the final product when doing so is a Figure object that holds the data, and all other information required to produce a well-formatted figure. This is discussed much more in the technical chapter on Plotly (Chapter I).

CHAPTER C

Working with Text

Most of your interactions with computer data occur through the process of entering or reading text. This text could be letters and words, numbers, symbols, or emoji, but the computer treats them all basically the same (as binary values). It is important to remember that the text on the screen, or that you enter via a keyboard, is just a representation of data—a pictorial indication of what the data is. The computer does not perform mathematical operations using numerals, and indeed, has no need for them. However, because humans are not typically adept in dealing directly with binary data, text is a useful intermediary between us and the machine. In this chapter, we will discuss how to use text as an expression of information in a way that humans and computers both can understand, how to send text to your program and have it give you text back, how manipulate it, and how to convert between text and numerical values.

C.0 WHAT IS TEXT DATA?

Text is just a collection of characters, which are individual symbols that humans recognize as representing letters, numbers, or other symbols,[1] as well as things you don't see, like spaces, new lines, and even literally nothing (the "NULL" character). Because data in a computer is binary, for your computer to understand text, some system must be devised that defines which binary numbers correspond to which characters. This is called an **binary encoding**. An early, widely used encoding is ASCII, which uses seven bits to identify capital and lower-case letters (although only those commonly used in the English language), numbers, a selection of symbols, and many non-displaying characters, for a total of 128 characters (Table C.1). Eventually, 128 characters was not enough, so many expanded encodings were developed. The one that has become dominant is UTF-8, which is backwards compatible with ASCII and uses up to four bytes to represent (at the time of writing) over a million characters. We will leave the details to other

[1] In some languages, characters can also represent syllables, things, concepts, and so on.

TABLE C.1 Codes for ASCII characters.

<table>
<tr><th rowspan="2">Digit 1</th><th colspan="16">Digit 2</th></tr>
<tr><th>0</th><th>1</th><th>2</th><th>3</th><th>4</th><th>5</th><th>6</th><th>7</th><th>8</th><th>9</th><th>A</th><th>B</th><th>C</th><th>D</th><th>E</th><th>F</th></tr>
<tr><td>0</td><td>NUL</td><td>SOH</td><td>STX</td><td>ETX</td><td>EOT</td><td>ENQ</td><td>ACK</td><td>BEL</td><td>BS</td><td>TAB</td><td>LF</td><td>VT</td><td>FF</td><td>CR</td><td>SO</td><td>SI</td></tr>
<tr><td>1</td><td>DLE</td><td>DC1</td><td>DC2</td><td>DC3</td><td>DC4</td><td>NAK</td><td>SYN</td><td>ETB</td><td>CAN</td><td>EM</td><td>SUB</td><td>ESC</td><td>FS</td><td>GS</td><td>RS</td><td>US</td></tr>
<tr><td>2</td><td>Space</td><td>!</td><td>”</td><td>#</td><td>$</td><td>%</td><td>&</td><td>’</td><td>(</td><td>)</td><td>*</td><td>+</td><td>,</td><td>-</td><td>.</td><td>/</td></tr>
<tr><td>3</td><td>0</td><td>1</td><td>2</td><td>3</td><td>4</td><td>5</td><td>6</td><td>7</td><td>8</td><td>9</td><td>:</td><td>;</td><td><</td><td>=</td><td>></td><td>?</td></tr>
<tr><td>4</td><td>@</td><td>A</td><td>B</td><td>C</td><td>D</td><td>E</td><td>F</td><td>G</td><td>H</td><td>I</td><td>J</td><td>K</td><td>L</td><td>M</td><td>N</td><td>O</td></tr>
<tr><td>5</td><td>P</td><td>Q</td><td>R</td><td>S</td><td>T</td><td>U</td><td>V</td><td>W</td><td>X</td><td>Y</td><td>Z</td><td>[</td><td>\</td><td>]</td><td>^</td><td>_</td></tr>
<tr><td>6</td><td>‘</td><td>a</td><td>b</td><td>c</td><td>d</td><td>e</td><td>f</td><td>g</td><td>h</td><td>i</td><td>j</td><td>k</td><td>l</td><td>m</td><td>n</td><td>o</td></tr>
<tr><td>7</td><td>p</td><td>q</td><td>r</td><td>s</td><td>t</td><td>u</td><td>v</td><td>w</td><td>x</td><td>y</td><td>z</td><td>{</td><td>|</td><td>}</td><td>~</td><td>DEL</td></tr>
</table>

Note: Three-letter characters are non-printable and are instructions for how to display or transmit characters.

resources because this is something that is mostly handled behind the scenes and so will not be critical for understanding this book. We have, however, included a library, `symbols()`, in the `codechembook` library at https://www.codingforchemistsbook.com/ that simplifies the inclusion of common characters in both command line and plot outputs. Also, because text has so many specialized functions associated with encoding and rendering characters, it is given its own **data type**: **strings**.

C.1 WORKING WITH TEXT IN PYTHON USING STRINGS

The Python standard library provides numerous functions for modifying, taking apart, and analyzing strings. Most string creation is done by coding string **literals** or by **functions** that read text from files, user inputs, or the internet. Here we present an overview of common string operations, and we refer you to the Python documentation for less-common ones.

C.1.0 CREATING STRINGS

Although you will probably most often create string literals by enclosing text between single (`'`) or double (`"`) quotes, there are many other ways to create strings.

C.1.0.0 Creating Strings from Other Data Types Using `str()`

Recall that there is a difference between the integer `1` and the string `'1'`. To turn the number 1 into the string, you can use the built-in `str()` function. This function actually works on any data type, although the output that you receive from it varies tremendously, particularly for collection data types. Particularly for floating point numbers, `str()` will tend to but not always give you the same number of digits of precision as the floating point number can provide.

C.1.0.1 How Strings Are Interpreted and Displayed

Because text data contains characters that represent ideas that often do not need to be drawn with "ink", Python needs a way to handle these characters. Consider the function of the Enter (or Return) key on your keyboard. We are very used to this creating a new line of text *and* having this represented as... well... a new line. However, within the data file for text, this is not the case. Instead, the text data is simply a uninterrupted set of characters with a special character used to indicate that there should be a new line. This "newline character" is represented by `\n` in Python. If we specify such a character in a string, we can see it is present:

```
In:  s6 = "One line \n Second line"
In:  s6
Out: "One line \n Second line"
```

However, if we ask to print the line, then the `\n` character will be interpreted as a new line:

```
In:  print(s6)
Out: 'One line
Out:   Second line'
```

There are a host of such special characters. For instance, tab is represented by `\t`. Additionally, there are a large number of characters we might wish to display, but which are not on a standard keyboard. For instance, in the US, the most common keyboard has no Greek characters, although we often use Greek letters in science. We can access such characters using the unicode designation: `\u+####`, where the # = any hexadecimal number. The unicode for the lowercase Greek delta is `u03B4` so we can use it in a string as follows:

```
In:  "\u03B4"
Out: 'δ'
```

Here, it is important to note that the capitalization of the "u" is important, and that the specification for u and U are different. This is taken up more in the chapter on text.

Also, because the `'` and `"` are used in the creation of a string, they are also special characters. This can create problems when trying to indicate quotations.

```
In:  'If you want to use a "double" quote to indicate quotations, then start ↩
         ↪the string with a single quote'
Out: 'If you want to use a "double" quote to indicate quotations, then start ↩
         ↪the string with a single quote'
```

```
In:  "If you want to use a 'single' quote to indicate quotations, then start ↩
         ↪the string with a double quote."
Out: "If you want to use a 'single' quote to indicate quotations, then start ↩
         ↪the string with a double quote."
```

The other is to use the \ character, which is also known as the "escape" character.

```
In:  "If you want to use \"double\" quotes to start the string and to ↩
        ↪indicate a quotation, then you have to use a backslash to escape ↩
        ↪the quotes used for quotations."
Out: If you want to use "double" quotes to start the string and to ↩
        ↪indicate a quotation, then you have to use a backslash to escape ↩
        ↪the quotes used for quotations.
```

C.1.0.2 Modifying How Strings Are Interpreted

We can control how Python interprets strings by preceding the ' or " characters that start the string creation with a letter. For this book, we will consider two: `r` and `f`.

First, *r-strings* are those for which interpretation of the characters is removed. We will get the characters as they are in the string.

```
In:  print(r"For r-strings, special characters like \n are not interpreted.")
Out: For r-strings, special characters like \n are not interpreted.
```

When we had a string with the `\n` character in a string before, printing that string resulted in a new line. However, in the r-string, it is not interpreted, and we see the newline character printed out.

This can be extremely useful when we want to write the string to be interpreted by something else... like a file.

Next, *f-strings* are those which allow direct interpretation of variables within them. We do this by adding an "f" in front of the string declaration, and then placing variables we want interpreted between curly brackets:

```
In:  print(f"A prior string we had was: {s1}")
Out: A prior string we had was: hi
```

Of course, as scientists, one of the most common uses of variables is going to be numbers, and so it is worth a separate discussion of how to format them.

C.1.1 Using Format Codes to Control the Display of Values in f-Strings

You will probably want to exercise more control over the appearance of the data inserted into f-strings. Handily, f-strings provide a means to do this by inserting a colon (:) after the variable name in the {} symbols, and after the colon, a special format code:

```
In:  f"The value of the variable concentration is {concentration:.6f}"
Out: 'The value of the variable concentration is 0.001000'
```

Here, we requested that the value of concentration be written out to six decimal places. An overview of common format codes is given in Table C.2.

TABLE C.2 Format Codes for Python f-strings

Format Code	Description
Strings	
{}	Default format for strings
{:<n>}	Left-align the string within a field of width *n*
{:>n}	Right-align the string within a field of width *n*
{:n⊠}	Center-align the string within a field of width *n*
Integers	
{d}	Decimal integer
{:b}	Binary format
{:X}	Hexadecimal format (uppercase)
Decimal Numbers	
{f}	Fixed-point number
{.nf}	Fixed-point number with *n* digits after the decimal point
{:,.nf}	Fixed-point number with *n* digits after the decimal point and comma as a thousand separator
Scientific Notation	
{e}	**e notation** (lowercase)
{E}	e notation (uppercase)
{.ne}	e notation with *n* digits after the decimal point

C.1.2 USING .format( ) TO INSERT VALUES INTO STRINGS DYNAMICALLY

An older and less readable way to insert variable values into strings is using the `.format()` **method**. You will see this a lot in older Python code because f-strings are relatively new. There are a number of ways to use this method:

```
In:  "The value of the variable concentration is {}".format(concentration)
Out: 'The value of the variable concentration is 0.001'
```

```
In:  "The value of the variable concentration is {0}".format(concentration)
Out: 'The value of the variable concentration is 0.001'
```

```
In:  "The value of the variable concentration is ↩
       ↪{concentration}".format(concentration = concentration)
Out: 'The value of the variable concentration is 0.001'
```

You can still use the format codes discussed above to format the text being inserted into the string:

```
In:  "The value of the variable concentration is {:.6f}".format(concentration)
Out: 'The value of the variable concentration is 0.001000'
```

C.1.3 Creating Strings Derived from Other Strings

Strings are immutable, and thus string modification methods must output new strings! Also indexing operations are available with them.[2]

[2] For a more thorough discussion on mutability and indexing, consult the chapter on data types: Chapter B

C.1.3.0 Concatenating Two Strings into One

Concatenating is the process of joining two strings to make one. The resulting string's length is the sum of the lengths of the two original strings. There are many ways in Python to concatenate strings, but the simplest is just to use the addition operator (+):

```
In: str1 = "Hello"
In: str2 = " World"
In: str3 = str1 + str2
In: print(str3)
Out: Hello World
```

C.1.3.1 Replacing a Subset of a String with a Different String Using .replace()

This method will look through a string (let's call it `parent_string`) and replace all instances of a shorter string (`child_string`) with a different string:

```
In: parent_string = "This book is the best book!"
In: parent_string.replace("book", "class")
Out: "This class is the best class!"
```

If you want to remove a substring entirely, simply use an empty string (`""`) for the second argument. If `child_string` is not found, then this method has no action. Remember, strings are not mutable, and thus this method produces a completely new string.

C.1.3.2 Changing Case

Python includes many standard **methods** to change the case (i.e., uppercase/lowercase) of words in strings. A few common ones are:

- `.upper()`: change all letters to uppercase.
- `.lower()`: change all letters to lowercase.
- `.capitalize()`: capitalize all words.
- `.swapcase()`: change capital letters to lower case and *vice versa*.
- `.title()`: force capitalization to follow title case conventions.

- `.casefold()`: like `.lower()` but useful for more characters than just the Latin alphabet.

C.1.4 TAKING APART A STRING TO CREATE MULTIPLE NEW STRINGS

Often, we have a string that has multiple substrings in it that we would like to extract. A simple example is breaking a sentence up into individual words. One common scenario in chemistry is reading text from a file or website that has numerical data presented in natural sentence form, like from simulation software. In this case, it is useful to break the string up into substrings and convert the relevant substrings into numerical data types.

C.1.4.0 Using a Specific String to Denote the Divisions Between Substrings with `.split()`

The example of splitting a string representing a sentence into individual words is a classic use case of `.split()`. This method searches a string for a given substring, and at each instance of that substring, divides the string into two pieces. After reaching the end of the string, the pieces are returned as a list of fragments. For the example of splitting a sentence into words, the natural substring is the space (`' '`).

```
In: parent_string = "This book is the best book!"
In: parent_string.split(' ')
Out: ['This', 'book', 'is', 'the', 'best', 'book!']
```

C.1.4.1 `.splitlines()`

Perhaps your string covers multiple lines by inclusion of a new line character (`\n`). You can return a list of one-line fragments of the string using `.splitlines()`:

```
In: multiline_string = "This book is the best book!\nBut there are other ↩
    ↪very good ones too!"
In: multiline_string.splitlines()
Out: ['This book is the best book!', 'But there are other very good ones too!']
```

This is particularly helpful when you get a large block of text all at once from a file or a website.

C.1.5 ANALYZING STRINGS TO DETERMINE SOME INFORMATION ABOUT THEIR TEXT

Sometimes, you don't actually want to change a string, just to know some information about it. There are many helpful methods to do this, a few of which are mentioned here.

C.1.5.0 Finding the First Instance of a Substring with `.index()`

This function searches the string for a substring and returns the index of the beginning of the substring:

```
In:  parent_string.index("book")
Out: 5
```

If the substring is not found, then a `ValueError` is raised.

C.1.5.1 `.find()`

The `.find()` is basically the same as `.index()`, with the only difference being that in `.find()`, if the substring is not found, then the value `-1` is returned. This can be useful if you want to straightforwardly handle not finding a string. Of course, you could handle the error thrown by `.index()`, but sometimes that would be more work.

C.1.5.2 Checking If the String can be Converted to a Different Type with `.is<type>()`

It is useful to check to see if you can do a type conversion before you actually try to do it. It is also useful to check in a comparison expression what the implied type of a string is (basically, "does this string represent a number?") There are a series of methods to do this:

```
In:  '18'.isnumeric()
Out: True
```

Note this evaluates to false if your number has a decimal point!

C.1.5.3 Determining If a Substring is in Another String with `StringA in StringB`

If you just want to know if a substring is in a string, you can use the `in` **keyword**:

```
In:  "book" in parent_string
Out: True
```

CHAPTER D

FILES AND FOLDERS

D.0 UNDERSTANDING THE FILE SYSTEM

Your computer is essentially useless without files and folders. Every program you use, website you visit, and piece of data that you collect, is stored in a file, which is stored in a folder, which is stored in another folder, and so on. This is true for apps on your phone as well, but your phone hides this fact from you better than your computer does. To work with experimental data, you will need to understand your file system.

D.0.1 ALL OF THE INFORMATION THAT YOU USE IS STORED IN FILES

Files make up essentially everything on your computer. Files are a container for data that makes the data easy to name, organize, copy, and transfer. They can contain almost anything: programs that you run, data that you analyze, documents that you read or write, videos of cats, or settings for your computer, to name just a few.

There's nothing special about a file; it is just an artificial construct that lets the computer keep all of these related pieces of data together. Your computer likely has tens to hundreds of gigabytes of data on it. There has to be a mechanism to figure out what data goes with what other data, and the first layer of this mechanism is the file. Imagine that every data point of your UV-vis spectrum that you just recorded was just *somewhere* different in your computer. In principle, the spectrum could still be recalled, but the work to find all of those individual pieces of data and put them together into a useful spectrum would be immense. Instead, why not invent a container for then and put them all in the container? And while you're at it, you might as well give that file a name so that you can remember what it in it or what it does. The container is a file, the name of the container is the filename.

It is worth pointing out that your computer also has files that you are not really intended to change or access. These can be files that are integral to how your computer works, and so changing them can lead to disastrous consequences. For this reason, most operating systems have a way hide these files (called "hidden files") so that they are not visible in your file browser. However, because there can be legitimate and useful reasons to access these files, most operating systems also have ways to temporarily or permanently unhide these files.

D.0.1.0 Files Have Names that (Hopefully) Describe Their Content and Extensions that Give Clues on What Program to Use to Open Them

Files have names that are text **strings**. You may have noticed that those names often take the format of `<name>.<extension>`, where `<name>` is something that you get to choose and `<extension>` is a seemingly arbitrary string of letters and numbers (or not so arbitary, like `.py`) that the program you're using appears to choose for you. For instance, you are probably used to seeing filenames like "Homework5.docx" for Microsoft Word files. Here, the filename is "Homework5" and the extension is "docx". Sometimes the extension of files are hidden by your computer. There is nothing magic about the extension, it is just there to help your operating system guess which program on your computer to use to interact with the file's contents. You can change the extension if you want, but it usually just makes your life harder.

D.0.2 Files Are Stored in Folders

The name "file system" is an allusion the old days of manila file folders and filing cabinets. These folders were used to store related files and had a tab on which you could write some description of the contents that should be found inside. This allusion is carried forward in a visual metaphor, where the folder icon is a stylized representation of a manila folder. This is basically what folders on your computer do: they group files that you think are logically related into collections. Folders (also often called directories) have names but have no real data associated with them, and they don't do anything by themselves. In most computers, no two files in the same folder can have the same name, and no two folders in a given sub-folder can have a same name. However, two files with the same name can quite easily be accommodated in two different folders.

Just as with files, there are folders that you either don't need to access or which accessing and changing would be dangerous for the continued operation of your computer. To help you avoid that outcome and to avoid unnecessarily clutter, operating systems often hide these folder. However, just as for files, you often can unhide them either temporarily or permanently.

Like files, folders can sometimes be hidden in your file system.

D.0.3 Folders Are an Organizational Hierarchy

Folders can contain other folders, which can in turn contain other folders, and so on. Each level of the folder hierarchy can also contain files, too. At some point, if you track this system to the top level of the hierarchy, you will find the "root" folder—the folder that is itself not contained in a folder. On a Unix-like system (including MacOS or Linux (or iOS or Android)), the root folder is actually referred to as "root" and is denoted by `/`.

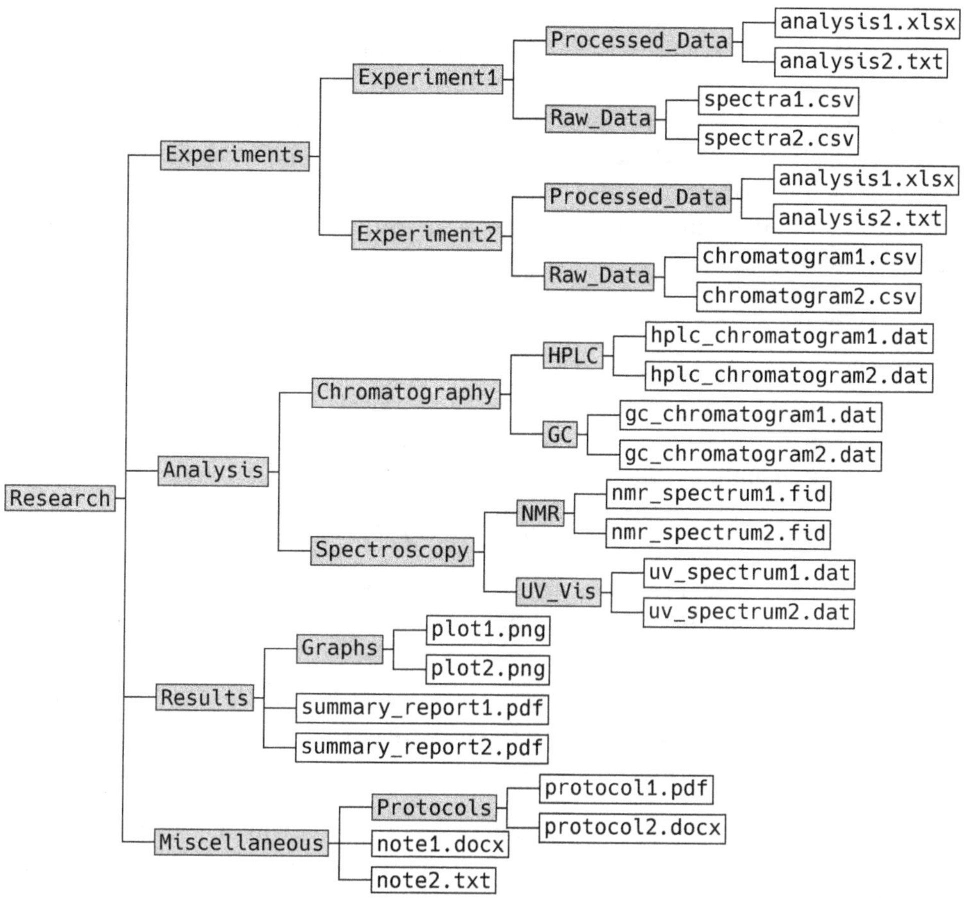

FIGURE D.1. View of a typical file structure that might look something like yours. Gray boxes denote folders and white boxes denote files.

On a Windows computer, the root folder that is denoted by a "drive letter," most often a "C", which means that the root folder is denoted `C:\`.

Figure D.1 presents a hypothetical piece of a file system. We can see that the main folder is called Research—this is not the root folder, just the highest level of the hierarchy that we are looking at. In the Research folder, there exist four more folders: Experiments, Analysis, Results, and Miscellaneous, and each of these folders contains more files and/or folders.

D.0.4 THE LOCATION OF A FILE OR FOLDER IN THE FILE SYSTEM IS CALLED A PATH

Within this hierarchical system of folders, sub-folders, and files, we need a method to *uniquely* identify any file or folder in the system. Because files in different folders or sub-folders in different folders can have the same name, it is not enough just to keep track of all of the names. In addition, there are billions of files on a computer—to have to search through all of them to find the right one would take a long time.

The unique location of a file or folder in a file system is called its **path**. The path contains the names of every folder that must be traversed to get from the root folder to the file or folder in question. Unhelpfully, Unix-like operating systems separate each subsequent sub-folder in the path with `/`, whereas Windows uses `\`. However, conceptually, both systems work the same way. So, for one of the authors (writing on a Mac) the path to the **Python interpreter** is:

```
/Library/Frameworks/Python.framework/Versions/3.12/bin/python3
```

where the first `/` denotes the root directory, `Library` is a folder in the root directory, `Frameworks` is a folder in the `Library` folder, and so on. Eventually, we get to `python3`, which is the actual file that is opened in order to run Python.

For the other author (writing on a PC) the path to the Python interpreter is:

```
C:\Users\Admin\AppData\Local\Microsoft\WindowsApps\python.exe
```

The operating system provides all of the facilities you need to move files or folders, delete them, copy them, and more, and many of these capabilities are also included in **libraries** in Python.

D.0.5 Absolute Paths Specify the Location of Any File or Folder in the File System

These paths are known as **absolute paths** because they start with the root folder. Absolute paths are the most specific way to specify the location of a file or folder, but because they are so specific, they are less flexible.

D.0.6 Relative Paths Specify the Location of a File or Folder with Respect to Some Other Folder

Sometimes, the exact specificity of an absolute path is undesirable. Let's take the example in Figure D.1. Suppose that this is the file structure required by the laboratory that you work in. You are going to write a script to work up data, and a co-worker expresses interest in using it. On your computer, the absolute path to "hplc_chromatogram.dat" may be

```
C:\Users\cjj\Research\Analysis\Chromatography\HPLC\hplc_
chromatogram.dat
```

but if you look at the location on your co-workers machine, you find it is

```
C:\Users\bjl\Research\Analysis\Chromatography\textbackslash
HPLC\hplc_chromatogram.dat
```

so the absolute paths are not the same, which means that your co-worker will need to change the path to use the script.

However, there is a solution! Looking at these two absolute paths, you can see that once you get to the folder `Research` the rest of the paths are the same. The structure of the entire file system is the same after `Research` folder. Thus, as long as you are thinking about the position of files from `Research` forward, the location will be the same on both of your computers. We would say that all the paths are the same "relative" to the `Research` folder, and this idea is known as a **relative path**.

In this case, the relative path from "hplc_chromatogram.dat" from "Research" is:

```
.\Analysis\Chromatography\hplc_chromatogram.dat
```

So, if you are producing a graph of this data, and you want to keep it in the "Graphs" folder (where you will store the plot), you can use the following relative path:

```
..\..\Analysis\Chromatography\hplc_chromatogram.dat
```

The first two usages of `..` move from the folder with the script ("Graphs") back to the "Research" folder, and then the path proceeds from there to the data.

If you write the script this way, then your co-worker will be able to use it directly, without any change.

Relative paths are not only useful when you are trying to specify paths on different computers, they're also useful shortcuts when you're just working on your own computer. Often, the file or folder that you want to specify is very close to the one in which you are working. Having to specify the absolute path for one or both files or folders would require a lot of typing, whereas the relative path is likely to be much shorter. In addition, if you move entire parts of your file system from one folder to another (let's say you're reorganizing your data, and move the hierarchy in Figure D.1 from a folder called "Active Projects" to a folder called "Finished Projects," then the absolute paths will change, but the relative path between them will not!

D.1 FILE SYSTEM METADATA PROVIDES ADDITIONAL INFORMATION BEYOND FILE AND FOLDER CONTENTS

You may notice when using Windows Explorer, the Mac Finder, or any number of tools in Linux, that you can see when a file was made, when it was last modified, what its size is, and many other attributes. This data is not actually part of the file—instead, it is collected and updated by the file system as **file metadata**. In addition to a map of the file hierarchy, the file system itself keeps a list of all of the properties of individual files and folders and updates this list every time a change is made to any file. Thus, if you can access this metadata, you can choose between two files, selecting that which was edited most recently.

D.1.0 METADATA IS NOT ALWAYS RELIABLE

Because the file metadata is maintained by the file system, it is not always reliable. For instance, if you copy a file onto a USB flash drive, and the flash drive is formatted a certain way, the date and time that the file was created may be lost. Once you copy that file onto another computer, from that computer's file system's perspective, it is a new file. File systems will try to reproduce each others' metadata, but it is not always possible. Thus, for critical applications (for instance, using the file created date to determine when a UV-vis spectrum was recorded), it is best not to rely on the metadata. Fortunately, most instrument or simulation software will note inside the files they produce when the file was made. If you make your own files in your own scripts, it is never a bad idea to do the same.

D.2 PYTHON PROVIDES A BUILT-IN LIBRARY TO WORK WITH THE FILE SYSTEM, BUT YOU SHOULD USUALLY USE `pathlib` INSTEAD

Python includes the `os` library which does provide many routines for interacting with the operating system and its file system, including routines to move, copy, delete, and rename files and folders. However, because Windows and Unix-like operating systems are so different, different routines must be provided to accomplish this. This means that, if you use the `os` library, you will need to write your code to account for either type of operating system, or accept that it will not be usable on one of the two. For instance, the use of relative paths discussed earlier in this chapter would only work across Mac and Windows if you had a way to change between \ and / characters as needed. This is either a lot of extra work, or a headache for potential users, and thus it is best to avoid this. The Pathlib library, discussed in Chapter E, alleviates most of these issues, for instance, automatically handling the change between \ and / when you move between operating systems.

CHAPTER E

PATHLIB

Needing to open, create, and modify files and folders on a computer is a common enough task that several libraries exit to handle these operations. In this book, we recommend using the library `pathlib`. Specifically, we suggest using the `Path` class.

Of course, the first question to address is: why do we need a library? Because files and folders and the paths to them can be represented by **strings**, then can't we always just supply a string when we are referring to files and folders? The answer is that we *can*; however, there are several functionalities that are missing from that approach. Learning to use `Path` will make code more general and will make handling files and folders more straightforward. A few examples of this are:

- The strings needed for Windows and Mac/Unix/Linux are different. In Windows, one separates folders and files with a backslash (\). In Mac/Unix/Linux, it is a forward slash (/). Thus, if we had a folder containing a file (file.fl), in Windows it is represented by "folder\file.fl", but on Mac/Unix/Linux, it is represented by "folder/file.fl". This means that if you write a code that uses strings for paths and might be used on different operating systems, then you will need to handle switching between these two representations. However, if you use `Path`, then this is handled automatically.

- When we look at a path to a file, we automatically break it down into components. For instance, when looking at "folder/file.fl", we can see there is a folder and a file and a file extension. It is common to call the folder the "parent." If you use strings to give file paths, then the string has no idea about these components. However, if you use `Path`, then you will have an object that includes ideas like parent and extension. Below, we will see how to access these components and how they can simplify file handling.

E.0 CREATING A `path` OBJECT

There are two ways to generate a `Path` object, both involve the function `Path( )`:

1. Supplying a string representing a path to a file or folder to the function `Path( )`. The string can be formatted using Windows or Mac/Unix/Linux conventions.

2. Supplying any number of strings separated by commas. The `Path( )` function will treat each of these as a component of the path, with them occurring in the order that they are supplied.

This creation of a path object requires a string be provided that contains the information for the path to the object. You can get this string by typing in the information, copying and pasting from your computer, or using native file dialogs, as introduced in Chapter 2. However, there is one other way: using Pathlibs `glob( )` function. This function will look inside of a specified folder and return all the files within it. You can supply an argument that is a string specifying a pattern that filenames must match to be returned and this can contain wildcards. for instance, `glob("*.csv")` will return all filenames that end in .csv—in other words all csv files, because the `'*'` character acts as a **wildcard**.

And there is a variant of this function, which will recursively glob—that is, it will pull all files from the current folder and within any folders contained within that folder, and so on. This function is `rglob( )`

E.1 ACCESSING COMPONENTS OF `path` OBJECTS

There is other functionality that you might find useful, for instance, using attributes of an existing Path **object**:

- **Access a filename.** If all you want is the filename (including the extension), you can use `<path object>.name`

- **Access the filename without the extension.** If you want the filename but not the extension, you can use `<path object>.stem`

- **Access the extension only.** If you only want to know the file extension (for instance, to know what kind of file it is), you can use `<path object>.suffix`

- **Access the path to the parent folder.** To find the folder that contains the file (the parent folder), you can use `<path object>.parent`

- **Access any folder in the path.** To get just a folder in the path, you can use <path **object**>.parents[n], where the parents method returns an **slicable object** containing the folders in the path, ordered from the folder containing the object to the root folder, and [n] selects the nth folder in this list. In this instance, counting starts at the parent folder (which contains the file) and goes until you get to the root directory (where the path starts). So, if you only want the parent folder, you can use <path **object**>.parents[0], whereas the root folder is obtained using <path **object**>.parents[-1]

E.2 MODIFYING EXISTING path OBJECTS

You can modify existing Path objects as follows:

- **Adding to an existing path**
  This is done using the "/" operator. For instance, one might have a folder with several files (file1.fl, file2.fl, etc.) and you want to create a Path object for each. You could supply the full path to each, or you can create a path to the folder, and then extend this for each file. If the folder is in a path such as folder1/folder2/folder3, then we can do the following:

```
from pathlib import Path
folder = Path("folder1/folder2/folder3")
f1 = folder/"file1.fl"
f2 = folder/"file2.fl"

```

  In this example, folder, f1, and f2 are all path objects. The folder object leads to the folder containing the files, and f1 and f2 lead to their respective files.
  In constructing paths in this way, the variable that precedes the \ character *must* hold a Path object. The object that follows the \ character can be a Path object or a **string**.

- **Changing the extension**
  To change the extension, you can use the with_suffix() method of the object. When you supply an extension, you must still use the dot, and the extension must be a string data type. So changing the extension of a file to be csv would use <path **object**>.with_suffix(".csv").

- **Renaming the file**
  To change the name of the file you can use the rename() method of the Path object. When you do that, you need to supply the path to where you want to have the renamed file. So you use <path **object**>.rename("<new_file_path>").

CHAPTER F

CONTROL STRUCTURES

It is often helpful to have a computer make your life easier or faster by doing math for you, or doing a set of tasks for you, as in the case of using arrays, but the real power of programming starts when the program can make its own decisions on what to do and when to do it without the programmer having to anticipate exactly what needs to happen in exactly what order. Consider how you interact with your computer—which does not know which application you will open at any given time. Being able to handle a wide-range of inputs in an arbitrary order allows us to move from writing simple scripts that perform a list of instructions to writing general purpose software that can be used by different users with different goals and data. In this section, we will first introduce you to the concepts underlying the process of decision-making in computer code, and then we will discuss specific control structures that you will use to lay out the decision-making process.

F.0 GETTING YES OR NO ANSWERS: COMPARISONS AND BOOLEAN LOGIC

Remembering that computers only actually operate based off of ones and zeros or True/False values, we first have to ask, how can they make decisions? How can we approach questions in which there may be more than two possible answers, or processes that could yield more than two outcomes?

A useful way to think about designing a computer decision-making process is to turn it into a series of questions like, "if some specific condition is met, then do these instructions." This breaks every decision down to a question with a Yes/No or True/False answer. Most of the decision-making structures used by computer programs have this basic question at their core. It is the job of programmers to design and order these questions to get the behavior that we want.

F.0.0 Comparison Statements Compare Two Values

Comparison statements are the questions that we design to give True/False answers to feed decision-making processes. Comparison statements will look familiar to you from earlier math classes, but they are not limited to only numbers, and there are more ways to compare data in programs than you learned in algebra or calculus.

Comparison statements usually involve two values and a comparison operation, taking the form

```
(Data) A <comparison operation> (Data) B
```

They can be used in assignment operations in much the same way as other mathematical statements, like

```
Result = (Data) A <comparison operation> (Data) B
```

where `Result` is a variable that will hold a boolean value. This usage implies that the outcome of a comparison operation is always either `True` or `False`.

Examples of comparison operations from mathematics, are "is greater than" ($>$), "is less than or equal to" ($\leq$), or even "is equal to" ($=$). Note that, in this context, we are not assigning something to be equal to something else, but rather asking *if* it is; sometimes, this is denoted "is equivalent to" to avoid this confusion. The symbols used to represent these operations are similarly familiar, but note that the computer has a hard time figuring out context and thus, in programming, == is typically used to denote "test if these two values are equal," whereas = is the assignment operator. The syntax for these statements is then conveniently natural: `A >= B` can be simply read as "A is greater than or equal to B?" where the comparison statement implicitly ends with a question mark. The response from the computer is then one of the two options: `True` or `False`. Like any other statement, if you type this code into the Python command prompt, the result is output to the standard output, usually the line after the command prompt.

One straightforward comparison operator with which you may not be familiar is "is not equal to", which in Python is denoted `!=` and simply behaves the opposite of "is equal to." Another is "is in the collection", which tests to see if a given value is in a given collection (list, set, tuple, and so on). This comparison returns `True` if the value is in the collection, and `False` if it is not.

F.0.1 Complex Comparisons can be Evaluated Using Boolean Logic

The basic comparison statements above only allow the comparison of two values, but we often find ourselves needing to come up with more complex tests. The solution to this problem is much like the one we know from evaluating complex arithmetic expressions (which also deal with two values at a time): we evaluate one comparison statement to produce a result, and replace that comparison statement with the result of the evaluation and continue until there is no comparison operator left. Thinking about arithmetic, we know that we can simplify $(1+2)\times(3+4)$ as follows:

$$(1+2)\times(3+4)$$
$$3\times(3+4)$$
$$3\times 7$$
$$21$$

Now, all we need is a form of arithmetic that works on boolean statements and we are set. Luckily for us, mathematicians have worked this out long ago.

Let's consider a contrived example: we have some solvents in the lab, but only two of them (methanol and water) can be used for a specific reaction. If we want our program to test if the solvent we have in our hands is one of these two, we (as humans) might think "solvent is equal to methanol or water." However, this will not work in Python (nor most other languages). The problem is that this single statement is actually three distinct comparisons and, when programming, we need to explicitly state all three comparisons. So, we can write "solvent is equal to methanol or solvent is equal to water." If you are struggling to see how this is three comparisons, then realize that "or" is also a comparison operator.

This is how boolean algebra is constructed. We only have two possible values, true and false, and we need to break down our work into simple statements that can return these values.

These two values can only be related so many ways: they could be the same and both true, they could be the same and both false, or one could be true and the other false. A number of different boolean operators can thus be devised to compare the two values, with a general form:

```
Result = Boolean Value A <boolean operator> Boolean Value B
```

The most common ones that you will encounter are `and` and `or`. Note that `and` evaluates to `True` only when both values in the expression are True, and `False` for all other cases (Table F.1). Meanwhile, `or` evaluates to `True` whenever *at least one* value is `True` (including when both are), as shown in Table F.2.

TABLE F.1 Truth table for `and`

A	B	A AND B
False	False	False
False	True	True
True	False	True
True	True	True

TABLE F.2 Truth table for `or`

A	B	A OR B
False	False	False
False	True	True
True	False	True
True	True	True

With one other operation, we can perform any comparison we want: `not`, which is an operation only on one value that converts it to the opposite value (`not True` evaluates to `False` and vice versa). These three operations allow us to generate arbitrarily complex comparison statements. For instance

```
In: pH > 7.0 and I > 1.0
```

returns `True` only when pH is greater than 7.0 *and* I is greater than 1.0, and `False` in all other circumstances. We can break this down to see how the statement is evaluated. Suppose `pH = 7.4` and `I = 2.2`:

$$7.4 > 7.0 \texttt{ and } 2.2 > 1.0$$
$$True \texttt{ and } 2.2 > 1.0$$
$$True \texttt{ and } True$$
$$True$$

Suppose instead that `I = 0.5`:

$$7.4 > 7.0 \texttt{ and } 0.5 > 1.0$$
$$True \texttt{ and } 0.5 > 1.0$$
$$True \texttt{ and } False$$
$$False$$

Meanwhile

```
In: pH > 7.0 or I > 1.0
```

returns `True` when pH is greater than 7.0 *or* I is greater than 1.0, as well as when pH is greater than 7.0 *and* I is greater than 1.0. For the last example:

$$7.4 > 7.0 \texttt{ or } 0.5 > 1.0$$
$$True \texttt{ or } 0.5 > 1.0$$
$$True \texttt{ or } False$$
$$True$$

It is important to control the order in which the individual operations are executed in complex comparison statements. There is a systematic set of rules for this, similar to algebra (remember PEMDAS?). In everyday use, it is good practice to use parentheses to make your statements clearer to a human reading your code and to avoid the possibility of errors, particularly as your statements become more complex. Suppose we add a condition to the first example so that we now want to do something not only if pH is greater than 7.0 and I is greater than 1.0, but also if pH is greater than 10.0 regardless of I. The code for this comparison statement would look like

```
In: (pH > 7.0 and I > 1.0) or pH > 10.0
```

We don't necessarily need the parentheses here, but it makes it much easier to figure out quickly what is going on.

F.1 CONDITIONALS: CHOOSING BETWEEN TWO OR MORE OPTIONS

Now that we know how to design tests, we are ready to use them to tell the computer how to make decisions and what to do after these decisions are made. There are several common decision structures, which can basically be broken down into two categories: conditionals, which execute different sets of instructions depending on the results of a comparison, and loops, which execute instructions over and over until a specific comparison statement is met. First, we will discuss specific implementations of the former.

F.1.0 `if`: DO SOMETHING OR DON'T DO IT

The simplest flow control statement that we have at our disposal is the `if` statement (sometimes called "if-then" statements): "`if <comparison statement evaluates to True>`, then do these instructions." If the comparison statement ends up being `False`, those instructions are not executed. Figure F.1 demonstrates this process in the form of a flowchart. In Python, this control structure is expressed as follows:

```
pH = float(input('What is the pH of your solution?'))

if (pH <= 2.0):
  print('Wear rubber gloves and a face shield, that is a strong acid!')

#<rest of code>
```

Note that the indentation following the `if` statement indicates the code to be executed if the comparison statement evaluates to `True` (sometimes called the 'then' part of an 'if-then' statement).

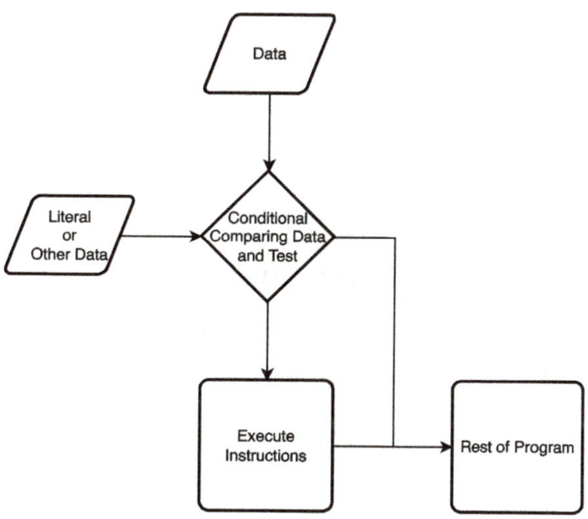

FIGURE F.1. Flowchart depicting the operation of an `if` statement.

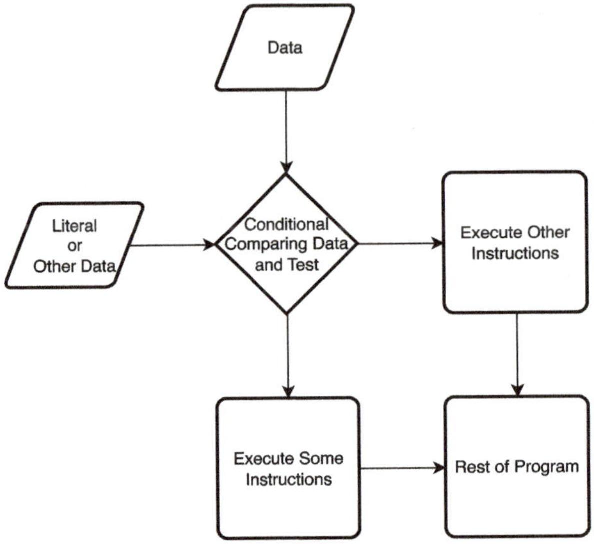

Figure F.2. Flowchart depicting the operation of an `if-else` statement.

F.1.1 `if-else`: Do One of Two Things

Sometimes we will want to run some other code if the comparison in the if statement ends up being false, as shown diagrammatically in Figure F.2. In this case, we can add an `else:` command at the end of the indented block of code that is supposed to be run if the comparison is true. The code for such a statement looks like this:

```
pH = float(input('What is the pH of your solution?'))

if (pH < 7.0):
  print('Your solution is acidic')
else:
  print('Your solution is basic')

#<rest of code>
```

Note here that the code to be executed if the comparison is `False` is also indented in its own block, while the rest of the code removes this indent.

F.1.2 `if-elif`: Do One of Many Things

We may still want to handle more possibilities than we can with just the `then:` and `else:` cases. The intuitive way to do this is just to put another `if` statement inside the code block of the `else:` case:

```
pH = float(input('What is the pH of your solution?'))

if (pH < 7.0):
  print('Your solution is acidic')
else:
  if (pH > 7.0):
    print('Your solution is basic')
  else:
    print('Your solution is neutral')

#<rest of code>
```

This is straightforward but not ideal:

```
pH = float(input('What is the pH of your solution?'))

if (pH < 3.0):
  print('Your solution is strongly acidic')
else:
  if (pH < 7.0):
    print('Your solution is acidic')
  else:
    if (pH == 7.0):
      print('Your solution is neutral')
    else:
      if (pH < 11.0):
        print('Your solution is basic')
      else:
        print('Your solution is strongly basic')

#<rest of code>
```

There are a few things to worry about here:

1. The code is harder than it needs to be to read. Your eyes have to search to find each ensuing comparison, you are running out of horizontal space on each ensuing line because of the need for white space, and it is generally ugly.

2. Although the code as written works, we have to be careful about the comparisons we choose. If, for whatever reason we had chosen a different order, we could see unexpected behavior.

3. The code assumes without checking that the user actually entered a number and that the number was in a reasonable range for pH values.

With regard to point 1, a shorthand notation exists to make the code cleaner: combining the `else` and the next `if` into `elif`. The flowchart for a `if-elif` statement is shown in Figure F.3.

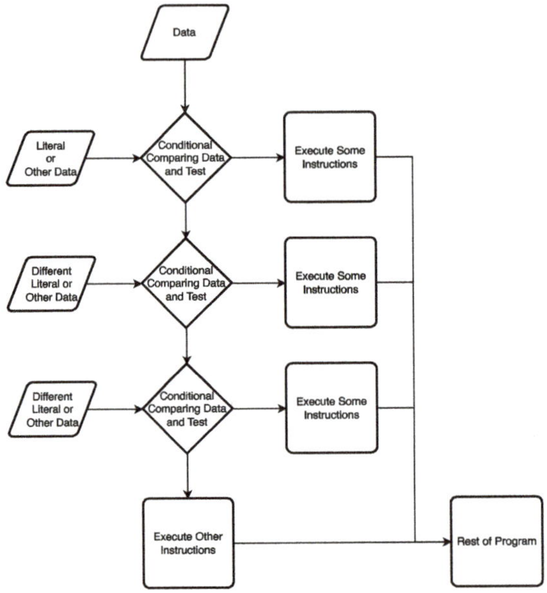

FIGURE F.3. Flowchart depicting the operation of an `if-elif` statement.

```
pH = float(input('What is the pH of your solution?'))

if (pH < 3.0):
  print('Your solution is strongly acidic')
elif (pH < 7.0):
  print('Your solution is acidic')
elif (pH == 7.0):
  print('Your solution is neutral')
elif (pH < 11.0):
  print('Your solution is basic')
else:
  print('Your solution is strongly basic')

#<rest of code>
```

This code certainly looks nicer and it is much quicker to figure out how the different cases are actually different.

F.1.3 POSSIBLE PITFALLS OF CONDITIONALS

With regard to point 2, consider if we simply reversed the order of the first two comparisons:

```
pH = float(input('What is the pH of your solution?'))

if (pH < 7.0):
  print('Your solution is acidic')
elif (pH < 3.0):
  print('Your solution is strongly acidic')
elif (pH == 7.0):
```

```
  print('Your solution is neutral')
elif (pH < 11.0):
  print('Your solution is basic')
else:
  print('Your solution is strongly basic')

#<rest of code>
```

This *looks* like it should do the same thing, but if the user inputs `2.0` when prompted, the response will be `'Your solution is acidic'` instead of the expected `'Your solution is strongly acidic.'` This example illustrates a common problem in nested `if` statements: the order in which the comparisons are executed can matter. It is best to be as specific as possible when creating your comparison statements. For instance, we can avoid the order of the statements mattering by comparing to specific ranges:

```
pH = float(input('What is the pH of your solution?'))

if (pH < 3.0):
  print('Your solution is strongly acidic')
elif (pH < 7.0 and pH >= 3.0):
  print('Your solution is acidic')
elif (pH == 7.0):
  print('Your solution is neutral')
elif (pH <= 11.0 and pH > 7.0):
  print('Your solution is basic')
elif (pH > 11.0):
  print('Your solution is strongly basic')

#<rest of code>
```

This code now has a specific range for each possible *numerical* value of pH and will execute as expected regardless of the order of the comparison statements.

In the last modification of the code, we also note that comparing the input to a range for each possibility also required removal of the `else` statement. This is also good practice for this kind of application of the `if-elif` construct, and it leaves the `else` outcome available to use as a potential way to simply handle user input errors. We could further modify the last example:

```
pH = float(input('What is the pH of your solution?'))

if (pH < 3.0) and pH >= -10.0:
  print('Your solution is strongly acidic')
elif (pH < 7.0 and pH >= 3.0):
  print('Your solution is acidic')
elif (pH == 7.0):
  print('Your solution is neutral')
elif (pH > 7.0 and pH <= 11.0):
  print('Your solution is basic')
elif (pH > 11.0 and pH <= 20):
  print('Your solution is strongly basic')
```

```
else:
  print('That is a crazy pH value.  Either you made a typo, or you should write a paper on that ASAP!')

#<rest of code>
```

F.1.4 When to Use `if-elif` Versus Several `if` Statements

It is important to realize that an if-else or if-elif statement will only execute *one* possible pathway: the first comparison statement that it comes to that evaluates to `True`. No other pathway after that will execute, even if its comparison statement also evaluates to `True`. Often this is the desired behavior, but not always. Suppose, for instance, that we wanted to identify the hazards associated with a particular chemical. If we only used an `if-elif` statement, the first hazard that we come across causes that statement to run and no other ones, subsequently continuing the program. However, chemicals can certainly have more than one hazard, and thus this would yield incorrect output of this program. In this circumstance, you should use a series of `if` statements: each hazard would be tested, and each one that the chemical satisfies will cause the code specific to that hazard to execute, as expected.

F.2 LOOPS: DOING THE SAME OPERATIONS OVER AND OVER

When humans do the same thing over and over again, they get bored and may make mistakes. Computers have no trouble doing things over and over and doing them perfectly each time. In fact, this is one of the things at which they excel.

Loops are the control structures that allow you to tell the computer to do repetitive tasks. It might be tempting to simply copy and paste the code we want to repeat until we get the desired number of repetitions, but there are a few flaws with this approach: (1) if you need to make a change to the repeated code, you have to change it in every instance, and (2) you have to know how many repetitions should run, and if that number changes, you have to change the code. It would be better, then to only tell the computer what to do in one repetition, and then either tell it how many repetitions to do, or how to decide for itself when to stop repeating. There are two basic types of loops to handle these tasks.

F.2.0 General Remarks on Loops

Regardless of the type of loop, there are a few common things we need to know. First (in Python), the loop is denoted by a statement that tells the computer how to determine how many times to run the loop, followed by a colon (:). Next, an indented block of code tells the computer what instructions to run for each repetition of the loop, where a repetition of the loop is known as an **iteration**. The computer will run these commands in

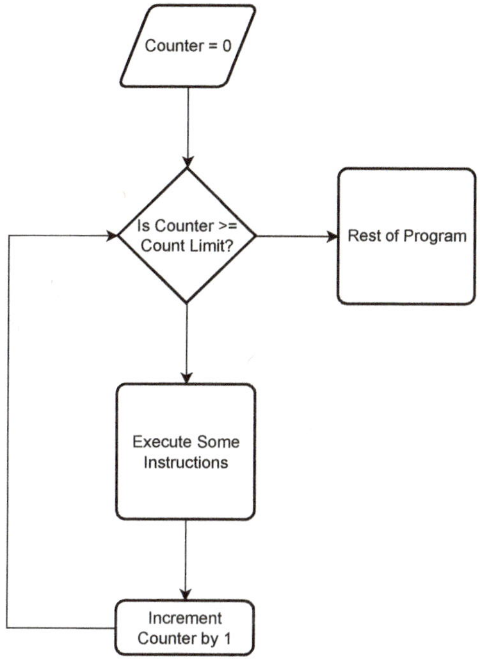

FIGURE F.4. Flowchart depicting the operation of a for loop.

order until there are no more lines of code at the same level of indentation, at which point it will decide whether to stop iterating and move on to the rest of your code, or it will go back to the first line of the indented black and start the next iteration. You will commonly find people saying "iterating through the loop" to describe the process of going from one iteration to the next.

Loops can utilize **nesting**, meaning that there is a loop within a loop. This can be useful for dealing with 2D or higher dimensional data such as images or matrices, in which each x and y pair of values must be addressed independently. For instance, in a case with two nested loops, for each iteration of the outer loop, the inner loop will run all of its iterations in a row. At the end of the iterations of the inner loop, the outer loop will move on to its next iteration and the inner loop will begin again. Always be careful with nested loops because the number of iterations of the nested loops grows as the product of the number of iterations of the individual loops!

In almost all cases, if you update variables defined outside of the loop in any given iteration, that variable will retain its value for the next iterations until you change it again.

F.2.1 `for` Loops: Do the Same Set of Instructions a Fixed Number of Times

For loops (Figure F.4) are a type of loop that iterates a set number of times. In Python (but not in many other languages), we do not directly tell the for loop how many times to iterate. Instead, we provide it with a variable of a type that is iterable,[1] and the for loop runs one time for each element of this iterable variable. Python provides a specific syntax to do this:

[1] See Chapter B for details.

```
for <temp variable holding element> in <collection>:
  <code to repeat>
```

What is happening here is that Python takes a look at your iterable variable to see how many elements are in it—from that it will determine the number of iterations. Then, the for loop will iterate one time for each element in the iterable variable, in the order that they arise in the iterable variable. For the purposes of the code to repeat, that element will be referred to as whatever variable name you give in the `<temp variable holding element` position of the for statement. If you perform an operation within an iteration that changes the value of this element, those changes are reflected in the iterable variable. When the loop moves from one iteration to the next, the value in this temporary variable is replaced by the value of the next element in the iterable variable.

```
for pH in pH_array:
    if pH > 7.0:
        print('Sample is basic')
    elif ph < 7.0:
        print('Sample is acidic')
    else:
        print('Sample is neutral')
```

Sometimes, we don't specifically want to iterate over the elements of an iterable like a list, but rather to do something a set number of times or to know which iteration we are doing. In this case, we can create an iterable that just contains the iteration number using the `range` type,[2] which produces an iterable containing the integers 1, 2, 3, ... We can then use this `range` variable as the iterable variable, and thus our variable holding the value of the iterable variable for each element is simply the iteration number.

[2] For more on this data type, see Chapter B.

```
for i in range(len(x)-1):
    delta_x[i] = x[i+1] - x[i]
```

Note that this is not necessarily the best way to do this!

F.2.1.0 `enumerate`: Get the Iteration Number and a List Element

Sometimes, in each iteration we want to know *both* the iteration number and the current element of the iterable variable. For this, we can use the `enumerate` type, which returns a tuple like (<iteration number>, <current element of iterable variable>) for each iteration. We could achieve similar results by simply iterating over a `range` and then using the current iteration number as the index of an iterable variable, but the `enumerate` type makes this simpler and more readable. Section 6.1.4 gives some example uses.

F.2.1.1 `zip`: Iterate Elements from Multiple Lists Together

Similarly, we sometimes would like to iterate over two or more different iterable variables, like the x- and y-coordinates of some data. We can do this with a `zip` type, which is similar to `enumerate` but it returns a tuple containing the current elements of two or more iterable variables. Again, we could achieve the same essential functionality by iterating over a `range` and indexing each iterable variable, this is easier. Section 4.1.12 gives some example uses.

F.2.2 `while` LOOPS: DO THE SAME SET OF INSTRUCTIONS REPEATEDLY UNTIL A CONDITION IS MET

For loops are great when there is a straightforward way to determine the number times a loop should iterate, and although this is generally preferable, it is not always possible. In these cases, we can use a `while` loop (Figure F.5) instead. A `while` loop statement contains a conditional test, and the loop will continue to iterate as long as this test evaluates to `True`. Notably, if the conditional test is not `True` when the while loop is first encountered, it will not run at all. If the conditional test becomes `True` in the middle of an iteration, that iteration will complete before the loop executes.

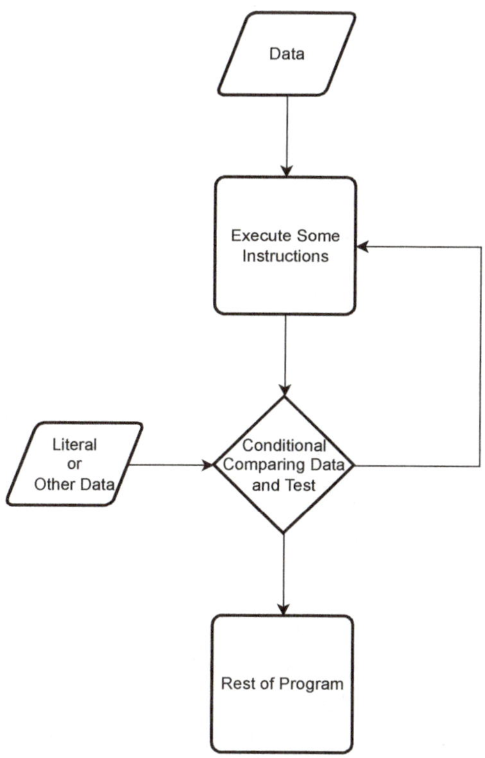

FIGURE F.5. Flowchart depicting the operation of a while loop.

```
data = []
stop = False
while stop is not True:
    pH = input('Enter the pH, or "stop" to stop.')
    if pH == "stop":
        stop = True
    else:
        data.append(float(pH))
```

Although useful, `while` loops have two major disadvantages. First, they are slower than `for` loops because Python doesn't have a way to know how many times they will iterate. Secondly, and more importantly, it is possible (easy, even) to create a `while` loop that *never* ends:

```
counter = 0

while counter == 0:
  print(counter)
```

Because `counter` never changes, each time an iteration ends, the conditional test `counter == 0` will always be true.

F.2.3 `with`: Letting Python Take Care of Boring Stuff for You

In coding, there are many things that you have to just do that don't actually move you towards a solution to a problem. To the extent that a language reduces the number of these things, it will likely make you more productive. The `with` structure is one instance of Python relieving you of tedium.

The most common scenario in which you will encounter `with` is reading or writing data from/to a file. A typical code that reads arbitrary data from a file looks like this:

```
file_to_read = open(<filename.txt>)
<code to read and parse input>
file_to_read.close()
```

Using `with`, this is somewhat simplified:

```
with open(<filename.txt>) as file_to_read:
  <code to read and parse input>
```

This may not seem much better, but there are two advantages here: first, Python will figure out when you need to close the file. This can be useful for codes that do more complex file I/O. Secondly, and more importantly, sometimes your code will create an error and stop executing before you have a chance to close the file. If this happens, the file could remain open, preventing another program from using or changing it, or it could become corrupted, causing the important data stored in it to be no longer usable. The `with` statement will notice when the code in its code block throws an error and will still close your file for you, preventing most of these unfortunate situations. Thus, it is almost always better to let `with` handle file opening and closing for you.

F.2.4 `try`: KEEP YOUR CODE EXECUTING EVEN AFTER AN ERROR OCCURS

Sometimes (often), your code will have one of two errors: a logic error, in which the code runs as intended but does something that you do not expect, and a runtime error, in which your code tries to do something improper, or that is impossible. The latter of these will raise an **exception**, which is what happens when Python basically doesn't know what to do. When exceptions occur, Python stops trying to execute the line of code that it is working on, and produces a **traceback** with a history of your code's execution and a terse description of the problem that caused it to stop. More details on errors in Python can be found in Section H, but for our purposes here, the most important one is that an exception is an object that contains descriptive information that helps you determine what the problem was.

Not all exceptions are equally alarming and you might wish that your code would keep running even if there was an error. Suppose that you ask a user to input a number that you intend to use to do math, but they instead input a letter. As soon as you try to do math with this letter, your program will raise an exception:

```
In: number = input('Please type a number to be squared: ')
Out: Please type a number to be squared: k
```

```
In: number ** 2
Out: Traceback (most recent call last):
Out:
Out:   Cell In[4], line 1
Out:
Out:     number **2
Out:
Out:
Out: TypeError: unsupported operand type(s) for ** or pow(): 'str' and 'int'
```

The last line tells you what the exception was, in this case, a **TypeError**. However, this is not a catastrophe, and certainly no reason that we would want our code to stop. It would be better to just tell the user that they messed up and ask them to try again. This way, we won't lose any of our previous work, and we won't require the user to start over again. We will also be able to give the user clear feedback on what they did wrong so that they won't (hopefully) do it again. The `try` statement is Python's way to make this easy for us.

The structure of a `try` statement is somewhat complex, and not all of it is required. In its simplest form, it looks something like:

```
number = input('Please type a number to be squared: ')
try:
    float(number)

except ValueError:
print('You did not type a valid number!)
```

The `try` statement is required. The code in the code **block** of the `try` statement (ideally one line) will be executed regardless. If this code executes without raising an exception, then the code in the `except` block is ignored. However, if the code in the `try` **block** raises a `ValueError` exception, then Python will execute the code in the `except` code block.

In cases like this, just handling the error is not very helpful. Yes, the code continued to execute, but if any other code later in the program needs *number* to be a number, it will also fail. It would be better to not just tell the user that they messed up, but also to allow them an opportunity to fix it. This can be done with a while loop:

```
input_okay = False
while input_okay == False:
    number = input('Please type a number to be squared: ')
    try:
        number = float(number)
    except ValueError:
        print(f'{number} is not a valid number type.  Please try again.')
    else:
        input_okay = True
```

Here, the `else` statement is intuitive: if no exception is raised, then everything went fine and we can exit the `while` loop.

In addition, `try` structures include a `finally` block, which will run regardless of what happens in the `try-except-else` blocks. This segment is useful for showing the results of the code in the `try` block or cleaning up opened files/streams, and so on. For instance, we could add the following to our last example:

```
    finally:
        print(f'The square of {number} is {number**2}.')
```

Adding this to the end will ensure that we output the final answer, and that it happens as part of the control structure.

CHAPTER G

FUNCTIONS

In many ways, functions are what give programming its power. Functions allow us to efficiently reuse code, minimizing the lines of code we need to write. This makes us more efficient coders and less likely to make mistakes.

Python has many built-in **functions** that we have already been using throughout this book. For instance, when we create a `float` from an `int` using `int()`, we are using the function `int()`.

However, there will be multiple times that you want to do something that Python does not have a built-in function for. As an example, Python does not have a built-in function to normalize spectroscopic data—a process that is regularly used when comparing spectra. However, Python *does* allow you to create your own function that will do this, and then call that function later on. Thus, to better understand functions and realize their true power, it is important to understand how to create and use your own.

G.0 CREATING AND USING FUNCTIONS

In Python, creating a function is called "defining" the function. A basic function definition looks like this:

```
def add_numbers(parameters):
    # do some things you want to do
    return # can can follow the word "return" with a value or not.
```

We begin the definition with the **keyword** `def`, which tells the Python interpreter that what follows is the definition of a function.

Immediately following `def`, we provide the name of the function. This name must be all one word, and must follow standard naming rules (see Section 0.1.4). As with variables, it is best to give these functions descriptive names. Because functions do things, it can be helpful to include a verb in the name.

Following the name, we have the parameters, held between parentheses. Parameters are pieces of data that the function needs (or can optionally use) to operate. You can have any number of parameters you want (including no parameters), and you separate them by commas. It is important to realize that there are a number of different types of parameters,[1] and we will discuss each of them the next section.

[1] You might see these called "**arguments.**" Parameters are arguments that are distinct but have closely related ideas. A brief disambiguation is this: parameters are the names of variables that the function expects to receive, whereas arguments are the values passed to these parameters when the function is **called**.

The first line ends with a colon, and this tells the interpreter that the series of operations that the function will specify follows on the next line(s). This is also considered a code **block**, and so the lines that follow the definition must be indented, with respect to the definition line. This block can contain any number of lines that have valid Python code, such as **control structures**. It could even include the definition of another function.

You end the function with a line that begins with the **keyword** `return`. The things that follow `return` will be passed back when the function finishes running. If you supply no values, then the function simply ends, without returning any values. Or, more properly, it returns the value `None`.[2]

[2] You can find this data type discussed in Chapter B.

Considering a very simple example can help us see how all this fits together. Functions are best used to gather together multi-step operations or to simplify the handling of operations that will require multiple inputs. One operation that that requires multiple inputs and is not uncommon in chemistry is the conversion of mass to number (or counts) of things (i.e., moles). This involves knowing the mass, the molecular mass, and also using the constant Avogardo's number. In Python, we might generate a function that accomplishes this task as follows:

```
def mass_to_counts (mass, MW):
    count = mass/MW*6.022*10**23
    return count
```

If we run this code, we will have now defined a function that will take a mass and a molecular mass and convert the grams into a number of atom or molecules. We say that `mass` and `MW` are parameters of the function. When we want to use the function—which is termed "**calling**" the function, we simply give the name of the function and then supply values (arguments) that will be passed to the function's parameters. The function performs a series of operations, and then returns a value. For instance, imagine we wanted to know how many molecules were in 100 g of water:

```
In: mass_to_number(100, 18.02)
Out: 3.341842397336293e+24
```

Here, we have given values of 100 for the mass, and 18.02 for the molecular weight. Importantly, we did not supply units, so we have to keep track of this. Also, the output is not formatted in a super convenient way, but this could be adjusted as described in the technical chapter on text (Chapter C).

In the above case, we used numbers directly, but we could also supply variables to the function:[3]

[3] Again, even though we have variables that seem to have units in the name, Python does not know this. Please be careful with keeping track of units. Here, it is done using comments, so that we can remind ourselves what we expect.

```
g = 100 # g
gmol = 18.02 # g/mol
mass_to_number(g, gmol)
```

```
Out: 3.341842397336293e+24
```

Also, we can choose to assign the returned value to a variable by using the assignment operator (=):

```
g = 100
gmol = 18.02
molecules = mass_to_number(g, gmol)
```

In this case, the returned value is not automatically output. In order to display the results, we can print the value of `molecules`.

```
In: print(molecules)
Out: 3.341842397336293e+24
```

This is the basic manner of using functions. However, there are also some advanced concepts on using functions, specifically with respect to the parameters, which are worth discussing. We discuss these next.

G.1 FUNCTION PARAMETERS

In Python, there are four types of parameters that can be used in a function.

G.1.0 POSITIONAL PARAMETERS

These are probably what most people first encounter, and the type that was demonstrated in our example above. The basic idea behind this is that Python assigns the values of the variables within the function based upon the position of the arguments being passed to the function. The first argument is assigned to the first function parameter, and so on. In our example above, Python "knew" that 100 goes with mass and 18.02 with MW because we gave the mass first and the MW value second.

If we changed the order of things, then we would get a very different answer:

```
In:  mass_to_number(18.02, 100)
Out: 1.0851644e+23
```

We get this answer because Python assigned 18.02 to the mass and 100 to the MW. You can verify this by doing the calculation yourself and take this as another friendly reminder that Python has no idea what units the numbers have—so please be cautious of this.

Thus, when using positional arguments, it is important that you understand the order that values need to be supplied in and then supply them in that order.

What if we don't want to have to remember the order? Or maybe we have a list of values that are not in the same order as the function expects? Our next parameter type solves this problem.

G.1.1 KEYWORD PARAMETERS

At their most basic, you can use parameters in the definition of the function as keywords to assign a value.[4] If we do this, then Python will assign the value to the parameter we want, even if we don't supply them in the correct order:

[4] Read the subsection on parameter order to see when this will not work.

```
In:  mass_to_number (MW = 18.02, mass = 100)
Out: 3.341842397336293e+24
```

Now, even though we have supplied values in a different order from which they were defined, we get the desired outcome. This is because we assigned values by keyword, which takes precedence over the ordering.

So, why wouldn't we always use the parameters in this keyword style? Although it is true that using keyword parameters means you don't have to know the order to specify them in, it has the disadvantage that you have to know the names of the parameters. Depending on the situation, one of these will be easier than the other, and so you will find yourself using both. In this book, we find ourselves using positional more often than keyword.

One thing to note is that we did not have to go out of our way in order to ensure these parameters were keyword parameters. Instead, being able to use them this way is an automatic result of the function definition. This may make it seem like positional parameters and keyword parameters are the same, but they are not quite. The difference comes in the places that positional and keyword parameters can occupy within the definition of a function. However, before we can truly appreciate this, we need to introduce the remaining two parameters. We will then return to the difference between positional and keyword parameters.

G.1.2 ARBITRARY POSITIONAL PARAMETERS: `*args`

These are positional parameters, but they can accept an arbitrary number of values.[5] They are designated by placing an asterix in front of a parameter name.

[5] As a reminder, when a function is called, the values passed to the parameters are called "arguments." This is why this parameter is called "*args." It is referencing the *use* (when they are called arguments) rather than the creation (when they are called parameters).

Imagine that you had a specific compound (such a glucose) that you wanted to weigh out an arbitrary number of times. You want to convert all these masses to the number of molecules per weighing and also get the mean value and standard deviation of these values. However, you don't know the number of times you are going to do the weighing. You could use the function above, just calling it multiple times, collecting the results, and the calculating the mean from them. However, you could create a new function that more naturally handles this situation.

```
import numpy as np
def masses_to_stats (*masses):
    MW = 180.156 # g/mol for glucose
    N_A = 6.022*10**23
    numbers = []
    for mass in masses:
        numbers.append(mass / MW * N_A)
    numbers = np.array(numbers)
    return [np.mean(numbers), np.std(numbers)]
```

Using this function, we can supply any number of masses we want:

```
>>> masses_to_stats (95.2, 95.1, 95.2)
```

```
Out: [3.181096753184277e+23, 1.9298848337330848e+20]
```

```
>>> masses_to_stats (44.8, 45.1, 45.2, 44.9)
```

```
Out: [1.5041963631519348e+23, 6.102831696411533e+20]
```

And we will then have the means and standard deviations recorded for two different sets of values that had different numbers of values.

What is happening is that Python is taking all of the values you gave the function, and putting them into `tuple` (see Chapter B) that the function assigns to the variable `masses`. Thus, when you write code to use *args you need to realize you are using a `tuple`.

This also suggests that you could write code that gathers masses in a `tuple` and pass them to a function with a normal positional argument, and this true. However, that requires more code *outside* of the function. If we know that we want to have a sequential type that we will iterate through, then it is better practice to implement this in the function. That way, if you need to call the function multiple times, you don't have to repeat the part of the code where you create a tuple first. Additionally, this can help you avoid unwanted scope problems associated with passing mutable lists to functions. This concept is discussed in more detail in the last section of this chapter.

G.1.3 ARBITRARY KEYWORD PARAMETERS: `**kwargs`

If one also has the ability to supply an arbitrary number of keyword arguments using the `**kwargs` parameter type. Doing so will build up a **dictionary** of key:value pairs based on these keywords.[6]

Perhaps the reason you were weighing out multiple samples was that you are working in a group and you want to understand the standard deviation in measurements that each team member made. You could calculate this while also keeping track of who did what. Perhaps you think it would be fun to know who's mass had the largest contribution to the standard deviation? This is, of course the same as who's mass is farthest from the mean:

[6] For more information on dictionaries, you can refer to Chapter B.

```
import statistics as stats
def person_mass_to_molecule (**trial):
    # make a list of values.
    masses = []
    for person, mass in trial.items():
        masses.append(mass)
    # now get the mean value and
    mean_mass = stats.mean(masses)
    diff = 0
    for person, mass in trial.items():
        if abs(mean-mass) > diff:
            diff = abs(mean-mass) # update the reference difference
            contributor = person
    return [mean_mass, stats.dev(numbers), contributor]
```

When calling this function, you can now give a series of key:value pairs that will be used in the function.

```
person_mass_to_stats(Ben = 14.3, Chris = 13.9, Stanley = 11.7, Shana = 12.3, Diana =
13.0)
```

```
[13.040000000000001, 1.080740486888504, 'Stanley']
```

The first number in the list is the mean value, the second is the standard deviation in values, and the third is the person who contributed most to the standard deviation.

G.1.4 Order of Parameters: Mixing and Matching Parameter Types

Above, we discussed the four different types of parameters that we can use in the definition of functions. We also showed their usage in isolation. We can also use multiple parameter types in a single function.

However, when using more than a single type of parameter, there are rules for the order they are supplied in. The basics are:

positional arguments → arbitrary positional arguments → keyword arguments → arbitrary keyword arguments.

The fact that this ordering is enforced shows how we can make a clear distinction between positional and keyword arguments. Specifically, if we have arbitrary positional arguments, then things that are to the right of them *must* be keyword arguments of some type, and parameters to the left *must* be positional. However, outside of this ordering, the distinction might be a bit academic.

For instance, let us return to the example where we were calculating the amount of glucose over multiple weighing trials. When we did this, we "hard-coded" the molecular mass into the function. But if we used a keyword instead, we can supply the molecular mass when we call the function *and* still have multiple measurements passed to an arbitrary positional parameters.

```
import statistics as stats
def masses_to_stats (*masses, MW):
    N_A = 6.022*10**23
    numbers = []
    for mass in masses:
        numbers.append(mass / MW * N_A)
    return [stats.mean(numbers), stats.stdev(numbers)]
```

Because we have added the MW parameter to the *right* of the arbitrary positional parameter, then the MW is a keyword argument, and we can assign values to it as described above.

Thus, if we had two measurements of water, we could have:

```
masses_to_stats(10.3, 11.1, MW = 18.02)
```

```
Out: [3.575771365149834e+23, 1.890431536650561e+22]
```

Whereas if we had three measurements of Fe_2O_3, we would have:

```
masses_to_stats(2.3, 2.1, 1.9, MW = 58.443)
```

```
Out: [2.1638519583183615e+22, 2.0608113888746285e+21]
```

G.1.5 Setting Default Values for Parameters

The final topic for definition of functions is to point out we can provide default values for parameters. These defaults are used in the case that no value is provided when the function is called. However, because they are used when no value is provided, this also means that there must be a way to handle the value being present or absent. Both arbitrary positional parameters and arbitrary keyword parameters handle this case naturally: they can always accept between zero and an arbitrary number of parameters. However, this is not the case for the positional and keyword arguments.

Instead, for both positional and keyword parameters, default values are specified using an assignment operator (=) when providing parameters in the function definition. However, assignment between argument and parameter depends on the number of parameters passed. So to make this fault tolerant, once a default value has been specified, all other positional or keyword arguments to the right of that one *must* also have default values provided.

```
def myFunc (p1 , p2="True", *p3, p4 = "three", **p5):
```

In this definition, `p1` is a positional parameter, but does not have a default value; `p2` is also a positional argument (because it is left of the arbitrary positional parameter), but it has a default value assigned. This means that all other positional or keyword parameters that follow also must have default values. Next, `*p3` is neither of these because it is an arbitrary positional parameter, and so cannot have defaults provided in the definition line; `p4` is a keyword argument (because it is to the right of the arbitrary positional argument) and given that there was a positional argument with a default value to the left of this, it also must have a default value assigned. Finally, the parameter `**p5` is a an arbitrary keyword argument and does not have defaults defined in the definition line.

G.2 SCOPE

A basic concept for using functions is this: if you define a variable within a function, you cannot access it outside the function—even if you have run the function already.

```
def mass_to_number (mass, MW):
    number = mass/MW*6.022*10**23
    return number

mass_to_number(100, 18.02)

print(number)
```

If you run this code, you will get an error:

```
Out:    NameError: name 'number' is not defined
```

Looking at the code we ran, you can see that we had defined `number`, and we even used it within the function with no problem (this was our first example in this chapter). But, this was done *within* the function. The error happens when we try to access this function *outside* the function.

On the other hand, if you have defined a variable outside of a function, you can access it inside the function.

```
units_i = "grams"
units_f = "molecules"

def mass_to_number (mass, MW):
    number = mass/MW*6.022*10**23
    return f'measured {mass} {units_i}, which is {number} {units_f}'

mass_to_number(100, 18.02)
```

Running this code, we get the following output:

```
Out:    'measured 100 grams, which is 3.341842397336293e+24 molecules'
```

Even though we defined the variables `units_i` and `units_f` outside the function, we were able to access them inside the function, which is how we returned the units in the string.

That is scope in a nutshell.

There are a few reasons to keep this in mind. First, it does control what variables you can and cannot access, and this is important to remember.

However, it also can be convenient because if you have two functions that will never call one another, you can use the same variable names within them, and that can really aid you in keeping variable names simple.

For example, when you are working with x- and y-data, you might be able to have multiple functions that use x and y as simple variable names. In the following example, we have two functions that work to either baseline correct data or normalize data when that data is held in a Python list.

```
from plotly.subplots import make_subplots

x = [0, 1, 2, 3, 4, 5, 6, 7, 8, 9, 10]
y = [1, 2, 3, 4, 5, 4, 2, -1, -2, -1, 0]


# baseline correct
def baseline_correct_data (y):
    low_value = min(y)
    corrected = []
    for v in y:
        corrected.append(v - low_value)
    return corrected

# normalize
def normalize_data (y):
    high_value = max(y)
    normalized = []
    for v in y:
        normalized.append(v/high_value)
    return normalized



norm_y = normalize_data(baseline_correct_data(copy.deepcopy(y)))

fig = make_subplots()
fig.add_scatter(x = x, y = y)
fig.add_scatter(x = x, y = norm_y)
fig.show("png")
```

Running this code will produce a default Plotly plot with two traces. The first is of the data that has not been altered, and the second is of data that has been corrected so that the lowest point is zero and then so that the highest point is one.

G.3 USE copy.deepcopy( ) TO PASS MUTABLE OBJECTS TO FUNCTIONS

This applies to one of the most used data types in Python: the list. So PLEASE read this section.

Above, we used `copy.deepcopy( )` when passing our list to the function. Here, we explain why. To understand this discussion, you need to understand **mutability**, as discussed in

Chapter B, and it will be useful to also understand **memory**, which is covered in Chapter A. So you may need to read that as well. However, it is extremely important to understand why we used `copy.deepcopy()` because misunderstanding it leads to a large number of errors in Python code.

The basic idea is this: when you pass values to a function, you are not actually giving the function that value, but the reference to that value in memory.

The reason that this is important is that if the object you are passing is mutable, then you can modify the values inside that object within the function, and these values will also change outside the function.

Let us see an example with a common mutable type: lists. If we repeat the above example but don't use `copy.deepcopy()`, we will obtain two traces that are right on top of one another (it may look like a single trace).

```
norm_y = normalize_data(baseline_correct_data(y)

fig = make_subplots()
fig.add_scatter(x = x, y = y)
fig.add_scatter(x = x, y = norm_y)
fig.show("png")
```

This is because both lists are the same... or rather they are pointing to exactly the same values. Again, this is explained in Chapter B.

This behavior is different from if you had an immutable object. The immutable analog to the list is the tuple, and if we repeat this with tuples, we get the following:

```
norm_y = normalize_data(baseline_correct_data(tuple(y)))

fig = make_subplots()
fig.add_scatter(x = x, y = y)
fig.add_scatter(x = x, y = norm_y)
fig.show("png")
```

Now, we again get two spectra that are different, the original and the baseline-corrected and normalized spectra. This is because we converted the list to the tuple and then operated on this immutable object.

The point here is that you have to be very careful in how you handle mutable objects and functions, and so it would be a good idea to keep which data types are mutable and which are immutable. An explanation of what data types are mutable and not mutable is covered in Chapter B.

It may also seem that this behavior is somehow at odds with our discussion of scope above; however, they are slightly different. In the discussion on scope, we said that if we create a new object within a function, we cannot access it outside the function. Here, we are saying that if we create a mutable object outside a function, we can still modify it within a function.

Most of the time when people get in trouble with this behavior is when they pass a list to a function and forget that it will be modified, so they end up with a list that is different from what they think they have. Of course, if you do realize this, you could exploit this behavior to your advantage. However, we think that, for the case of this book, best practice is to never pass a mutable object to a function.

In the case of a list, we have two options: we can convert the `list` to a `tuple` (as was done above), or we can use the `copy.deepcopy()` function to create a new list that is then passed to the function, as was done in our original example.

If you use the `copy.deepcopy()`, we are technically passing a mutable object to the function, but it is a new object with no reference to it outside the function. It is created as it is passed, so we will never have the opportunity to use it outside the function.

Therefore, we recommend the use of `copy.deepcopy()`. The reason is that it can handle mutable objects that hold other mutable objects. What `copy.deepcopy()` does is to make a new copy of very item *and sub-item* in the object you feed it.[7] It will do this for however many levels you have. If you have a series of 100 nested lists, it will copy them all.

[7] The technical term is that it does this recursively.

For instance, imagine that we had a list that contained another list. If we turn the outer list into a tuple, the tuple still contains the mutable lists within it. So, if we do something to alter those lists, **they will be altered outside the function.**

G.4 RECURSION

There is nothing that prevents you from calling a function from within itself. This is called "recursion," and although it may sound strange, it can be a very powerful technique. As the function executes, if it reaches another call to itself, it will pause the current progress through the function and start up another iteration from the beginning. Like a loop, care must be taken to ensure that this does not happen infinitely by including some code that determines the appropriate time to no longer call the function again. At that point, the last function call will complete, then the one before that, and the one before that, and so on until all of the nested function calls are complete.

Recursion is neat but is never necessary to solve a problem. It can make solutions easier to understand and more readable. Recursion particularly shines when an algorithm needs to iterate some number of times that is both not known at the time of coding and is difficult to express simply using conditional expressions. An example of this is computing the isotopic abundances of molecules of arbitrary chemical formula—here, the number of each atom, and the number of different atoms, is unlikely to be known to the programmer, and a recursive algorithm is likely to be much simpler and less prone to error than a series of nested `while` loops.

It is unlikely that most readers of this book will run into a recursion problem in the context of the material of the book, but it is a programming concept that is good to know about and worth reading more about if your interest is piqued.

CHAPTER H

Common Errors

Here is a fact worth coming to terms with: a significant amount of your time spent coding will be spent debugging your code. This means you will spend time looking for mistakes in your code. In general, there will be two kinds. First, you will write code that the Python interpreter cannot use. There might be many reasons for this, such as improper **syntax**, missing **white space**, or trying to use a bit of data that simply does not exist. These will produce errors that Python will tell you about. Second, you will write code that will run, but will give you wrong results. This might be because you forgot to account for the units you are working in or perhaps simply implemented the task incorrectly.

The second problem is much harder to provide general diagnoses of. We simply point out that is *always* a good idea to run your program using a test case or two, in order to verify the results you are getting are correct. The first case can be simpler to address because Python will often tell you what went wrong. Nevertheless, the first few times you encounter the messages Python gives, you might need help understanding them. This chapter attempts to give a bit of background about what the most common types of errors are, and why they are happening.

H.0 GENERAL ADVICE

Although we will address several of the most common errors below, there are so many possible errors we cannot explain them all. Thus, it is worth a brief word on how to troubleshoot errors in general.

Historically (*i.e.*, since roughly 2010), one of the best ways to solve a problem was to copy and paste the error message directly into Google, perform a search, and then read any of the Stack Overflow or Stack Exchange pages that come up. These sites are devoted to discussion of programming concepts and trouble shooting. Sometimes they have very detailed answers, but often they also have fairly straightforward answers. They will also often provide links to other discussions related to the error. By reading these discussions, you can learn a lot about the specific issue you are encountering and general programming concepts.

It is also worth mentioning that since 2023, LLM-based AI tools (like ChatGPT, Claude, Gemini, Co-Pilot, etc.) can provide answers to problems you are encountering. Again, a good approach is to directly paste the error message into the AI tool. However, you can also directly paste in the code that gave the error and ask the AI tool to explain what is going wrong. You can also ask the AI tool for solutions to the problem you are having.

A few words of caution are in order here. First, although AI tools can be useful, it can also supply answers that are simply wrong. When you are new to coding, it can be hard to know if this is the case—although if you try them and they simply don't run, that is clear enough. However, even if they do run, make sure to take the time to understand what is happening in the code. One way to do this is to also ask the AI tool to explain what when wrong and why, as well as how the solution it provided solved this problem. Taking the time to do so ensures that the solution provided by the AI tool is doing what you want and also helps you learn something about programming. Of course, you can also always fall back to the use of Stack Overflow and Stack Exchange for further explanation if you find those from the AI tool lacking.

Although this above approach will work in general, it is still worth examining a few problems that are commonly encountered. Understanding these might save you a bit of time down the road.

H.1 SYNTAX ERRORS

Syntax describes the "grammar" of the programming language. It controls how the interpreter understands what you are asking of it. Thus, when you fail to use proper grammar, it cannot work.

There are a number of potential issues, but the ones that are most common to Python are discussed next.

H.1.1 INCORRECT INDENTATION

In Python, white space can have meaning. Specifically, the white space at the start of a line controls what belongs to a code **block**. All code that is in the same block must have exactly the same amount of leading white space. Some people use spaces for this, some use tabs. In this book, we have used tabs, with a width = four spaces.

If you make an indentation error, then your code will likely not run. Instead, when you try to run your code, you will get a message like "`expected an indented block`."

H.1.2 UNMATCHED BRACKETS AND PARENTHESES

Python uses things like square brackets, curly brackets, and parentheses in the declaration of collections like lists, dictionaries, sets, and so on. A common mistake is to open one such bracket and then not close it. This is very common when you nesting these collections, like in a list of lists. When you do so, you will likely get an error telling you this:

```
"'[' was never closed"
```

In the slightly less common case that you added too many closing brackets, you will get a message that things are unmatched:

```
"unmatched ']'"
```

However, we also use parentheses in calling functions. When we call a function, we pass arguments between parentheses. If you open these, but do not close them, you will get:

```
"incomplete input"
```

If you try to close one you haven't opened, you will again get an "unmatched" error.

H.1.3 Missing Colon

The beginning of a new code block usually ends with a colon. For example, if statements end with a colon, as do the opening line of a for loop. If you are missing this, the Python interpreter will not run, and you will get an error message:

```
"expected ':'"
```

H.1.4 Missing Comma

If you create a collection, you will separate items in this collection with commas. If you forget the comma, the interpreter will not run and you will get an error message:

```
"invalid syntax. Perhaps you forgot a comma?"
```

H.2 Segmentation Fault

In general, segmentation faults are encountered when you try to access a bit of memory that cannot be accessed, or you try to access a specific value that no longer exists. In Python, it is very rare to encounter a true segmentation fault, but there are a number of other errors that are closely related, which you might encounter:

TypeError: This occurs when you are attempting to perform an action that cannot be performed on the data type. For instance, if you try to convert a **list** to a H.0string, you will get such an error.

IndexError: This occurs when you attempt to access a value at an **index** that does not exist. For instance, imagine you had the list `mylist = [1, "purple", True]`. If you tried to access the index 3, you would get such an error. Here, `mylist[3]` will yield `'list index out of range'`. In our experience, this is most commonly encountered when you are using a counter to advance elements in a list, and then advance it too far. For instance, you could be using `enumerate` with a list, but then accessing elements one value ahead (so-called "look ahead"). If you don't take care, you will automatically try to access indecies that do not exist.

KeyError: These happen when using mapping data types, like dictionaries, but then attempt to use a key that does not exist. In our experience, this happens most often when you are building up a dictionary and then attempt to access a value that you have not yet added.

NameError: This occurs when you try to use a variable that has not yet been created. There are probably three ways you will encounter this error. First, you may have a typo—using a name that you did not intend to—in either the declaration of the variable or the usage of it. Second, you may being trying to access a variable that doesn't exist due to scope. Perhaps the variable was created in a function, but you are trying to access the variable outside the function. Finally, you may have previously had a perfectly good code, and then cut and pasted parts, accidently moving the usage of a variable to before it was declared.

H.3 OVERFLOW

This is the name assigned to the problem of trying to generate a number that is too large for Python. As noted in the discussion of data types there is no limit for integers, but there is for floats. Thus, if you try to generate a float that is too large or too small for Python, you will get this error. These errors are why you might encounter "strange" units in computational chemistry. For instance, units can be used that allow avoidance of Planck's constant. Because this constant is very small, calculations that might involve the inverse square larger can lead to such errors.

H.4 USING MUTABLE OBJECTS INCORRECTLY

We conclude this chapter with a discussion of something that is not, strictly speaking, an error. But it is encountered commonly enough that we feel it should be mentioned here. The basic problem is that one accidentally changes the value of a mutable object. This idea was discussed in our the chapter on data types (Chapter B) and again at the end of the chapter on function (Chapter G, so please read that section). We mention it here, because it is a commonly encountered problem.

CHAPTER 1

PLOTLY

If you are doing science, you are going to collect data. And if you are collecting data, you are going to want to plot it. In Python, you have many options for libraries that can plot data. In this chapter, we will explore how to do so for one of the most common ones: Plotly.

We wish to note up front that no library can perform every task that you will desire in terms of plotting. There are always some specialty plots that libraries may not handle. However, Plotly offers a fairly substantial range of plot types and offers interactivity with these plots as a natural result of plotting.

We also wish to note that we are going to teach you a somewhat non-standard approach to plotting in Plotly. Much of the information that is present online focuses on how to plot single plots in a simple manner and often use an approach called "Plotly express." Although this is a nice approach for making figures that have only a single plot, if you want to make more complex figures that include many subplots, you will find this challenging. We wish to give you a tool that is more flexible and can be used to create any sort of complex plot you may choose to make.

One last thing to acknowledge: we are focusing on how to construct plots and control many of the aspects of plotting, but we are not focusing on the theory of why you might want to change these aspects. If you are curious to understand this, there are many good books on graphic design and the design of data visualizations. Here, focus is not on the why, but on the how.

1.0 THE BASICS OF PLOTTING USING PLOTLY

The construction of a figure showing data can be broken down into five elementary steps. Assuming you already have data you want to plot, these steps are:

1. **Create a Figure object that holds the plotting region(s) you will need.** You might think of the **figure object** as the canvas you are "drawing" on. Such a canvas might

contain a single plot area, or it might have several plotting areas. In technical terms, we need to specify the number of "subplots," as well as their arrangement—how many rows and columns of these subplots we will want to use.

2. **Add trace object(s) to your figure.** As scientists, we are conversant with many different means of representing data. We use bars, scatterplots, line plots, histograms, box and whisker plots, and so on. A key aspect of constructing a plot is to choose the means to represent it. When doing so, we might also choose things like the color of lines and the shape of markers, among others. These are parameters that control the appearance of our data representation. The choice of representation and the style of the representation are all held in the Plotly trace object. The purpose of making these representations is to add them to our figure, so we will also create and add them in a single step. It is useful to note that the order in which we add traces is the order in which they are drawn, so the last trace added will be on top.

3. **Specify the representation of axes.** Even if we do not explicitly show axes, data is always plotted against axes. However, these axes are not part of the data. Thus, we should consider how to format them (including adding titles) separately from formatting data representations.

4. **Design on an overall appearance for the figure.** This addresses questions of appearance that are separate from data and axes, and includes questions like the color of the background, the font, the overall size, and so on. Although it might seem a bit strange to specify these late in the construction of a figure, when using Plotly, it is much easier to do it at this stage.

5. **Display/save the figure.** Plotly allows us to create interactive figures, static figures, and to save files. This is how we will ultimately use our plots.

We can demonstrate these steps in a simple example. We wish to note that the following example is not the absolute simplest means to generate a figure using Plotly; however, it strikes a balance between being simple and being robust to customization. Also, the steps do not need to be done in the exact order presented. The first and last steps need to remain in their positions, but the other steps could move around.

```python
from plotly.subplots import make_subplots
import numpy as np

x_data = np.linspace(0, 10, 50)
sin_data = np.sin(x_data)

# Step 1: Create a Figure object that holds the plotting region(s) you will need
sinplot = make_subplots()

# Step 2: Add trace object(s) to your figure
sinplot.add_scatter(x = x_data, y = sin_data, mode = "markers")
```

```python

# Step 3: Specify the representation of axes
sinplot.update_xaxes(title = "x")
sinplot.update_yaxes(title = "sin(x)")

# Step 4: Design on an overall appearance for the figure
sinplot.update_layout(template = "simple_white" ) # we can used a set of namedtemplates

# Step 5. Display/save the figure
sinplot.write_image ("sin.svg")
sinplot.show ('png')
sinplot.show ('browser')
```

Running this code in Spyder will save a figure (as an .svg) to your **working directory**; it will also show a .png version of the figure in the "Plots" window of the Spyder IDE, and it will open a web browser that contains an interactive version of the plot (you can point to data, zoom in, and so on). In all cases, the plot should look similar to Figure I.1:

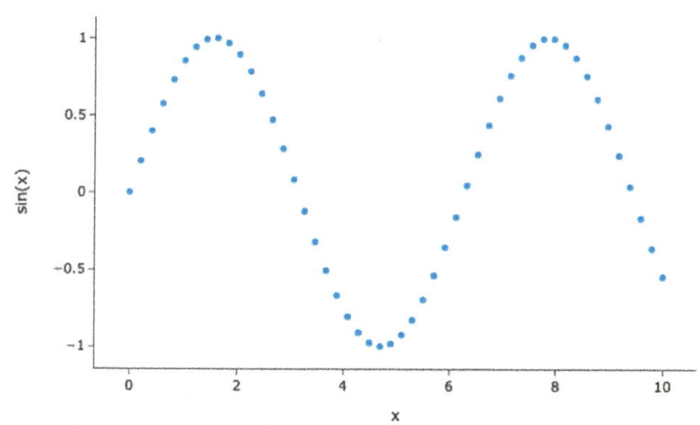

FIGURE I.1. Results of running the code above. This is a simple scatter plot using many of the Plotly defaults.

If we compare the code to our elementary steps, we can see the following:

1. `make_subplots()` is the function that creates our Figure object. Though we haven't done so in the above example, we can use this function to specify the number of subplots we want and how they are layed out. Here we do not, but below we will consider an example where we do. If we don't do so, we will just have a single subplot. By assigning the output of this function to a variable (in this case `sinplot`), we can refer to the Figure object by this variable name.

2. `add_scatter()` is the **method** that creates the traces from our data and then adds them to our Figure object. This function creates a scatterplot-like representation. If we wanted a different representation, then we could a different function, like `add_bar()`. Within these function calls, we have the option to specify the appearance of this representation. In this case, we have told it to display the data using markers.

As will be discussed below, other options would be to show just lines or markers and lines both. We can also change the color of the markers, the size of the markers, and so on. The data representations is then added to subplot(s) we wish. Because we only have a single subplot, we do not need to specify the specific subplot in this case.

3. `update_xaxes()` and `update_yaxes()` are also methods of the figure we created and are used to specify the formatting of the x- and y-axes in our figure. In this case, we specify a title for each.

4. `update_layout` is another method of the Figure object we created, and is used to control the general formatting of the figure (i.e., the formatting that is not for data or axes). We can assign a template (here: `simple_white`) or alter specific traits (see next example plot).

5. `write_image()` is a method of the Figure object that produces a file of the figure. We control the format of the file (i.e., svg, png, jpg, or others) simply by specifying this in the name we wish to save the file. Also, `show()` is a method of the Figure object used to display an image and we can display either static images (first usage in our example) or interactive figures (second usage in our example). Note that when outputting static images (svg, png, and so on), Plotly relies upon another library. You have the option of either kaleido (default) or orca. If you do not have these libraries, you will be unable to generate such images. The "browser" option, however, should always work.

The above example shows the very basics of generating a Plotly plot. However, there is much more we can do in Plotly. We can create multiple subplots (step 1), represent data in different ways (step 2), put traces on specific subplots (step 2), format specific axes (step 3), fine-tune the layout (step 4), and control the output (step 5). An example code that does all of these is shown below, and many more example plots are shown at https://www.codingforchemistsbook.com/.

```
import numpy as np
from plotly.subplots import make_subplots

#Obtain data.  In this case, it is a simple sine and cosine set.
x_data = np.linspace(0, 10, 50) # create 50 equally spaced points between 0 and 10
sin_data = np.sin(x_data) # calculate the sine of the above points.
cos_data = np.cos(x_data) # calculate the cosine of the above points.

# Step 1. Specify the figure layout. In this case, we will have two subplots.
sincosplot = make_subplots(rows = 1, cols = 2,
                           horizontal_spacing=0.12
                           )
```

```
# Step 2. Specify the data representation.  A line and scatter plot
sincosplot.add_scatter(x = x_data, y = sin_data,
                       mode = "lines", # show as a line
                       line = dict(color = "#FF00FF", width = 3), # format the line
                       name = "sine", # name of the representation
                       )

sincosplot.add_scatter(x = x_data, y = cos_data,
                  mode = "markers", # show using markers
                  marker = dict(color = "darkcyan", symbol = "diamond"), # format markers
                  name = "cosine",  # name of the representation
                  row = 1, col = 2, # specify the subplot we want this in
                  )

# Step 3. Format the axes
y_format = dict(zeroline = True, # set up a dictionary we can use multiple times
                zerolinecolor = "grey",
                )

sincosplot.update_xaxes(title = "x-specified", # text for the axis
                        linecolor = "grey", # change the color of the axis line
                        range = [0,10],
                        showline = False, # don't show the line at bottom
                        ticks = "inside"
                        )

sincosplot.update_yaxes(y_format, # start with the formatting from the above dictionary
                        title = "intensity", # text for the axis
                        color = "#FF00FF", # change the color for the text
                        col = 1, row = 1 # specify which subplot to apply this to
                        )

sincosplot.update_yaxes(y_format, # start with the formatting from the above dictionary
                        color = "darkcyan", # change the color for the text
                        col = 2, row = 1 # specify which subplot to apply this to
                        )

# Step 4. Format the figure using update_layout()
text = dict(family = 'times', # it can be convienent to make dictionaries to pass along
            size = 18, # size in pixels
            color = "grey" # this will change the color for font on the axis labels and title.
            )

sincosplot.update_layout(template = "simple_white",
                         font = text,
                         title = "Example with subplots"
                         ) # alter the font from the template

# Step 5. Save/Show the figure
sincosplot.write_image("sin-cos.svg")
sincosplot.show('png', width = 1000, height = 500) # change the size of the figure
sincosplot.show('browser')
```

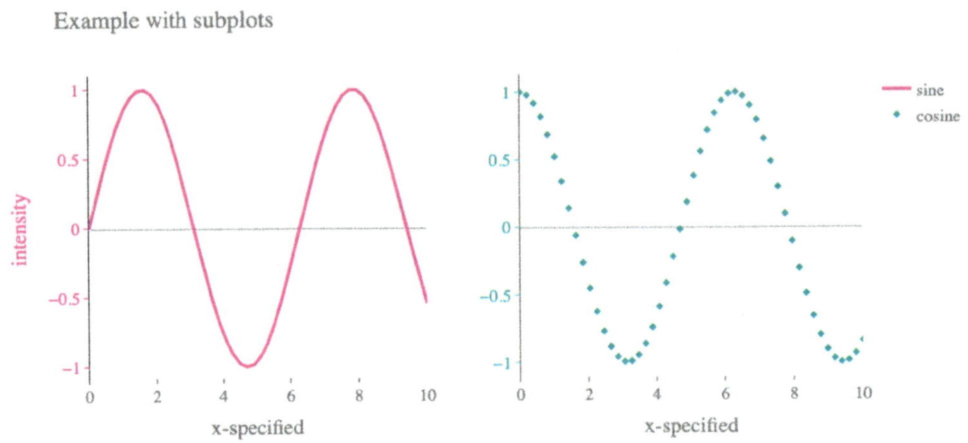

FIGURE I.2. Result from running the above code. This is a more complex example, showing the use of subplots and non-default formatting.

Running this code should produce an image like that shown in Figure I.2.

Comparing the outputs shown in Figure I.1 to Figure I.2 illustrates how we can control many aspects of plotting. We are not recommending you use this particular style, simply showing what can be changed.

There are many more changes that could be made. For instance, it might look odd to you that the numbers on the y-axis have different numbers of decimal places. This can be adjusted using the `tickformat` **keyword argument** when updating axes. For the rest of this chapter, we will explore the various ways to control each of our five elementary plotting steps in Plotly. This discussion is not meant to be exhaustive, but to provide information on what we think will be the most common tasks and help provide a framework for understanding Plotly. Complete information can be found in the Plotly documentation.

I.1 CONTROLLING FORMATTING DURING PLOT CONSTRUCTION

Because our aim is to make as general a tool as possible, we are going to construct every figure as a figure with subplots, using the `make_subplots()` function. In addition, we wish to leverage a general objected-oriented approach, where we are always creating and manipulating different parts of the Plotly Figure object. For this reason, once we make a figure, we will continue to modify it using **methods** of that Figure object, including methods like `add_scatter()` and `update_layout()`. In what follows, let us consider some of the options available during each of the elementary steps of plotting:

STEP 1. CREATING THE FIGURE OBJECT AND SPECIFYING SUBPLOT LAYOUT

The `make_subplots()` function creates two-dimensional grid of subplots (rows and columns) and then assigns this to a Figure object, and we can then pass this Figure object to a variable.

The `make_subplots()` function accepts a large number of **keyword arguments**. Complete documentation can be found on the Plotly website, but we can summarize the most relevant here by the behavior they control:

- `rows` and `cols`: integer values that control the number of rows and columns, respectively. If you do not specify a value for these keywords, the default for each is one. Thus, if you call `make_subplots()` without any arguments, you will return a figure that has a 1×1 grid.

- `specs`: a list of lists used to specify more advanced layouts, such as if subplots should span multiple rows or columns. The outer list should contain a number of lists equal to the number of rows in the figure. The inner lists contain a number items equal to the number of columns in the figure. Thus, each item corresponds to a subplot directly. The items in the inner list are either the word `None` (specifying the **data type** `nonetype`), or a **dictionary**. If `None` is used, then the subplot will not exist and you cannot add traces to it. If it is a dictionary, then you can leave it empty or add information to it. For this book, we will consider two keywords for these dictionaries:

  - `rowspan` and `columnspan`: integers that specify how many rows or columns the subplot will span. If not included, the subplot will span a single row and column. The spanning behavior is such that the row and column numbers specify the range of the span starting from the top left of the subplot. This means that the span extends downwards (`rowspan`) and to the right (`columnspan`).

Perhaps the best way to think about the impact of `rowspan` and `columnspan` is to think of them controlling the size of the subplot, in units of the subplot grid. So, you still add traces to the row and column corresponding to the subplot (i.e., the top left position). You can also add images to the subplots that are overlapped by the subplot. As an example of how to use `specs`, imagine we wanted to have a 3×4 grid, where there was a subplot that spanned 2×2 and was centered on the bottom middle. The specs would look like this:

```
specs = [ # this outer list contains three lists... one for each row
        [{},{},{},{}], # each inner lists will one entry per column
        [{},{'colspan':2, 'rowspan':2},{},{}],
        [{},{},{},{}],
       ]
```

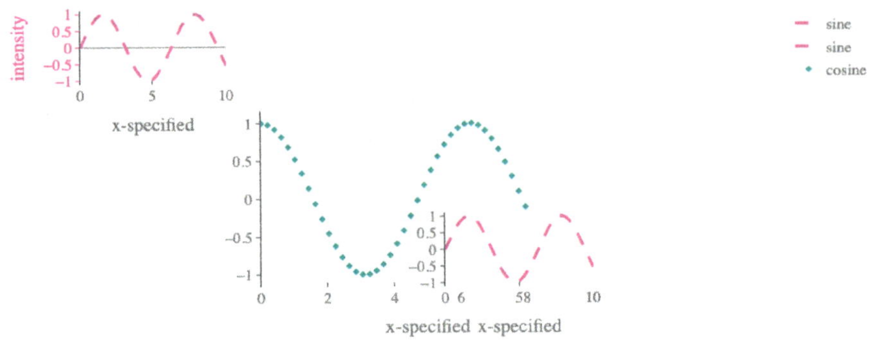

FIGURE I.3. Plot that results from using the `specs` values shown above. This illustrates how the subplot grid is organized in Plotly.

In this example, all subplots occupy a single grid space, except for the subplot at row = 2, column = 2. This spans two rows and two columns. Now, imagine we added our `sin_data` from Figure I.2 to two traces added to two subplots at (`row = 1, col = 1`) and `row = 3, col = 3`) and the `cos_data` to a trace in the subplot at (`row = 2, col = 2`). The traces showing `sin_data` will each occupy a single grid space, and the trace showing `cos_data` will span a 2×2 block. Also, the cos trace subplot will overlap with the trace at (`row = 3, col = 3`). The output would look like Figure I.3.

- `vertical_spacing` and `horizontal_spacing`: float values between zero and one that control the spacing between rows and columns, respectively. The value specifies the fraction of horizontal or vertical space on the plot that is devoted to separation between subplots. You can think of this as the internal margins. Thus, if you had two rows and used `vertical_spacing = 0.5`, then half of the availible vertical space would be devoted to separation between the plots. If you set the vertical spacing value too large, then there will not be sufficient room to hold the figure, and you will receive an error that there is insufficient room. A good rule of thumb is to keep the spacing less than 0.5/# of rows. The same holds for cols. The default values are 0.2/# of cols and 0.3/# of rows, respectively.

- `shared_yaxes` or `shared_xaxes`: boolean value that controls if axes titles and tick numbers are shared. If set to `True`, title and tick numbers appear on only the leftmost y-axes and the bottom-most x-axes. If set to `False` (default), all axes bear numbers and titles (if supplied).

- `subplot_titles`: list of strings that specifies the titles for the subplots. You must pass all of the titles you want as a list and, if you use `subplot_titles`, then you *must* have an entry for each subplot. Use empty strings (`''`) for subplots you don't want titles to display. When counting the subplots in the list, you count across rows first, before moving to the next row.

- `column_widths` and `row_heights`: list of floats that control the relative widths of columns or heights of rows, respectively. The length of the list needs to be equal to the number of columns or rows, respectively. You can use any numbers you like, as they will be normalized. Thus, if you had to rows, and you wanted them to have heights in a ratio of two to three, you could use `row_heights = [4, 6]`.

STEP 2. CREATING TRACE OBJECTS AND ADDING THEM TO FIGURE OBJECTS

In Plotly, the representations of data we use are called "traces," and there are many types availible. Herein, we focus on the 2D plots that we think you will most commonly want to plot. You can read the Plotly documentation for more 2D plots, and all 3D plots, as well as help on creating animations for any of these.

We will create representations and add them to Figure objects, using a methods of the Figure object. The **methods** for the most common traces are as follows:

- `add_scatter()`: This is used to create scatter plots (where data is represented by markers) as well as line plots (where data is represented by lines). The keyword arguments used for this method are:
  - `x` and `y`: list or array floats. These are the x-axis and y-axis data, respectively.
  - `mode`: string specifying the display mode. Options include: `'lines'` (only lines), `'markers'` (only markers), and `'lines+markers'` (markers connected by lines).
  - `name`: string that provides a name for the trace, used in legends and hover labels (labels that appear when the cursor is over the trace in an interactive plot).
  - `marker`: dictionary specifying marker properties. Common keys are:
    * `color`: specify using Plotly conventions, as described in Figure I.4.
    * `size`: integer specifying marker size in pixels.
    * `symbol`: string specifying the shape for the marker symbol. There are a large number of options but ones that are common are: `'circle'` (a filled circle, which is the default value), `'square'` (a filled square), `'diamond'` (a filled diamond), `'cross'` (plus sign), and `'x'` (a × shape). For many of these, you can add the suffix `'-open'` to make them non-filled. For instance, `'circle-open'` produces a non-filled circle.
  - `line`: dictionary specifying line properties. You can use the common line keys for Plotly as shown in Figure I.4.

  - `showlegend`: boolean value indicating whether the trace should appear in legends.

  - `fill`: string specifying fill mode. Options include: `'tozeroy'` (will fill until the bottom of the y-axis) and `'tonexty'` (will fill below, until the next trace is encountered).

  - `fillcolor`: color of the fill specified using the Plotly conventions shown in Figure I.4.

- `add_bar()`: This method produces a bar chart, which is often used to show relative magnitude between data that is both quantitative and categorical. Keyword arguments for the bar trace are:

  - `x` and `y`: list or array. One of these (the numerical axis) should hold floats or ints, while the other (the categorical axis) could also hold floats or ints (specifying the bar locations) or it can hold strings (providing labels to be associated with the bars). Which is the numerical axis and which is the categorical axis will depend on the value of the keyword argument: `orientation`.

  - `orientation`: string specifying the orientation of the bars. Options include `'v'` (vertical) and `'h'` (horizontal).

  - `marker`: dictionary specifying marker properties. Common keys for this include:

    * `color`: color of the bars, specified using the Plotly conventions shown in Figure I.4.

    * `line` dictionary specifying line properties. You can use the common line keys for Plotly as shown in Figure I.4.

  - `name`: string that is the trace, used in legends and hover labels.

  - `text`: list or array of strings that will serve as the text labels for each bar.

  - `width`: float specifying the width of the bars along the categorical axis (as a fraction of the plot dimensions).

  - `showlegend`: boolean value indicating whether the trace should appear in legends.

  - `textposition`: string specifying the position of text labels. Options include `'inside'` (text inside bar) and `'outside'` (text outside bar).

- `textfont`: dictionary specifying font properties for text labels. Font is specified as shown in Figure I.4.
- `bargap`: float specifying the gap between bars in a group (as a fraction of the total plot dimensions). *NOTE: This must be specified in the* `update_layout()` *method of the Figure object.*
- `bargroupgap`: number specifying the gap between bar groups (as a fraction of the total plot dimensions). *NOTE: This must be specified in the* `update_layout()` *method of the Figure object.*

- `add_histogram()`: This method is used to construct a histogram, often used to show shape of distributions of data.
  - `x`: list or array of floats from which the histogram is constructed.
  - `histnorm`: string specifying the normalization mode for the histogram. Options include `'count'` (display the count of data points in each bin), `'percent'` (display the percentage of data points in each bin), and `'density'` (display the probability density, normalized such that the integral of the histogram is one).
  - `name`: string that is the name of the trace, used in legends and hover labels.
  - `marker`: dictionary specifying marker properties. Common keys for this include:
    * `color`: color of the bars, specified using the Plotly conventions shown in Figure I.4.
    * `line` dictionary specifying line properties. You can use the common line keywords for Plotly as shown in Figure I.4.
  - `orientation`: string specifying the orientation of the histogram bars. Options include `'v'` (default, vertical bars) and `'h'` (horizontal bars).
  - `xbins`: dictionary specifying the settings for the bins on the x-axis. Keys include:
    * `start`: float specifying starting value of the bins.
    * `end`: float specifying ending value of the bins.
    * `size`: float specifying the size of the bins.

  - `ybins`: dictionary specifying the settings for the bins on the y-axis (when orientation is given the value `'h'`). Same as the above dictionary for the `xbins`.

- `add_box( )`: This method creates a box and whisker plot, used to succinctly summarize statistical information about a set of data—typically showing the median, mean, inner quartile range, and outliers.

  - `y`: list or array of floats used to create the box plot.
  - `x`: list or array of strings that provide categorical labels or groupings for the data.
  - `name`: string that is the name of the trace, used in legends and hover labels.
  - `marker`: dictionary specifying marker properties. Common keys are:
    * `color`: color of the boxes and outliers, specified using the Plotly conventions, as shown in Figure I.4.
    * `outliercolor`: color of the outlier points, specified using the Plotly conventions, as shown in Figure I.4.
    * `line`: dictionary specifying line properties. You can use the common line keywords for Plotly as shown in Figure I.4.
  - `boxpoints`: string or array-like object specifying which data points should be shown as individual points. Options include: `'outliers'` (show outliers as individual points) and `'all'` (show all data points as individual points).
  - `boxmean`: string specifying whether to show a mean line inside the box. Options include `'false'` (yes, this is string, not a boolean and this is also the default value, which results in not showing a mean line), `'sd'` (show the standard deviation as a line inside the box), `'mean'` (show only the mean as a line inside the box), and `'sd+mean'` (show both mean and standard deviations).
  - `meanline`: dictionary specifying properties of the mean line when using `'mean '` or `'mean+sd'`. You can use the common line keywords for Plotly as shown in Figure I.4.
  - `orientation`: string specifying the orientation of the box plot. Options include `'v'` (default value, vertical boxes) and `'h'` (horizontal boxes).
  - `markeroutliercolor`: color of the outlier points, specified using the Plotly conventions as shown in Figure I.4.

- `add_pie()`: method that creates a pie chart, often used to show part to whole relationships. Although often frowned upon in science, they do have their place.
  - `labels`: list or array of strings that function as labels for the pie chart sectors.
  - `values`: list or array of floats corresponding to the sizes of the pie chart slices.
  - `hole`: float between zero and one specifying the size of the center hole in the pie chart. This creates a donut chart when hole is greater than zero. The default value is zero.
  - `textinfo`: string or array-like object specifying the information to display on the chart. Options include `'label'` (display labels for slices), `'text'` (display the values of slices as text), and `'percent'` (display percentages). Any of these can be joined together with a "+" to display multiple kinds. For instance, `'percent+label'` displays both the percentages and labels.
  - `direction`: string specifying the direction in which the sectors are drawn. Options are `'clockwise'` (default) and `'counterclockwise'`.
  - `sort`: boolean value indicating whether to sort the sectors by size.
  - `rotation`: float specifying the rotation angle of the first slice in degrees. The default value is zero, and the first slice starts at the top of the chart.
  - `textposition`: string specifying the position of text labels. Options include `'inside'` (text labels inside the slices) and `'outside'` (text labels outside the slices). If neither is specified, Plotly will try to make the best choice.
  - `marker`: dictionary specifying marker properties for the sectors. Common keys include:
    * `color`: specified using the Plotly conventions, as shown in Figure I.4.
    * `line`: dictionary specifying line formatting. See Figure I.4 for options.
  - `showlegend`: boolean value indicating whether the trace should appear in legends.
- `add_contour()`: This method creates a 2D contour plot. Three-dimensional plots can be very challenging to interpret, and so a 2D rendering of the contours of a 3D plot is often a better choice.

- `z`: 2D array of floats. These are the values to be contoured.
- `x` and `y`: list or array floats or strings specifying the labels or categories for the x-axis and y-axis, respectively.
- `ncontours`: integer specifying the number of contour levels to be displayed.
- `contours`: dictionary specifying the contour lines. Commonly used keys include:
  * `coloring`: string specifying how contour representation is colored. Options include fill (color the contour surfaces) and lines (color only the contour lines).
  * `showlines`: boolean value indicating whether contour lines should be displayed.
  * `start`, `end`, and `size`: Each of these accept floats that specify the range and step size of contour levels, respectively. If not specified, then Plotly guesses at the values. Can be used to provide more control than when using `ncountours` (see above).
  * `colorscale`: string or list specifying the colorscale to use for coloring the contour plot. The string refers to a named colorscale (see Figure I.4 for a list), while the list can specify the exact colors to be used, following the Plotly conventions shown in Figure I.4.
- `showlegend`: boolean value indicating whether the trace should appear in legends.
- `showscale`: boolean value indicating whether a color scale should be displayed alongside the contour plot.
- `line`: dictionary specifying line properties for the contour lines. You can use the same line specifications shown in Figure I.4.

- `add_heatmap( )`: Method that creates a heatmap. This plot can be used to show the projection of 3D surfaces in 2D, but can also be used to show correlations between variables.
  - `z`: 2D array of floats that are the values to be displayed in the heatmap.
  - `x` and `y`: list or array of strings specifying the labels or categories for the x-axis or y-axis, respectively.

- `colorscale`: string or list specifying the colorscale to use for coloring the heatmap. The strings correspond to named color scales, and the list contains colors (specified using the Plotly convention shown in Figure I.4) to be stepped through.

- `showlegend`: boolean value indicating whether the trace should appear in legends.

Adding data traces to subplots

All of the above methods both create the trace object and add it to our Figure object. Because we created the Figure object using `make_subplots()`, we will naturally have subplots we can add the traces to. In order to specify the exact subplot you wish to add the trace to, we use two keyword arguments: `row` and `col`. These are used to specify the location of the subplot you wish to add the trace to. The default values for these are `None`, but they must be specified if you are using a grid of subplots. If you do not supply them, the trace will be added to the subplot at `row` = 1, `col` = 1. This subplot is the top left most subplot, that is, rows count top by bottom starting at one and columns count left to right starting at one.

You can assign multiple traces to a single subplot. If you do so, they will be drawn in the order in which they were added. Thus, the last trace you add will be drawn on top of all the other traces.

That is it for assigning traces to the subplots.

STEP 3. FORMATTING AXES

If you want to change the formatting of your axes from the default, you need to use the `update_xaxes()` and `update_yaxes()` methods of the Figure object. When using these, you can specify a particular subplot to update using the `row` and `col` keyword. Specifying these will update the axes for *only* the corresponding subplot. If you do not specify anything for these keywords, then axes for *all* subplots will be updated.

When specifying the formatting, you have two options. First, you can use each of the **keyword argument**s described below. Alternatively, you can gather these keyword arguments and their values into a dictionary, and then pass this dictionary as the `first` **positional argument**. The former might be quicker when you are formatting a single axis; however, the latter will be superior if you are trying to apply the same formatting multiple times—for instance, across multiple figures.

Common keyword arguments for the `update_xaxes()` and `update_yaxes()` methods are:

- `title`: string that will be used as the axis title.

- **`range`**: list of floats that provide the lower and upper limits for the axis. The default behavior is to not use such a list and Plotly takes a reasonable guess at the range, given the data.

LINE

keyword	value
`width`	**float:** controls the thickness of the line in undefined relative units. Default = 1
`opacity`	**float:** from 0 (totally transparent) to 1 (totally opaque)
`dash`	**string:** name of a style, including: "solid", "dot", "dash", "longdash"
`color`	**string:** any of the valid Plotly specifications for color

FONT

keyword	value
`size`	**float:** determines the size of the font in units of pixels
`opacity`	**float:** from 0 (totally transparent) to 1 (totally opaque)
`family`	**string:** name of a font such as: "arial", "helvetica", or "times"
`color`	**string:** any of the valid Plotly specifications for color

COLOR

Color in plotly is specified in one of four ways:

'named' colors	hexidecimal value	tuple of RGB values	tuple of RGBA values
"red"	"#FF0000"	(255, 0 , 0)	(255, 0, 0, 1)
"blue"	"#1E90FF"	(113, 255, 116)	(255, 0, 0, 0.75)
"darkcyan"	"#228B22"	(0, 32, 94)	(255, 0, 0, 0.5)
"gray"	"#B8860B"	(255, 181, 181)	(255, 0, 0, 0.25)
and 174 more names in the Plotly documentation	you can find values online or in standard color pickers on your computer.	specifies values of the red, green, and blue color channels. You can values online, or using a standard color picker.	like RGB, but with an additional value for alpha, which ranges from 0 (transparent) to 1 (opaque)

FIGURE I.4. Keys and values for Line and Font formatting, as well as how color is specified in Plotly.

- **type**: string that specifies the type of axis (`'linear'` is the default), but other options are `'log'` (logarithmic), `'date'` (for dates), and `'category'` (for categorical).
- `showgrid` and `showgrid_minor`: boolean value that, when set to `True`, shows gridlines at the major or minor tick marks, respectively. If you do not want gridlines, then set these to `False`.
- `line`: dictionary controls the appearance of the line that represents the axis. You can use the values shown in Figure I.4.
- `showline`: boolean that, when set to `True`, results in the axis line being shown. This is the default value.
- `zeroline`: boolean that, when set to `True`, results in a line being drawn at the value of zero. The default value is `False` and this line is not drawn. The color of this line can

be controlled using the keyword `zerolinecolor` and using the standard Plotly color designations, as shown in Figure I.4.

- `ticks`: string that control the position of the tick marks. Accepted values are `'inside'`, `'outside'`, `'both'`, or an empty string (`''`). If an empty string is used, then no tick marks are drawn.

- `tickvals`: list of floats that specify where the values at which tickmarks are drawn. If no list is specified, Plotly takes a guess at reasonable tickmark values.

- `ticktext`: list of strings or floats that specify exactly what text appears at the tickmarks. If no such list is specified, Plotly will default to something reasonable (i.e., numbers). If you want to use your own custom labels, specified in a list, then you need to also specify `tickvals` as well as include `tickmode = 'array'`

- `showticklabels`: boolean that, when set to `True`, results in the values at the ticks being displayed.

- `tickformat`: string that specifies the formatting for numbers at the tick marks. The formatting of numbers was discussed in our chapter on strings.

- `tickprefix` and `ticksuffix`: strings that specify text to be added before or after the tick labels, respectively. For instance, you could add a $ before the number, or a % after the numbers.

STEP 4. FORMATTING THE REST OF THE FIGURE

There are two kinds of formatting that you might like to do to your figure:

1. Add an element that is not yet present. For instance, you might want to add text to the plot to annotate some features. Or you might want to add an image to your plot.

2. For the elements that are already present, or apply some global changes. For instance, you might want to change the background color for the plotting area, or change the margins that are used, or perhaps change the default font size of the figure.

These two actions require different approaches, and so we discuss them separately.

Adding elements to the plot

For the purposes of this book, you will be either adding text-based annotations or images.

For text-based annotations, you can use the `add_annotation()` method of the Figure object. This method accepts a number of keyword arguments, including

- `text`: String, the text to be displayed in the annotation. This can be an empty string. You can also supply formatting using HTML within the string.
- `font` dictionary, containing key-value pairs for formatting text, as shown in Figure I.4.
- `align`: string with possible values of `left`, `center`, and `right`. Specifies the alignment of text.
- `bgcolor`: string that specifies the color of the text box background. Values set as shown in Figure I.4.
- `bordercolor`: string that specifies the color of the text box border. Values set as shown in Figure I.4.
- `showarrow`: boolean that states if an arrow should be drawn from the annotation position to the text. Default is `True`, so you need to be aware that the default behavior is to have a line.
- `xref` and `yref`, `axref` and `ayref`: string with possible values `paper`, `x`, or `y`. The first two specify the positional coordinates for the annotation are relative to the plot area or the axes coordinates. If there is an arrow, this will be where the arrow starts. If there is no arrow, then it is the position for text alignment. The second two specify which axis to use as the reference. If you have multiple subplots, you will have multiple axes of each type, and this is why you might need to specify which to use.
- `x` and `y`, `ax` and `ay`: The first two specifies the positions of annotation. If there is an arrow drawn, this is where the arrow will start. If not, it species where the text will be aligned on. If there is an arrow drawn, the second two specify where the arrow will end, as an *offset* along the x and y axes, in units of *pixels*. If the reference is set to `paper`, then you use a float. If it is set to `x` or `y`, then you use a number (if the axis is numerical) or a label (if the axis is categorical).
- `arrowsize`: float that specifies the scaling factor applied to the arrow. Default is one.
- `arrowhead`: int that specifies the style of arrow to use. The options are 0 (no arrowhead, so a plain line), 1 (triangle for arrowhead), 3 (open triangular arrowhead), 4 (a T-junction), 6 (filled circle), 7 (open circle), and 8 (x-shape).
- `arrowwidth`: float that specifies the width of the arrow.
- `arrowcolor`: string that specifies the color, as shown in Figure I.4.

If you want to add an image to the figure, then you will either be adding a **scalable vector graphic** image or a **bitmap** image. Both of these require the use the `add_layout_image()` method. The manner in which scalable vector graphics are added is detailed in Chapter 4, so this will not be repeated here.

If you want to add a bitmap, you can use the same basic approach; however, the `source` keyword argument needs to be given an object containing the information for a bitmap. The simplest way to do this is to use the Pillow library (often imported using `import PIL`) to open up an image as a Pillow Image object, and then supply this to the `source` keyword argument. For instance:

`source = PIL.Image.open(<image_path>)`, where `<image_path>` is the **path** to the image on your computer. If you want to know how to use Pillow, we have chapter on that as well: Chapter 16.

Formatting elements that are already present using `update_layout()`

When you want to format standard elements of a plot that are not an axis or the data representation (trace), you will use the `update_layout()` method. This accepts keyword arguments that let you change various aspects of the plot. We are going to suggest that you always use the `template` keyword, and then follow up its usage with any other keywords you want to change. This approach allows you to work from a reasonable start. The common keywords for `update_layout()` are:

- `template`: string that specifies a pre-made template from the Plotly library. Examples are `'plotly'`, `'seaborn'`, `'simple_white'`, and more. Of course, one you specify these, you can continue to change them by using any of the keywords below or in the Plotly documentation.

- `row` and `col`: integers that specify the row and/or column to which you wish to apply the specified formatting. If you do not use these keyword arguments, then the Layout will be applied to all subplots. If you do, then the Layout will only be applied to the specified subplot(s).

- `margin`: `dict` that controls the size of the outer margins of the plot—that is the distances from the edges of the figure. The value provided is a dictionary, with keys `l`, `r`, `t`, `b` (for left, right, top, and bottom, respectively). Each can be assigned an integer value, which specifies the size of the margin in units of pixels. So, if you wanted to have margins that are 50 pixels on each side, you would use: `margin = dict(l = 50, r = 50, t = 50, b = 50)` to do so.[1]

- `paper_bgcolor`: controls the color of the entire figure. So, not only the plot areas, but also all margins will have this color. Use the color specifications shown in Figure I.4.

- `plot_bgcolor`: controls the color of only the plotting areas. The color outside the plotting areas (e.g., margins) will be unaffected. Use the color specifications shown in Figure I.4.

[1] If you want to control the inner margins (between subplots in a figure with multiple subplots) then refer to the `specs` keyword associated with `make_subplots` (Step 1), specifically the `vertical_spacing` and `horizontal_spacing` keywords.

- `title`: string that will be used as the title of the plot.
- `height` and `width`: floats that specify the height and width of the page, respectively. Values are in units of pixels. The page is the object that holds all the subplots, so you are adjusting the entire canvas for the figure. If no values are defined, then Plotly will simply render the plot to fit the window it is being drawn in. You can also specify sizes when rendering and saving the plot (see next section) and so you don't have to specify a size here.
- `font`: dictionary that controls the appearance of font in the figure. The keywords that can be included in this dictionary summarized in Figure I.4.
- `legend`: This allows you to control the appearance of the legend (sometimes called key) of your plots. This accepts a dictionary. Commonly used keywords for the dictionary are:
  - `font`: dictionary that controls the appearance of font in the figure. The keywords that can be included in this dictionary summarized in Figure I.4.
  - `title`: string that will be used at the title of the legend.
  - `x` and `y`: floats between zero and one, which specify where the legend is placed on the plot area. These are relative to the min and max of the axes and so 0.5 is the middle.
  - `xanchor` and `yanchor`: strings that tell Ploty where the placement should be anchored to on the legend. Options are `'left'`, `'right'`, and `'center'` (for `xanchor`) and `'top'`, `'middle'`, `'bottom'` (for `yanchor`).
  - `borderwidth`: float that controls the line around the legend. Setting this to zero will result in no line being drawn.
  - `bordercolor`: You can use this to control the color of a border drawn around the legend. Colors are specified using the approach shown in Figure I.4.
  - `bgcolor`: You can use this to control the background color of the legend. Colors are specified using the approach shown in Figure I.4.
- `hovermode`: string that controls how hover information is displayed in interactive plots. Common options include `'x'`, `'y'`, `'closest'`, and `'x unified'`. These control whether the hover text is shown for points closest to the cursor or across entire vertical/horizontal slices. To format the appearance of hover text, use the `hoverlabel`

dictionary. If you do not want any hover functionality, set the keyword `hoverinfo` for individual traces to `'skip'` or disable it using `fig.update_traces(hoverinfo='skip')`.

- `annotations`: dictionary that allows you to add text elements to the plot. You can use the same set of options shown for the `add_annotation()` method, as shown above, but the keyword arguments serve as keys in this dictionary.

- `shapes`: dictionary that lets you draw shapes on your plot. Common keywords are:

  - **`type`**: Specifies the type of shape to be drawn. Options include: `'line'`, `'rect'`, and `'circ'`.
  - `x0`, `x1`, `y0`, and `x1`: float that the starting and ending points along x- and y-axes.
  - `fillcolor`: you can specify color using Plotly conventions shown in Figure I.4
  - `line`: dictionary that controls the appearance of lines, as shown in Figure I.4
  - `opacity`: float between zero (completely transparent) and one (completely opaque).
  - `layer`: string that specifies if the shape is drawn below or above other elements on the plot. Values are `'above'` or `'below'`.

- `images`: a list of dictionaries (one per image) that specifies and formats an image to include in your plot area. Common keywords of the dictionaries are:

  - `source`: string or variable that points to the image to be displayed. Although there are several options, we suggest creating an object that is a base64-encoded image data string, and passing this to the source. A discussion of image encoding is given in the chapter on images in Python.
  - `x` and `y`: floats that are the x- and y-coordinates of the left and top edges of the image. These are given in plot coordinates.
  - `xanchor` and `yanchor`: strings that specify the reference position for x and y. Options are `'left'` and `'right'` for the former and `'top'` and `'bottom'` for the latter.
  - `sizex` and `ysize`: floats that specify the width and height of the image. These are given in plot coordinates.

- `sizing`: string that specifies the image should be sized and/or stretched. Options include `'contain'` (keep aspect ratio) and `'stretch'` (stretch image).

- `opacity`: float between zero (completely transparent) and one (completely opaque).

One further note on usage: for keywords that accept dictionaries, we find it simpler and cleaner to construct the dictionary outside the `update_layout()` method call, and then pass the dictionary's variable to the correct keyword.

STEP 5. OUTPUTTING THE FIGURE BY SAVING OR DISPLAYING

For this book, we will use two different output types to generate three different kinds of figure output. The methods are:

- `write_image()`: This method is used to save a static image of your plot, such as you might include in a paper, report, or slide deck. The function has one positional argument, which is the filename. This accepts a string that can provide information on where to save and the file to save. Using the extension of the file, you can specify the file type, and options include `'png'`, `'jpg'`, `'svg'`, and `'pdf'`. However, you can also specify keyword arguments, and the three you are most likely to use are:

  - `width` and `height`: integer that specify the final image width and height (respectively) in pixels. Because you can call `write_image()` method multiple times for the same Figure object, you can output multiple files in different sizes for different use cases.

  - `engine`: string that specifies the engine you wish to use to write the image. In this book we have tried to ensure you only have a single engine and so Plotly will default to using it. However, if you have multiple engines installed, you may want to specify one or the other. Options are `'kaleido'` and `'orca'`.

- `show()`: This method is used to generate either static or interactive images. They can be displayed in browsers or other settings. It accepts a positional argument, which controls the display mode, and options are `'div'`, `'cdn'`, `'svg'`, `'png'`, and `'browser'`. The last of these will open the plot in a web browser window. We are going to suggest that you primarily use `'svg'`, `'png'`, and `'browser'`. The first two will output to the Sypder plot window and provide a static record of the plots you have made, and the last one will provide interactive plots that exist until you close the browser window.

I.2 UNDERSTANDING THE STRUCTURE OF PLOTLY FIGURE OBJECTS

As you become more familiar with the Plotly library, it will probably be helpful to understand the structure of the Plotly Figure object. Fortunately, you can gain access to this structure by printing the object. For instance, let us consider printing the `sincosplot` Figure object we created near the beginning of this chapter and which was shown in Figure I.2.

```
In: print(sincosplot)
```

```
Figure({
    'data': [{'line': {'color': '#FF00FF', 'width': 3},
              'mode': 'lines',
              'name': 'sine',
              'type': 'scatter',
              'x': array([ 0.        ,  0.20408163,  0.40816327,  0.6122449 ,  0.81632653,
                           1.02040816,  1.2244898 ,  1.42857143,  1.63265306,  1.83673469,
                           2.04081633,  2.24489796,  2.44897959,  2.65306122,  2.85714286,
                           3.06122449,  3.26530612,  3.46938776,  3.67346939,  3.87755102,
                           4.08163265,  4.28571429,  4.48979592,  4.69387755,  4.89795918,
                           5.10204082,  5.30612245,  5.51020408,  5.71428571,  5.91836735,
                           6.12244898,  6.32653061,  6.53061224,  6.73469388,  6.93877551,
                           7.14285714,  7.34693878,  7.55102041,  7.75510204,  7.95918367,
                           8.16326531,  8.36734694,  8.57142857,  8.7755102 ,  8.97959184,
                           9.18367347,  9.3877551 ,  9.59183673,  9.79591837, 10.        ]),
              'y': array([ 0.        ,  0.20266794,  0.39692415,  0.57470604,  0.72863478,
                           0.85232157,  0.94063279,  0.98990308,  0.99808748,  0.96484631,
                           0.89155923,  0.78126802,  0.63855032,  0.46932961,  0.2806294 ,
                           0.08028167, -0.12339814, -0.32195632, -0.50715171, -0.67129779,
                          -0.80758169, -0.91034694, -0.97532829, -0.99982867, -0.9828312 ,
                          -0.92504137, -0.82885774, -0.6982724 , -0.53870529, -0.35677924,
                          -0.16004509,  0.04333173,  0.24491007,  0.43632343,  0.6096272 ,
                           0.75762842,  0.8741843 ,  0.9544572 ,  0.99511539,  0.99447137,
                           0.95255185,  0.8710967 ,  0.75348673,  0.60460332,  0.43062587,
                           0.23877532,  0.0370144 , -0.16628279, -0.36267843, -0.54402111])},
             {'marker': {'color': 'darkcyan', 'symbol': 'diamond'},
              'mode': 'markers',
              'name': 'cosine',
              'type': 'scatter',
              'x': array([ 0.        ,  0.20408163,  0.40816327,  0.6122449 ,  0.81632653,
                           1.02040816,  1.2244898 ,  1.42857143,  1.63265306,  1.83673469,
                           2.04081633,  2.24489796,  2.44897959,  2.65306122,  2.85714286,
                           3.06122449,  3.26530612,  3.46938776,  3.67346939,  3.87755102,
                           4.08163265,  4.28571429,  4.48979592,  4.69387755,  4.89795918,
                           5.10204082,  5.30612245,  5.51020408,  5.71428571,  5.91836735,
                           6.12244898,  6.32653061,  6.53061224,  6.73469388,  6.93877551,
                           7.14285714,  7.34693878,  7.55102041,  7.75510204,  7.95918367,
                           8.16326531,  8.36734694,  8.57142857,  8.7755102 ,  8.97959184,
                           9.18367347,  9.3877551 ,  9.59183673,  9.79591837, 10.        ]),
              'xaxis': 'x2',
              'y': array([ 1.        ,  0.97924752,  0.91785141,  0.81835992,  0.68490244,
                           0.52301811,  0.33942593,  0.1417459 , -0.0618173 , -0.26281476,
                          -0.45290412, -0.6241957 , -0.76958007, -0.88302305, -0.9598162 ,
                          -0.99677222, -0.99235724, -0.94675453, -0.8618568 , -0.74118774,
                          -0.58975572, -0.41384591, -0.22075945, -0.01851037,  0.18450698,
                           0.37986637,  0.55945943,  0.71583215,  0.84249428,  0.93418872,
                           0.98710971,  0.99906074,  0.9695458 ,  0.8997899 ,  0.79268826,
                           0.65268613,  0.48559429,  0.29834787,  0.09871854, -0.10500809,
                          -0.30437638, -0.49111153, -0.65746312, -0.79652673, -0.90253053,
                          -0.97107484, -0.99931473, -0.98607811, -0.93191435, -0.83907153]),
              'yaxis': 'y2'}],
    'layout': {'font': {'color': 'grey', 'family': 'times', 'size': 18},
               'template': '...',
               'title': {'text': 'Example with subplots'},
               'xaxis': {'anchor': 'y',
                         'domain': [0.0, 0.44],
                         'linecolor': 'grey',
                         'range': [0, 10],
                         'showline': False,
                         'ticks': 'inside',
                         'title': {'text': 'x-specified'}},
```

```
                    'xaxis2': {'anchor': 'y2',
                               'domain': [0.56, 1.0],
                               'linecolor': 'grey',
                               'range': [0, 10],
                               'showline': False,
                               'ticks': 'inside',
                               'title': {'text': 'x-specified'}},
                    'yaxis': {'anchor': 'x',
                              'color': '#FF00FF',
                              'domain': [0.0, 1.0],
                              'title': {'text': 'intensity'},
                              'zeroline': True,
                              'zerolinecolor': 'grey'},
                    'yaxis2': {'anchor': 'x2',
                               'color': 'darkcyan',
                               'domain': [0.0, 1.0],
                               'zeroline': True,
                               'zerolinecolor': 'grey'}}
})
```

The very first thing to note is that the Figure object *appears* to hold information inside of dictionaries. This is *sort of* true. However, the Figure object is its own data structure, and so you cannot always interact with it in the way you would a dictionary. However, for the most part, trying to interact with it as a dictionary is a reasonable approach to start with.

The first object is given the `data` keyword, and holds information on the data. This object is associated with a list, where there is one entry for each trace. These entries are also dictionary-like, and contain a number of keywords specifying what and how the trace will be rendered. In our example, the first entry is of the type "scatter" and has information on the name, the mode for the display ("lines") and specification for the way in which these lines will be drawn. However, there is also information on the x- and y-data, where the data is directly given. This is super convenient because we can always pull the data that we are interested in directly from a figure! For instance, `sincosplot['data'][0]['x']` would output the x-data for the first entry in the data dictionary within the sincosplot Figure object.

One other critical piece of information in the `data` part of the Figure object occurs when data is referenced to a specific subplot. For instance, the marker version of the data is shown in the first row, second column. The axes on the subplots are numbered, and this subplot has the axes `x2` and `y2`. You can see that the second entry in the data structure of the Figure object includes these designations. So, this is a way to know where data is plotted in a complex figure. However, there is even more: we can *change* where a trace is drawn by changing these designations directly. The axes on the first subplot are `x` and `y`. So we can use the commands `sincosplot['data'][0]['xaxis'] = 'x2'` and `sincosplot['data'][0]['yaxis'] = 'y2'` to change the axes to which the data is assigned and move the data that was plotted on the first subplot to the second subplot.

The other top-level container within the Figure object is `layout`. Layout is another dictionary-like object, and it contains information on the styling of the plot, including information on the font, the template, the title of the plot. Additionally, this is where the identity of the axes is stored, including their format. Again, we can go in and directly change any of these.

Despite all of the above utility, **there is one thing you cannot do: you cannot simply add a trace object**. That is, you cannot create a dictionary that looks like a well-formatted trace object (with type, data, axes, and so on) and add it to the data structure. If you do, you will find that the apparent list is actually a tuple, and so we cannot add or remove things from it. At the best, we can add only subsets and permutations of the current data object. Thus, the only way to add a trace is to use the `add_<trace type>()` methods of the Figure object, as described above.

I.3 OTHER TASKS

There are many other tasks that you might like to do when working with Plotly, and we will outline just a few here.

I.3.0 MAKING INSETS

It is common to desire to make one figure inside another figure—an inset. There are two ways to accomplish this directly in Plotly, although neither are super straightforward:

- You can create a large number of subplots, create the make figure to span all these subplots, and then just add the inset to the subplot that is in the place you want it. For instance, you could make subplots that are given by `rows = 4, cols = 4`. You could then place a subplot in `row = 1, col = 1` and have it span four rows and columns. This figure will occupy the entire figure canvas. Then you could add another subplot in `row = 4, col = 1` and it will sit in the top right corner of the plotting canvas—the top right corner of the subplot spanning the entire figure canvas.

- You can add multiple axes and then use the `domain` keyword for the axes, to control where they appear on the plot. We will not go into more detail here, but the usage can be found in Plotly's online documentation.

I.4 THE CODECHEMBOOK PLOTLY WRAPPERS

As a final note, we have also written wrappers that produce Plotly images quickly, focusing on the types of plots that we, as experimental chemists, often make. These are included in the `quickPlots` module of our codechembook library, and there are many examples of its usage throughout the Narrative chapters, starting with Chapter 6. The ones we think you will use most are: `quickScatter()`, `quickBar()`, `quickHist()`, and `plotFit()`.

GLOSSARY

absolute path: The series of folders, starting at the root directory (folder), that must be traversed to locate a specific directory or file on a computer. (See pages 56, 58)
alias: A new name for an object that already has a name. Often used when importing Python libraries, where it can be used to provide a shorter name for an already named object. e.g., `import numpy as np`. In this example, `np` is the alias. (See pages 29, 65, 281)
analytical: In the context of models, this means that there is a solution that can be written down as a mathematical equation. This is in contrast to those solutions requiring numerical approaches. (See page 187)
argument: Data that is passed to a function when it is called. The data is either operated on, or it specifies the manner in which the function should operate. In Python, there are four types of arguments: positional arguments, keyword arguments, arbitrary positional arguments, and arbitrary keyword arguments. Data passed as an argument in a function call is assigned to the corresponding parameter specified in the function definition. (See pages 8, 16, 19, 30, 35, 55, 56, 102, 103, 104, 130, 187, 189, 203, 227, 265, 281, 343, 355, 358)
assign: The operation of associating an object with a variable name. (See pages 12, 30)
attribute: Data associated with an object. The value of the attribute can be accessed by naming the object, using a period, and then stating the name of the attribute desired. e.g., <object>.<attribute>. (See pages 55, 82, 87, 88, 104, 129, 130, 135, 153, 188, 302, 344)
base 64 encoding: A form of binary encoding that is based on 64 bits, or $2^6 possible values$. (See page 100)
binary encoding: A form of storing data that is not intended for humans to read. This differs from plain text data storage, often used when you intend humans to read the data. Binary encoded data typically uses less space than plain text to represent the same data. (See pages 126, 286, 336, 369)
binary format: The details of how data is translated when using binary encoding. (See pages 100, 280)
bitmap: One approach to encoding image data. In a bitmap, the image is stored by breaking it into a grid of pixels, and then assigning colors to each of these pixels. (See pages 309, 437)
block: A portion of Python code that shares the same indentation level. Often used to indicate that the code all belongs to the same control structure. (See pages 104, 169, 172, 281, 402, 404, 416)
boolean: A data type that has only two possible values: `True` and `False`. (See pages 54, 145)
broadcast: The ability of Numpy to expand a smaller Numpy array so that it has the same shape as a larger Numpy array, allowing for element-wise operations without explicitly duplicating values. (See pages 33, 34)

call: Invoking some code from some other code. For instance, if one utilizes a function, we say we have "called" the function, or that we invoked a function call. (See pages 8, 10, 18, 19, 55, 66, 86, 103, 261, 404)
class: A collection of Python code that defines a complex, self-contained object composed of data, methods that transform the data, and attributes that provide information about the data. A class can be thought of a template: it does not hold data, but has instructions on how to create an instance of the class that will hold data. (See pages 53, 55, 58, 66, 82, 98, 104, 150, 151, 152, 238, 337)
codechembook: Python library written by the authors of this book to accomplish tasks common to experimental chemists. (See pages 59, 126, 149)
collection: An object that contains multiple other objects but that can be referred to with a single variable name. (See pages 79, 82, 99, 102, 145, 148)
comment: A portion of code that is not meant to be executed, but is used to document the code. Comments are used to improve the readability and understandability of code. (See pages 11, 19, 40, 41, 62, 104)
comparison operation: A statement that uses operators (<, >, ==, <=, >=, !=, `not`, `and`, `or`) and evaluates to a boolean value. (See pages 147, 345)
complex: A numerical data type used to represent complex numbers.
concatenate: Join two collections together. (See pages 77, 83, 87)
console: Also sometimes called the command line. An interface by which a user can enter text-based commands that the computer will execute. This is done in an interactive manner, so that each command is executed as it is entered. (See pages 6, 14, 16, 18, 132, 272)
control structure: A statement that allows code to execute in a defined order other than simply line by line. (See pages 83, 168, 172, 281, 346, 349)
creation: Making a new instance of an object. (See page 30)
csv: Stands for "comma separated variable." Or sometimes the much more general "character separated variable." This is a format for structuring data in plain text files. A character (called a delimiter) is chosen that is placed between pieces of data. Then, one can simply separate out the values by breaking on this delimiter. This character is often a comma, but other ones are also used, such as tabs, spaces, colons, semicolons, or others. (See pages 51, 60, 65, 76, 86, 280, 322)
data type: Classification of the values that objects can hold. For instance, Python makes a distinction between text (strings) and numbers. Within number, Python makes a distinction between numbers with decimal places (floats) and numbers without decimal places (ints). (See pages 10, 16, 29, 30, 32, 54, 55, 77, 82, 102, 104, 108, 129, 133, 145, 337, 341, 370, 425)
delimited: A format of human-readable data whereby different values are separated by a delimiter. (See page 60)
delimiter: When data is delimited, this is the character that is used to separate the values. Often the delimiter is a particular characters of text, such as a comma, semicolon, tab, or space. Often this strategy is used to create data files in plain text. (See pages 60, 66)

dictionary: A collection of values in which each value is paired with a key that serves as a mnemonic label that identifies it and makes it easy to access programmatically. (See pages 82, 87, 108, 110, 152, 407, 425)
docstring: A multiline comment that directly follows the definition line for a function. This typically gives information regarding the function's purpose, the parameters/arguments for the function, and what the function returns. Some IDEs will use these docstrings to dynamically provide information on the function, when you include it in your code. (See pages 104, 105, 261)
document: The process of providing a guide for understanding and using code. This can be done by using comments and docstrings, but one can also point to online resources, or other means of documentation. (See pages 40, 62)
e notation: A notation used to represent numbers with decimal places. It is similar to scientific notation, but 10^X is replaced by eX. Thus, the e is a placeholder for "ten to the power of. (See pages 11, 125, 136, 285, 373)
error: The result of Python attempting to execute instructions that cannot correctly complete. The error can happen for many different reasons, such as incorrect syntax or missing values. (See pages 21, 133)
exclusive: Does not include the specified object. Often used when discussing indexing—especially when slicing. In such a case, a range of indices are given and the final index in the specified range is excluded. The opposite of inclusive. (See pages 80, 102)
executable instructions: Code that include valid commands/expressions. (See page 11)
expression: Code that evaluates to a value. (See pages 6, 12, 30, 54, 60, 90, 125, 337)
extension: Part of a filename, often hidden by an operating system, that follows a period. It is used to specify the type of data in a file and is used by a computer to know what programs can open the file. For instance, Microsoft Word files often end in .docx. In this case, "docx" is the extension, and it is used to tell the computer to open this file in Microsoft Word when you double click on it. (See pages 56, 57, 87, 128)
f-string: A way to create strings in Python in which the result obtained by evaluating an expression can be included in the string. This is done by placing an "f" before the string starts, and then adding the code to be executed between curly brackets ({}). (See pages 124, 303)
figure object: The object that holds all data for a Plotly figure. (See pages 61, 77, 87, 419)
file structure: The arrangement of files and directories in a computer's file system. (See pages 56, 57)
file system: A framework for defining the relationships of directories and files on a computer. File structures are implemented within this framework. (See pages 51, 56, 59, 170)
float: The name of the data type for working with numbers with decimals within Python. (See page 133)
for: A keyword that indicates that there will be iteration through the elements in a collection. (See pages 86, 97, 102, 133, 156, 249, 284, 290, 314)
function: A reusable block of code that performs a specific task and can be called by its name. Functions can accept inputs (called parameters), execute statements, and return

results to the code that calls it. (See pages 8, 12, 16, 18, 19, 28, 30, 32, 36, 52, 53, 54, 55, 60, 79, 81, 82, 86, 101, 102, 103, 104, 122, 123, 128, 148, 149, 171, 174, 187, 189, 203, 207, 227, 228, 237, 238, 265, 286, 337, 342, 345, 347, 360, 364, 370)

generator: An object that holds instructions for creating data, rather than the data itself. (See pages 102, 128, 133)

hard coding: When values are specified directly in the code, rather than being obtained from somewhere else (i.e., an external file or user input). (See pages 20, 122)

header: In a .csv file, text that appears on lines before the data rows. It is often used to provide important information about the data and/or to label columns to give the user information about the organization of the data. (See pages 51, 66, 87)

hexadecimal: A way to represent a base-16 number system. Binary can be represented using 1 and 0. Base 10 uses 0, 1, 2, 3, 4, 5, 6, 7, 8, and 9. Base 16 uses these same digits, but adds: a, b, c, d, e, and f. (See pages 126, 297, 311, 336)

histogram: A representation of data in which data is divided into bins with lower and upper limits and the counts or frequency of the data falling between these limits is drawn using bars in a chart. (See page 220)

HTML Stands for hypertext markup language, which is a means to specify how text, images, and other data can be represented on a screen. This is used widely for webpages. (See page 101)

immutability: A property of certain data types that, once created, their value cannot be altered. (See page 99)

import: The means by which to add functionality to Python. Once a class, module, or function is imported, you can use it in your code. (See pages 86, 168, 178, 281)

in: The keyword for an operator that evaluates to a boolean depending on whether an object exists within a collection. (See pages 83, 103)

in place: A behavior in which the values within an object are changed without re-naming the object or assigning it to a different variable name. (See page 78)

inclusive: Includes the specified object. Often used when discussing indexing—especially when slicing. In such a case, a range of indices are given and the first index in the specified range is included. The opposite of exclusive. (See pages 35, 80, 102)

index: An integer that specifies the position of an element within an sliceable collection. The first position has, by definition, the index 0. (See pages 79, 81, 99, 123, 147, 358, 417)

inner array: When the elements of a Numpy array are also arrays, the contained (or nested) arrays are called "inner arrays." The containing array is called an "outer array." (See pages 33, 80)

instance: A specific object holding data that has been created from a class. (See pages 30, 55, 65, 66, 123)

instantiation: The creation of an instance of an object. (See page 30)

int: The name of the data type for working with integers within Python.

Integrated Development Environment (IDE) A computer program that assists a programmer in creating code. Typically it includes a text editor with syntax highlighting, a console, and tools to aid debugging. (See page 67)

iterable: A property of an collection whereby the elements can be processed one by one. For ordered collections, the elements are processed in order. (See pages 103, 148, 208, 301)
iterate: Execute a block of code a single time in a loop. The iteration number is advanced each time a loop iterates. (See pages 76, 148)
iteration: A given round of execution of a code block in a loop. Often a specific iteration is identified using integers. So, one speaks of "the 4th iteration", for instance, or "iteration number 4". (See pages 83, 396)
key: Together with a value, this forms a key/value pair for a dictionary. One can access the value by referencing the key. (See pages 82, 87, 97, 108, 364)
keyword: Word used in a statement or expression with special meaning that causes an action. Examples in Python include `import`, `for`, and `in`. Keywords cannot be used as variable names. Alternately, keyword refers to the name of a keyword argument in a function. (See pages 19, 87, 103, 145, 203, 302, 376, 403, 404)
keyword argument: Arguments passed to functions that are not required for the function to be called. They must appear after positional arguments, but their order does not matter. Data is passed to a keyword argument by assigning the data to the corresponding keyword in an assignment statement within the function call. Here, the use of "keyword" is distinct from reserved keywords above. (See pages 9, 20, 61, 66, 86, 101, 127, 130, 149, 178, 188, 227, 228, 265, 302, 343, 424, 425, 433)
library: A external set of Python files that define new classes, functions, and so on that can be imported to add new capability to code. (See pages 29, 42, 55, 56, 57, 59, 60, 65, 82, 98, 101, 120, 126, 127, 129, 132, 144, 149, 150, 151, 152, 166, 168, 170, 187, 202, 218, 224, 258, 281)
list: A slicable, mutable collection data type that can hold any type of object. (See pages 77, 78, 79, 81, 83, 86, 99, 102, 103, 128, 129, 145, 146, 147, 417)
list comprehension: (See page 208)
literal: A fixed value that is directly written into the code to represent data. Sometimes this is called "hard coding." This approach stands in contrast to the use of variables. (See pages 12, 34, 104, 169, 370)
lmfit: A Python library that functions as a wrapper for many behaviors of another library: Scipy. This library is often used to fit the return of functions to data. (See page 149)
loop: A series of instructions that are repeated. (See pages 76, 83, 86, 97, 102, 133, 145, 156, 208, 249, 284, 290, 349)
mask: A Numpy array of True or False values that can be used to select elements from another array. It must have the same shape as the array it is applied to. (See pages 146, 147, 206, 208)
memory: Hardware in a computer that can temporarily store data. Variables identify the location of specific data in memory. (See pages 12, 13, 51, 99, 281, 337, 346, 412)
method: A function that is contained in an object and that acts on that object. Thus, one must call the function by first listing the object: <object>.method(). (See pages 55, 58, 66, 77, 78, 82, 98, 100, 104, 127, 128, 129, 151, 202, 239, 312, 373, 421, 424, 427)

model: A parameterized quantitative mathematical prediction of the response of a phenomenon to an independent quantity. For example, a linear equation ($y = m{\cdot}x + b$) is a model of linear behavior. (See page 129)
module: A Python file (i.e., a file ending in .py) that is typically part of a Python library. (See pages 53, 65, 104, 149, 187, 227, 237, 286)
mutability: A property of certain data types that, once created, their value can be altered. This can lead to unexpected behavior when multiple variable names point to the same mutable object. (See page 411)
nesting: The process of containing one collection of objects in another collection of objects. (See pages 33, 133, 397)
newline character: A text symbol that is often not rendered. Instead, it is used to mark where a line should end, and a new line begin. This is the behavior that is expected from the "Enter" or "Return" key on your keyboard. In Python, this is specified by the string: \n. (See page 17)
numerical: In the context of models, this means that there is not solution that can be written down as a single mathematical equation. Instead, all functions involved in the solution are approximated by a series of numerical values. This is in contrast to solutions that can be solved with analytical approaches. (See page 187)
numerical precision: The idea that not all numbers can be represented exactly in a computer. This is due to the fact that there are an infinite amount of numbers, but only a finite amount of computer memory. (See pages 7, 9, 35)
Numpy array: An ordered, sliceable collection of values. All values must be of the same data type. Numpy arrays allow for vectorization and broadcasting. (See pages 30, 33, 34, 35, 36, 42, 51, 54, 55, 77, 79, 99, 103, 123, 147, 155, 188, 227, 228, 281)
Numpy: A library that provides classes and functions for working with Numpy arrays. (See pages 29, 30, 42, 60, 68, 171)
object: An instance of a class. This means that the object will have specific data, attributes, and methods. (See pages 44, 53, 54, 55, 56, 60, 65, 78, 82, 98, 99, 102, 103, 108, 128, 129, 132, 148, 151, 155, 170, 188, 208, 239, 337, 358, 384, 385)
object oriented: A programming approach by which common data transformations and queries are built into the objects themselves (as methods and attributes, respectively), rather than separately programmed. Thus, solutions to problems become a sequence of requests for the objects to change themselves, rather than a sequence of original code to change their values. This allows the programmer to focus on the solution more than the specific implementation of algorithms.
operator: A keyword or reserved symbol in code that instructs the computer to perform a specific action on one or more objects. Examples include mathematical operations (+, −, *, / **) or assignment to variables (=). (See pages 6, 7, 12, 30, 54, 77, 79, 147)
outer array: When an array has multiple dimensions, this is represented by nesting arrays. An array that contains arrays is an outer array, while the arrays inside that array are called inner arrays. (See page 33)

parameter: In a function definition, a name of a variable that exists only in the function that takes on the value of data passed as an argument in the function call. (See pages 18, 19, 104, 150, 151)
parsing: The act of processing a collection to identify and extract a subset of its data. For instance, pulling words from a sentence can be called "parsing" the sentence. (See page 52)
pass: The act of transferring data from the code that calls a function to the code that specifies the behavior of the function. The data is passed using arguments defined by the function. (See pages 19, 30)
path: The text-based representation of a location in a computers file structure. Paths can be absolute or relative. (See pages 51, 59, 60, 81, 170, 281, 380, 437)
Pathlib: A library that provides classes that represent paths and methods for manipulating them. (See pages 56, 58, 60)
plain text: A representation of data that is readable by humans. Plain text files can be opened and read in text editors. (See pages 280, 322, 324)
Plotly: A library used to create visualizations of data. (See pages 52, 61, 173)
positional argument: Arguments that are required to be passed to functions and which must be included in a function call in the order specified by the function definition. This is different from keyword arguments. (See pages 9, 60, 65, 78, 81, 122, 127, 128, 148, 149, 158, 228, 238, 239, 261, 265, 281, 304, 356, 433)
Python interpreter: The program that accepts Python code and executes the instructions contained in that code. (See pages 6, 18, 38, 60, 380)
Pythonic: Describes code that adheres to the stylistic conventions of the Python language. Typically refers to code that strives for simplicity and readability. (See page 209)
Pythonpath: A list of locations in the computer's file structure in which Python will attempt to find files associated with libraries to import. (See page 171)
refactoring: The process of breaking a script down into a series of functions that group the code into logical tasks. Refactoring code makes it easier to reuse parts of it for other solutions. (See page 261)
relative path: The address to a specific directory or file on a computer, relative to another directory or file. (See pages 58, 381)
result object: The object that is returned from the fitting operation performed in lmfit. (See page 149)
return: A keyword that instructs Python to end execution of a function and pass zero or more values back to the place in the code that it was called from. In other words, it is the result of the evaluation of a function call. (See pages 10, 18, 55, 60, 81, 103, 104, 109, 122, 129, 148, 188, 189, 345, 347)
root: The lowest level directory in the file system. All other directories and files are contained in this directory and its various levels of subdirectories. The root directory is not contained in any other folder. (See page 56)
scalable vector graphic: A method to represent graphics using equations and other instructions that can be interpreted when the graphic is visualized, as opposed to specifying the graphic in terms of the appearance of each pixel in an array of pixels (the latter

is used in bitmaps. Because information is not specified by pixel, the graphic is easily resized without losing resolution, and in some cases can result in dramatically smaller image files. (See pages 50, 100, 309, 437)
script: A plain text file that contains Python code that can be executed using the Python interpreter. (See pages 38, 41, 172, 174, 207, 225, 237, 239, 258)
shape: A tuple containing the same number of elements as the dimensions of a nested Numpy array. Each element of the tuple specifies the number of elements held by each array layer, working from the outermost array inward. (See pages 33, 34)
shell: A specific session of a Python interpreter in a console. (See page 6)
slicable: A property of an object that indicates that it can be indexed. (See page 385)
slice: A subgroup of items selected from a iterable collection. This subgroup is specified using indices. (See pages 80, 84, 88, 99, 316, 351)
SMILES A text-based representation of the structure of molecules and complexes. SMILES stands for Simplified Molecular Input Line Entry System. (See pages 97, 98)
statement: A command that either assigns a value to a variable or changes the execution order of the code. (See pages 12, 84, 104, 171, 281)
string: Technical name for the text **data type**. (See pages 16, 54, 56, 66, 79, 86, 87, 98, 99, 104, 123, 124, 125, 127, 128, 133, 362, 370, 383, 385)
syntax: The rules that govern how computer code needs to be structured. Different languages define different syntax. (See pages 5, 6, 7, 8, 151, 209, 328, 415)
syntax highlighting: A feature of many text editors and IDEs that formats the different elements of code using different font properties. For instance, one might use different colors for numbers, functions, comments, and strings. (See page 11)
traceback: Textual output generated in the console upon an error in code execution. The text attempts to give a record of what code was executed leading up to the error and a terse description of why the error occurred. (See page 33)
tuple: A data type that is a collection that can hold other objects. Similar to a list, but a tuple is immutable. (See pages 99, 100, 148)
unicode: One common binary encoding framework that allows code to specifies characters to be displayed when rendering text. This allows you to display emojis. (See pages 126, 133)
unordered: A property of a collection which means that if it is iterated through, the order of iteration is not guaranteed to be the same each time. (See page 97)
unpacked: The behavior by which the elements within a collection are assigned to a number of variables equal to the number of elements in the collection. (See page 148)
uri binary stream: A data stream that is encoded using Uniform Resource Identifier (URI). (See page 100)
value: An object stored in a dictionary. Values can be accessed using the associated key. (See pages 82, 97, 108, 364)
variable: An object with a user-defined name that identifies the location in memory that holds data. Variables allow the data to be referenced or manipulated throughout a program using the name. (See pages 10, 12, 14, 16, 19, 30, 32, 37, 42, 54, 60, 65, 78, 82, 83, 86, 102, 122, 124, 148, 336, 342)

vectorization: A behavior of Numpy arrays whereby an operator applied to the array is applied to each item in the collection automatically. (See pages 31, 37, 148)
vectorized: A property of an instruction that specifies that the calculation should be performed using vectorization. (See page 205)
version control: The practice of organizing different copies of the same basic code in a way that makes it easy to identify the changes from one to another. Good version control practices make it easy to identify which copy of a code is the one that behaves the way you want. (See pages 39, 322)
view: A way to access a only a subset of a Numpy array. The view can be assigned to a new variable (which points to the elements of the original array), rather than making a copy of them to a new object. Thus, changes made to the view result in the same changes being made to the original array. (See page 146)
white space: Tab, space, newline, or any character which does not have a visual representation in text. In Python, white space is used to designate which lines of code belong to a common code block, for instance. (See pages 18, 84, 415)
wildcard: When matching patterns in strings, a wildcard can be used to specify that any combination of characters will be considered a match. Typically denoted by an asterisk (*). (See pages 86, 128, 384)
working directory: The directory in which the Python interpreter is currently running. This directory is included in the PythonPATH Files in the working directory can be accessed by Python without specifying their paths. (See pages 58, 170, 171, 421)
wrapper: A function that serves as an interface for another function to make it easier to use or to extend its capabilities. (See page 206)